Wild Relatives of Cultivated Plants in India

Anurudh Kumar Singh

Wild Relatives of Cultivated Plants in India

A Reservoir of Alternative Genetic Resources and More

Springer

Anurudh Kumar Singh
National Genebank Curator & Head,
Division of Germplasm Conservation
National Bureau of Plant Genetic
Resources, Indian Council
of Agricultural Research
Delhi, India

ISBN 978-981-10-5115-9 ISBN 978-981-10-5116-6 (eBook)
DOI 10.1007/978-981-10-5116-6

Library of Congress Control Number: 2017943483

Printed on acid-free paper

This Springer imprint is published by Springer Nature
The registered company is Springer Nature Singapore Pte Ltd.
The registered company address is: 152 Beach Road, #21-01/04 Gateway East, Singapore 189721, Singapore

Preface

The narrow genetic base of most crops/cultivated species has made them increasingly vulnerable to various stresses on one side and has limited the genetic variability on the other. This has restricted the capabilities to engineer new cultivars to mitigate the challenges of biotic and abiotic stresses and those created by climate change. This scenario of limited genetic diversity demands search for new and novel variability outside the gene pool of crop/cultivated species. In this regard, the immediate and distant wild ancestors of these cultivated species can be the first genetic resource in search for new genes and alleles that have helped them survive fighting against all possible odds and vagaries of nature for millenniums. They have been found with a reservoir of genes in several crops to overcome disaster situations in a number of crops like rice, wheat, maize, potato, tomato, etc., establishing their credentials in saving the crop species by genetically improving them with incorporation of genes, providing greater resilience and productivity. However, despite such established credentials, the global data on crop wild relative's germplasm collection, conservation, and evaluation in search of new genes and alleles has been found with large gaps. Consequently, their use in crop improvement has not received desired attention both at national and international levels. This situation demands creation of a national inventory on wild relatives of crop/cultivated species with basic information and potential to generate interest among researchers and plant breeders to make use of them in crop improvement and their further economic exploitation.

India is floristically one of the megacenters of biodiversity and the center, or one of the centers, of origin of a large number (215) of crops. Obviously, for this reason, it has a reservoir of crop wild relative's (>900) genetic diversity. This has been created in most plant species, because of the vast climatic diversity offered by the Indian Subcontinent. However, in many cases it has not been accessed because of various reasons and still lay untapped. The present book is an effort of producing an inventory of >950 wild species related to crops/cultivated species with basic information on their habitat, distribution, desirable features, and phylogenetic relationship with the cultivated species and the threat created to their existence, to facilitate their conservation and use. To promote the activation of necessary steps in this direction, it also discusses the strategies that can be utilized for improved information on their phylogenetic relationships with cultivated species, collection, conservation, and breeding strategies that can be adapted for their effective and efficient use.

The author would like to acknowledge the help extended by various peers and colleagues in the preparation of the book by perusing the part of the manuscript related to their expertise, providing comments, and sharing information, namely, Drs. SN Nigam, former principal breeder, ICRISAT, Patancheru; YS Nerkar, former VC, MPKV, Rahuri; AK Roy, project coordinator, IGFRI, Jhansi; SR Pandravada, NBPGR RS, Hyderabad; JC Rana, head of the Division of Germplasm Evaluation; Veena Gupta, Division of Germplasm Conservation; and K. Pradheep, Division of Plant Exploration and Germplasm Collection, NBPGR, New Delhi.

Thanks are also due to Springer for accepting the project proposal and providing the opportunity and support for this book project. In this regard, the author is particularly grateful to the Life Science Team and Miss Manjusha Nalamolu, production editor (Books), and her associates for the assistance extended in the preparation and production of the book.

It is hoped that researchers in general and those interested in wild plant genetic resources would find the information compiled useful for the identification of the gaps in our knowledge and the researchable issues, to promote their conservation and use, whereas the plant breeders and biotechnologists involved in genetic enhancement and/or crop improvement are expected to benefit from the information compiled, in promoting the use of wild relatives of crop species in crop improvement.

Delhi, India Anurudh Kumar Singh

Contents

1 Introduction 1
References 8

Part I Wild Relatives

2 Defining a Wild Relative 13
References 17

3 Importance of Wild Relatives as Genetic Resource and Otherwise 19
3.1 Introduction 19
3.2 Wild Species as Source of Genetic Enhancement to Broaden Crop Genetic Base 21
3.3 Wild Species as Source of Genetic Variability for Crop Improvement 23
3.3.1 Disease and Insect Pest Resistance 23
3.3.2 Abiotic Stress Tolerance 25
3.3.3 Yield Traits 25
3.3.4 Quality Traits 26
3.3.5 Earliness and Adaptation 26
3.3.6 Modes of Reproduction 26
3.3.7 Miscellaneous Traits 27
3.3.8 Quantitative Traits 27
3.4 Resource for Other Economic Uses 28
3.4.1 Source of Alternative Food Crops 28
3.4.2 Sources of Ethano-medicinal and Floriculture Crops 29
3.4.3 Sources of Renewable Bioenergy 30
3.4.4 Source of Restricting Soil Erosion, Soil Reclamation, and Soil Fertility Improvement 31

3.4.5 Securing Ecological Benefits Through Carbon Sequestration, Phytoremediation, Promoting Ecology, and Environment ... 32
3.5 Perspective ... 33
References ... 33

Part II Wild Relatives Distribution and Diversity in Different Crop Groups

4 Cereals ... 41
4.1 Introduction ... 41
4.2 Wild Species Found in India ... 44
4.3 Perspective ... 51
References ... 52

5 Grain Legumes ... 53
5.1 Introduction ... 53
5.2 Wild Species Found in India ... 55
5.3 Perspective ... 61
References ... 61

6 Oilseeds ... 63
6.1 Introduction ... 63
6.2 Wild Species Found in India ... 65
6.3 Perspective ... 67
References ... 68

7 Fiber Crops ... 69
7.1 Introduction ... 69
7.2 Wild Species Found in India ... 71
7.3 Perspective ... 75
References ... 76

8 Forage Crops ... 77
8.1 Introduction ... 77
8.2 Wild Species Found in India ... 79
8.3 Perspective ... 83
References ... 84

9 Vegetables ... 85
9.1 Introduction ... 85
9.2 Wild Species Found in India ... 89
9.2.1 Fruity Vegetable ... 89
9.2.2 Leafy Vegetables ... 97
9.2.3 Root, Tuber, and Bulb Vegetables ... 99
9.3 Perspective ... 105
References ... 106

10 Fruits and Nuts 109
10.1 Introduction 109
10.2 Wild Species Found in India 113
10.2.1 Temperate Fruits 113
10.2.2 Tropical and Subtropical Fruits 117
10.2.3 Arid and Semi-arid Fruits 126
10.3 Perspective 131
References 132

11 Spices and Condiments 137
11.1 Introduction 137
11.2 Wild Species Found in India 141
11.3 Perspective 152
References 153

12 Commercial Crops 155
12.1 Introduction 155
12.2 Wild Species Found in India 157
12.3 Perspective 161
References 162

13 Medicinal and Aromatic Plants 165
13.1 Introduction 165
13.2 Wild Species Found in India 166
13.3 Perspective 175
References 176

14 Floriculture Crops 177
14.1 Introduction 177
14.2 Wild Species Found in India 179
14.3 Perspective 193
References 194

15 Agroforestry 197
15.1 Introduction 197
15.2 Wild Species Found in India 199
15.3 Perspective 204
References 205

16 Cottage Industry Crops and Others 207
16.1 Introduction 207
16.2 Wild Species Found in India 208
16.3 Perspective 214
References 215

Part III Conservation of Wild Relatives

17 Classification of Wild Species to Facilitate Conservation and Gene Transfer 219
17.1 Introduction 219
17.2 Experimental Taxonomy 220
17.2.1 Biochemical and Molecular Biosystematics 222
17.3 Phylogenetic Classification 224
17.4 Classification of Species (Genetic Resources) for Better Management and Use 225
17.4.1 Gene Pool Concept 225
17.4.2 Primary Gene Pool (GP-1) 226
17.4.3 Secondary Gene Pool (GP-2) 226
17.4.4 Tertiary Gene Pool (GP-3) 227
17.4.5 Factors Contributing to Use Gene Pools 229
17.5 Perspective 230
References 230

18 Collection Strategies 233
18.1 Introduction 233
18.2 Collection 234
18.2.1 Planning 235
18.2.2 Selection of Populations/Sites 236
18.2.3 Number of Samples to Be Collected 238
18.2.4 Minimum Number of Plants to Be Sampled 239
18.2.5 Sample Size and Seeds per Plant 240
18.2.6 Sampling Approach 240
18.2.7 Seed Collecting 241
18.2.8 Information Documentation 243
18.2.9 Post Collection Care 243
18.3 Perspective 245
References 247

19 Conservation Strategies 249
19.1 Introduction 249
19.2 Probable Conservation Approaches 250
19.2.1 By Regulatory Mechanisms 250
19.2.2 By Promoting Cultivation Practices 251
19.2.3 By Conventional Conservation Strategies 251
19.3 Perspective 258
References 259

20 Breeding Strategies for Use of Wild Relatives 261
20.1 Introduction 261
20.2 Characterization and Evaluation of Wild Species 262
20.3 Identification of Potentially Valuable Species 262
20.4 Prioritization of Species 263

20.5 Breeding Options for Interspecific Gene Transfer 263
20.5.1 Conventional Cytogenetic Manipulations 263
20.5.2 Biotechnological Approaches Overcoming Hybridization Barriers 265
20.5.3 Asexual Biotechnological Approaches 266
20.5.4 Other Methods 268
20.6 Merit and Demerits of Various Breeding Approaches 268
20.6.1 Conventional Breeding 268
20.6.2 Sexual Biotechnological Breeding 268
20.6.3 Asexual Biotechnological Breeding/Genetic Modification/Transformation 269
20.7 General Steps of Interspecific Gene Transfer 269
20.8 Perspective 270
References 271

21 Future Perspectives 273
21.1 Introduction 273
21.2 Most Are Threatened: Action Plan for Protection/Conservation of Important Ones 274
21.3 Awareness Development Regarding the Importance of Wild Species 275
21.4 Building Comprehensive Collection of Wild Relatives of Indian Origin/Indigenous to India 275
21.5 Continued Conservation 276
21.6 Promotion of Biotechnological Approaches in Combination of Conventional Breeding 277
21.7 Developing Basic Information on Phylogeny and Genomics 277
21.8 Creation and Phenotyping of Purpose-Driven Hybrid Population Sets 278
21.9 Establishing a Predictive Network of Genotype–Phenotype Associations 278
21.10 Deployment of Identified Phenotypes into Crop Breeding Pipelines 279
21.11 Overcoming Other Problems 279
References 280

Erratum E1

Glossary 283

Appendices 291
Appendix I 291
Appendix II 297
References 303
Appendix III 304

About the Author

Dr. Anurudh Kumar Singh was educated at Aligarh Muslim University (1962–1966); Patna University (1969–1972); the Max Planck Institute, Koln, Germany (1986–1987); and Cornell University, Ithaca, USA (1994–1995). He has been interested in the wild relatives of crop species since the beginning of his career. For his Ph.D., he worked on their collection and conservation and conducted cytogenetic studies to trace the phylogenetic relationships between wild species related to cucurbit vegetable crop species. More significantly, from 1978 to 1997, while at the International Crops Research Institute for the Semi-Arid Tropics (ICRISAT), he was involved in a project on utilizing wild *Arachis* to genetically improve cultivated groundnut and successfully incorporated foliar disease resistance and other desirable traits from several wild diploid *Arachis* species. Later, he became involved in the collection and conservation of groundnut germplasm, including wild *Arachis* species. From 1998 to 2008, as a National Gene Bank curator and head of the Germplasm Conservation Division at the National Bureau of Plant Genetic Resources (NBPGR), New Delhi, he was responsible for the ex situ conservation of crop germplasm with special emphasis on wild relatives of cultivated species. As member secretary of the Germplasm Registration Committee, Indian Council of Agricultural Research (ICAR), he registered a number of wild relatives of crops; and as member secretary of the Task Force for the Protection of Plant Varieties and Farmers' Rights (PPV & FR), he was involved in the identification of agrobiodiversity hotspots in India and production of the first inventory of cultivated species and their wild relatives. For these contributions he received the "The CGIAR Chairman's Excellence in Science Award" for "Outstanding Locally Recruited Scientist" (1997). He was awarded a fellowship of the National Academy of Sciences, India (2001), and of the National Academy of Agricultural Sciences, India (2004). He has more than 210 scientific publications to his credit, including 4 books.

1 Introduction

The improvement of crop plants has progressed from the time of domestication 10,000 to 14,000 years back. The harvesting-sowing-harvesting cycle over generations with selection pressure for yield and yield-related traits and quality have transformed the low-yielding shattering plants into present-day high-yielding cultivars. However, this process of selection in a definite direction toward grain or economic yield or in yield-related traits and quality has resulted in narrowing down of the genetic base of these crop species due to funneling toward a selected set of genes (Tanksley and McCouch 1997). It has caused loss of genetic and allelic diversity for many other traits, particularly related with their potential to stand against various natural biotic and abiotic stresses, and for micro-nutritional quality. Consequent to this, often genetic diversity available in the cultivated species gene pool has been found limited for many stress factors and micro-nutritional factors.

The need for improvement in crop productivity in food crops has been ever increasing, particularly in thickly populated countries like India, to meet the ever-growing demand of increasing population to ensure national food and nutritional security. In addition, genetic improvement is also needed to overcome the various yield-reducing stress factors, created by the change in intensive agricultural practices/pattern and climate. Therefore, when the genetic diversity available in the cultivated species gene pool is found inadequate, it demands for search and use of genetic diversity beyond the cultivated species gene pool. This process may start with immediate ancestral and other wild and weedy relatives of crop species and may be carried to other living entities using the recently developed parasexual biotechnological approaches that can facilitate transfer of genetic material from distant alien species. However, the concerns of biosafety and technological expertise required under biotechnological approaches limit the use of other living organisms. Moreover, most of the estimated genetic diversity in relation to a cultivated species is confined to the related wild species. For example, in tomato, 95% of genetic diversity of the total generic gene pool of *Lycopersicon/Solanum* is confined to wild species (Tanksley and McCouch 1997). Therefore, use of wild relatives in genetic improvement of cultivated species should be a priority. This would demand

A.K. Singh, *Wild Relatives of Cultivated Plants in India*,
DOI 10.1007/978-981-10-5116-6_1

collation of information on the wild relatives of cultivated species with regard to their total number, distribution, potential genetic value, and threats to their existence. The efforts in this direction will facilitate their sustainable conservation and use, which is the need of present.

The richness of crop wild relatives with genes conferring resistance to biotic and abiotic stress is well recognized and documented. Many of alien genes have been transferred into their cultivated counterparts through wide hybridization with or without application of biotechnological techniques and mutagenesis. Recently, it has also been revealed that the wild relatives can also be source of genes related with yield and nutrition quality. Apomictic genes are another asset of genes in many crop relatives that deserve attention. The past and the present contributions of wild species in crop improvement have led to the realization that the so-called inferior wild species are highly superior in conserving many of desirable genes. They can be a reservoir of genetic resource in search of new genes for breeding elite plant varieties. With the availability of molecular tools, desirable genes and polygenes in wild donor species can be mapped and introgressed utilizing tightly linked markers, mitigating the constraint of linkage drag, etc. Also, genes from multiple sources of cultivated and wild taxa can be pyramided. Thus, characterization and evaluation of crop wild relatives belonging to primary, secondary, and tertiary gene pools and utilization of their novel genes are one of the leading strategies in today's plant-breeding efforts to design the cultivars to meet the challenges of ever-growing population and climate change. It is obvious that some wide hybridizations between cultivated and wild species will never be easy and may involve near-impossible constraints such as complete or partial sterility. In such cases, gene cloning and gene discovery, complemented by transgenic breeding, will hopefully pave the way for success. To date, utilization of wild relatives through traditional and molecular breeding has been demonstrated in many field and horticultural crops.

Based on richness of floristic diversity and species endemism, India is among the 17 mega-biodiversity centers of the world. It contains 3 of the 34 biodiversity hotspots identified at global level, namely, the Himalayas, the Eastern Indo-Myanmar region, and the Western Ghats and Sri Lankan region (Conservation International 2005). Being situated at the tri-junction of three realms of global biogeographical zones, namely, Afrotropical, Indo-Malayan, and Palaearctic, and the product of a collision of a Gondwana landmass from Southern Hemisphere with Euro-Asian plate during continental drift, it offers components of biodiversity and characteristic features of all three realms and of both Northern and Southern Hemispheres, thereby making the subcontinent very rich in diverse and unique biological diversity. It ranks tenth in the world and fourth in Asia in plant diversity (Khoshoo 1995).

The Indian flora represents nearly 12% of the global floral diversity. The floristically rich, India has about 148 endemic genera belonging to over 47 families of higher plants (Nayar 1980, 1996). As per Botanical Survey of India (BSI), India has 47,513 plant species (Singh and Dash 2014), including species of bacteria. Of these, about 20,141 taxa are of angiosperms with 17,926 species belonging to 2991 genera and 251 families, representing approximately 7% of the described species in the world (Karthikeyan 2009). There are about 5725 endemic species to India, representing 33.5% of the Indian flora (Nayar 1996). As per latest survey, this has increased to 6000 (Goyal and Arora 2009). Of the 5725 endemic species listed by Nayar (1996), the largest number of about 3471 species are found in Himalayas, followed by 2051 species confined to peninsular region and around 239 species to Andaman and Nicobar Islands. However, as per recent estimation about 4045 taxa belonging to 975 genera in 155 families are considered as strict endemics to the present Indian political boundary (Arisdason and Lakshminarasimhan 2015). Nayar (1996) identified 3 megacenters and 25 microcenters of endemic plants. The three megacenters of endemism in India are the Himalayas (both Western and the Eastern), the Northeastern Indo-Myanmar region, and the Western Ghats. The Central Himalayas not only includes the fringe endemic species of Eastern and Western Himalayas, but is a crucible of speciation. The moist evergreen forests of the Western Ghats, due to varied topography and microclimatic regimes, have some areas of active speciation and show several vicarious species (Ramesh et al. 1992). The dominant plant families of the Indian Subcontinent are family Poaceae represented by 263 genera and 1291 species, followed by Orchidaceae (184/1229), Leguminosae (173/1192), Asteraceae (167/1052), Rubiaceae (113/616), Cyperaceae (38/545), Euphorbiaceae (84/528), Acanthaceae (92/510), Rosaceae (40/492), and Lamiaceae (72/454). A total of 42 families have more than 100 species each.

Agriculturally, the Indian Subcontinent is one of the 12 megacenters of crop diversity (Zeven and Zhukovsky 1975). Also, it figures as the second most important among the Vavilovian centers of crop plants origin and diversity, after China (Vavilov 1926). The Indian Subcontinent is the center of domestication of 117 from Indian mainland plus 55 from Indo-Malayan center, totaling 172 (Vavilov 1935); 24 to India plus 73, including Southeast Asia (Harlan 1975); and 166 (Zeven and de Wet 1982) crop species. In addition, it has contributed in the diversification of many introduced crop/cultivated species, which include food crops, forage grasses, spices and condiments, ornamentals, medicinal and aromatic plants, forest trees, etc. As per the recent estimate, 800 (Goyal and Arora 2009) to 811 (Singh et al. 2013) crop plants are cultivated in India. This includes 172 and/or 166 plant species earlier reported under Hindustani center (one of the eight Vavilovian centers of origin and diversity) and many more crop/plant species introduced from various parts of the world. With regard to wild relatives of these crop species, a preliminary document listing 334 was published by Arora and Nayar (1984). Since then, more surveys and research have gone through in the last 30 years. These studies have changed picture about the number of species available as wild relatives of crop species, raising the number to 912 (Singh et al. 2013) and regarding their distribution, genetic diversity, and potential value. Therefore, there is a need for compiling updated information regarding the wild relatives of crop species, particularly on their distribution, characteristic features, and genetic potential from the genetic resources' point of view with improved illustrations, and also information on threat to their existence to facilitate protection and conservation and to promote use in genetic improvement of cultivated species. This will include highlighting the future research efforts required in search for new genes and alleles in the crop wild relatives' genetic diversity through systematic evaluation.

Under these circumstances, the wild relatives of crop plants, particularly those which are part of undisturbed natural habitats and have not been put to human selection pressure, form a reservoir of genes that can be searched to help fill the void of lack of genetic diversity. These rare untapped wild genetic resources can help further genetic improvement of crop species with incorporation of genes conferring greater resilience against yield-reducing factors, particularly biotic and abiotic stresses, and, thereby, help meeting various challenges, contributing to sustainability of agriculture. It could be the same way, like wild species, which have sustained themselves across the changes of time and space, surviving longer than the oldest cultivated variety, without any human assistance. Also, besides sustainability, it could help increase productivity. Thus, as source of inbuilt genetic resistance against various stresses, they offer a treasure of genes for crop improvement program. The wild gene pool includes the naturally occurring wild and weedy types and protected, semi-protected, and domesticated diversity (Hoyt 1988). Many of these have contributed significantly in genetic improvement of the world's major food crops such as rice, wheat, tomato, potato, etc. For example, in rice, resistance to grassy stunt virus and grass hopper has been introduced from wild *Oryza nivara* (Central India) and Pitambi rice (Silent Valley, Kerala) and resistance to brown plant hopper from *O. officinalis*. *Oryza granulata, O. officinalis*, *O. nivara,* and *O. sativa* var. *spontanea* as donors have significantly helped in the reduction of incidence of grassy stunt virus, bacterial blight, brown plant hopper, and green leaf hopper in the cultivated *O. sativa* (Khush 1977). Similarly, in often cross-pollinated pigeonpea, the species *Cajanus cajanifolius* (=*Atylosia cajanifolia*) endemic to Koraput region of India has contributed to the higher pod and seed yield of the cultivated *C. cajan* (Singh 2013). Therefore, the wild species offer a reservoir of potentially valuable genes to support crop improvement program through wide hybridization.

For these reasons, in crop improvement, the use of wild species has established credentials and has been successfully exploited in many of crop species (Maxted and Kell 2009; Brumlop et al. 2013; Singh et al. 1990). The potato and tomato are excellent examples of crop plants that have been extensively improved by controlled introgression of genes from primitive cultivars, wild species, and other forms of exotic germplasm. The use of wild species of potato dates back to 1840 Irish famine caused by late blight (Salaman 1949). This forced the use of late blight-resistant wild *Solanum demissum* for containment of late blight. As per one of the estimates in potato, 49% of the varieties grown around the world had genes from the wild species by the late 1970s (Ross 1979). The gene of this species continues to be effective in some areas, and currently 40% of total area of the most popular potato cultivars in the United States has *S. demissum* gene in their ancestry (National Potato Council 2003). Therefore, the wide hybridization involving wild species provides an excellent opportunity to tap the useful genetic diversity available in wild and weedy relatives of crop plants, using appropriate conventional and/or modern techniques.

The first step for exploitation of wild relatives in crop improvement would be survey, inventory, and documentation of the wild relatives available in India with information on their origin and distribution, unique/essential features, and potentially valuable genetic features for specific use. Generation of information/knowledge on their phylogenetic relationship with cultivated species shall facilitate selection of appropriate breeding strategies for introgression of desired genes, thereby their use in crop improvement. This is the principal endeavor of the present book. Application of unique and imaginative breeding procedures would help incorporation of desirable features, facilitating exploitation of full potential of crop/cultivated species. Utilization of wild species is basically designed to targeted introgression of desired genes into cultivated species from the alternative genetic resources, i.e., other than the cultivated species genetic variability.

Wide hybridization helps in generation of new combinations of variability from which the selection can be effected for the target recombinants, incorporating desirable features. Therefore, in addition to introgression of specific gene(s) for induction of inbuilt resistance against various biotic and abiotic stresses, increasing yield and quality, it helps in broadening the genetic base of crop species. Thus, genetic enhancement provides the breeders an endless choice of recombinants not only in relation to resistance to biotic and abiotic stresses but also to other traits such as yield and quality for which the potential has been recently reported in wild species of several crops. However, the interspecific breeding program may not be easy to implement and face several constraints, such as hybridization barriers, hybrid sterility, linkage drag, etc. It would need a multidisciplinary team of scientists specializing in genetic resources, cytogenetics, biotechnology, breeding, plant physiology and pathology, etc. This would help evaluate wild relatives, establish hybrids (embryo rescue), and execute cytogenetic manipulations, genetic transformation, and selection of desirable recombinants to support controlled introgression of desirable genes from wild species.

At the global level, there are many success stories on introgression of genes from wild species in crops like wheat, rice, cotton, potato, tomato, etc. (Brar 2005; Reem and Toby 2007; Rick 1979; Ross 1979; Stalker 1980; Singh et al. 1990). At the national level, there are a few success stories in case of rice, wheat, groundnut, brassica's, pigeonpea, *Vigna*, tomato, etc. (Singh et al. 2015; Singh and Nigam 2016). However, there is still a considerable scope for further intensification of wide hybridization in crops like rice, wheat, maize, sorghum, pearl millet, black gram, green gram, pigeonpea, chickpea, groundnut, tomato, potato, chili, brinjal, *Cucumis*, okra, etc. that are globally important and extending the benefits of wide hybridization to new crops. The present book has made a sincere effort in this regard with compilation of the information available on wild relatives of cultivated plant species, particularly of crop/cultivated plants indigenous to India to bring them into scientific domain. This shall generate desired interest in collection, conservation, characterization, and evaluation of wild relatives to facilitate their use in genetic improvement of cultivated species through conventional interspecific breeding and biotechnological techniques, including recombinant DNA technologies.

Besides contributing to genetic improvement of their cultivated counterparts, the wild relatives of crop/cultivated species have been playing enormously important roles in basic scientific studies in the defining/deciphering plant genomes. They have contributed immensely in resolving several fundamental questions, particularly those related to the origin, evolution, phylogenetic relationships, cytological status, and inheritance of genes of an array of crop plants. Since the advent of molecular markers, they have been frequently used in genetic studies. Their involvement in molecular mapping has facilitated the development of mapping populations with optimum polymorphism to construct saturated maps. This has helped in illuminating the organization, reorganization, and functional aspects of genes and genomes. Some of them are contributing enormously as model plant species to the elucidation and amelioration of the genomes of crop plant species. Also, they have helped in understanding the possible threats of gene flow from genetically modified plants developed using the recombinant DNA technologies, adversely affecting the biodiversity due to their invasive nature.

In several other cases, it is by now well established that many wild relatives of crops/cultivated species deserve serious attention for domestication. This is especially true for exploitation of many in herbal industry for their phytomedicinal and nutraceutical properties and renewable bioenergy production. In addition, many more can be harnessed for their environmental and other benefits such as in phytoremediation of ecology and environment, restriction of soil erosion, soil reclamation by improving soil fertility, etc.

Despite the realization of the above benefits of wild relatives of crop/cultivated species, one of the obstacles in its usage is their poor representation in germplasm collections for detailed scientific study. Maxted and Kell (2009) estimated that only 2–6% of international germplasm collections are of crop wild relatives. Although collections for a few crops and their wild relatives are large, wild relatives of many crops have been poorly collected or nearly ignored in crop groups, such as forage crops, ornamental crops, medicinal and aromatic plants, etc., and genera such as *Vigna*, *Trichosanthes*, *Lectuca*, *Trigonella*, *Allium*, *Dioscorea*, etc. to the extent that many lack well-identified wild relatives. A comprehensive documentation of basic information on the wild relatives of all the major field and horticultural crops and agroforestry/industrial trees that are either native or naturalized to India with significant variability shall help bring them into global scientific community domain and identify gaps. This shall promote systematic attempts on their collection, characterization, evaluation, and conservation to promote wider use, which has been the prime objective and driving force behind the publication of this book.

Vavilov, the Russian botanist, was the first to realize the importance of wild relatives of the crop plants in crop improvement in the early twentieth century (Vavilov 1926). Since then, there have been several national and global initiatives to further emphasize their importance to promote collection, characterization, conservation, and use. It is hoped that the present effort with respect to the wild relatives of crop species found in India shall stimulate:

1. Their collection, to build up a comprehensive collection of wild relatives covering spatial and ecological range.
2. Phenotypic characterization and evaluation of wild relatives, which has been one of the most significant constraints in promotion of their conservation and use

(generation of useful phenotypic information has been challenging for crop wild relatives, because of greater genetic diversity, absence of standard descriptors unabling description of total spectrum of variability, and, as valuable traits, even simple characteristics such as fruit size and color which are often masked in wild phenotypes).
3. Generation of full genome sequences for crop wild relatives, as genomic tools and technologies that have become available in recent past are increasingly powerful and inexpensive.
4. Development of databases as the amount of genetic and phenotypic data increases (Heywood and Dulloo 2006). The integration of databases with overall information system on plant genetic resources with direct links to source of availability shall be some of the future steps critical for the assessment of potential value, conservation, and precise use of wild relatives in breeding programs.

Future step may also include creation of pre-breeding populations that are made available for direct incorporation into conventional breeding programs through participatory breeding networks. This approach shall also support and promote the global multilateral systems of genetic resource exchange for the world's important food crops and their wild relatives in a transparent manner, ensuring ownership, fair and equitable benefit sharing, and intellectual property rights (IPR) issues.

Present effort of information documentation on wild relatives of crop/cultivated species shall also encourage demand for genomic and molecular taxonomic studies. These studies will facilitate the use of certain modern plant-breeding technologies, including recombinant DNA technologies to overcome some of the constraints of interspecific breeding and promote precise breeding for direct incorporation of specific natural desirable genes. For example, the use of conventional methods in overcoming the linkage drag problem, which leads of transfer both desirable and undesirable genes, is a long drawn and cumbersome process. However, it can be overcome relatively easily by using improved techniques of genome analysis, marker-assisted selection, next-generation sequencing and transcriptomics, etc. The crop wild relatives could be exploited using recombinant DNA technology known as cis-genesis. This involves efficient gene isolation methods such as map-based cloning and allele mining, opening new avenues for the use of cloned natural indigenous genes. It will accelerate the process of gene transfer, avoiding incorporation of toxic components of wild species. These can also enable the possibility of marker-free plant transformation without bacterial antibiotic gene as selection markers, leading to creation of clean cisgenic plants without transgenes (Jacobsen and Nataraja 2008).

Lastly, but not the least, the fact that is being comprehensively documented in the book is the number of wild relatives of crops threatened due to habitat loss, fragmentation, and environmental degradation. They are also being threatened by infrastructure development, agricultural intensification, and the climate change. On the other hand, despite being targeted under biodiversity conservation in many regions, there is little or no in situ conservation of these resources, because strategies, such as establishment of national parks or protection of other areas through alternative conservation strategies/facilities for this purpose, are independent, unrelated to the preservation of crop genetic resources. Additionally, crop wild relatives also face genetic risks from factors, such as introgression from cultivated forms, or, in the case of medicinally or pharmaceutically useful species, from direct overharvesting. Therefore, prioritization of the most threatened crop wild relatives is needed. Collaborations between local people, researchers, and various national and international organizations responsible for conservation of biodiversity need to be built up with consensus about the numerous benefits of protecting crop wild relatives both in situ and ex situ.

References

Arisdason W, Lakshminarasimhan P (2015) Status of plant diversity in India: an overview, ENVIS Center of Floral Diversity. Hosted by Botanical Survey of India, Kolkata

Arora RK, Nayar ER (1984) Wild relatives of crop plants of India. National Bureau of Plant Genetic Resources (NBPGR). Kapoor Art Press, New Delhi

Brar DS (2005) Broadening the gene pool of rice through introgression from wild species. In: Toriyama K, Heong KL, Hardy B (eds) Rice is life: scientific perspective for 21 century. Proceeding of the world rice research conference, Tokyo and Tsukuba, Japan, 4–7 Nov 2004. IRRI, Phillipines, pp 157–160

Brumlop S, Reichenbecher W, Tappeser B, Finckh MR (2013) What is the SMARTest way to breed plants and increase agrobiodiversity? Euphytica 194:53–66

Conservation International (2005) Biodiversity Hot Spots, 1919 M Street, NW, Suite 600, Washington, DC 20036. (202)912-1000, fax: (202)912-1030. Updated 2nd May 2005. www.conservation.org

Goyal AK, Arora S (2009) Chapter 1. Overview of biodiversity status, trends and hreats. In: India's Fourth National Report to the Convention on Biological Diversity. Ministry of Environment and Forest, Government of India, New Delhi

Harlan JR (1975) Crops and man, 2nd edn. American Society of Agronomy, Madison

Heywood VH, Dulloo ME (2006) In situ conservation of wild plant species – a critical global review of good practices, Technical Bulletin No.11. International Plant Genetic Resources Institute (IPGRI), Rome

Hoyt E (1988) *Conserving the Wild Relatives of Crops*. International Board of Plant Genetic Resources (IPBGR); International Union for the Conservation of Nature (IUCN); World Wildlife Fund (WWF)

Jacobsen E, Nataraja KN (2008) Cisgenics - facilitating the second green revolution in India by improved traditional plant breeding. Curr Sci 94:1365–1366

Karthikeyan S (2009) Flowering plants of India in 19th and 21st centuries – a comparison. In: Krishnan S, Bhat DJ (eds) Plant and fungal biodiversity and bioprospecting. Goa University, Goa, pp 19–30

Khoshoo TN (1995) Census of India's biodiversity: task ahead. Curr Sci 69:14–17

Khush GS (1977) Disease and insect resistance in rice. Adv Agron 29:265–341

Maxted N, Kell SP (2009) Establishment of a global network for the *In Situ* conservation of crop wild relatives: status and needs. FAO, Rome

National Potato Council (2003) Potato statistics: seed certification, approved acerage. Varieties Planted and Production,Washington, DC

Nayar MP (1980) Endemism and pattern of distribution of endemic genera (angiosperm). J Econ Taxon Bot 1:99–110

Nayar MP (1996) Hot spots of endemic plants of India, Nepal and Bhutan. Tropical Botanic Garden and Research Institute, Thiruvananthapuram

Ramesh BR, Pascal JP, De Franceschi D (1992) Studies of endemism as a criterion for conservation in the Western Ghats. In: Proc of Consultation Workshop on Biodiversity Conservation in the Western Ghats (Nov. 1992, Bangalore) WWF, India.

Reem H, Toby H (2007) The use of wild relatives in crop improvement: a survey of developments over the last 20 years. Euphytica 156:1–13

Rick CM (1979) Potential improvement of tomatoes by controlled introgression of genes from wild species. In: Proc. Conf Broadening Gentic Base of Crops, Wageningen, 1978, Pudoc. Wageningen, The Netherlands. pp 167–173

Ross H (1979) Wild species and primitive cultivars as ancestors of potato varieties. In: Proc. Conf Broadening Gentic Base of Crops, Wageningen, 1978, Pudoc. Wageningen, pp 237–245

Salaman RN (1949) The history of the social influence of the Potato. Cambridge University Press, London/New York

Singh AK (2013) Probable agricultural biodiversity heritage sites in India: XVI. The Koraput region. Asian Agri-History 17(2):97–122

Singh P, Dash SS (2014) Plant discoveries 2013 – New Genera, species and new records. Botanical Survey of India, Kolkata

Singh AK, Nigam SN (2016) *Arachis* gene pools and genetic improvement in groundnut. In: Rajpal VR, Rao SM, Raina SN (eds) Gene pool diversity and crop improvement. Springer, Cham, pp 17–77

Singh AK, Moss JP, Smartt J (1990) Ploidy manipulations in interspecific gene transfer. Adv Agron 43:199–240

Singh AK, Rana RS, Mal B, Singh B, Agrawal RC (2013) Cultivated plants and their wild relatives in India– an inventory. Protection of Plant Varieties and Farmers' Rights Authority, New Delhi

Singh RB, Renu KC, Singh AK, Gopala Krishnan S, Singh Nagendra K, Prabhu KV, Singh Ashok K, Bansal KC, Mahadevappa M (2015) Chapter 1 crop science. In: Singh RB (ed) 100 years of agricultural sciences in India. National Academy of Agricultural Sciences, New Delhi, pp 1–165

Stalker HT (1980) Utilization of wild species for crop improvement. Adv Agron 43:199–241

Tanksley SD, McCouch SR (1997) Seed banks and molecular maps: unlocking genetic potential from the wild. Science 277:1063–1066

Vavilov NI (1926) Centers of origin of cultivated plants. Tr po Prikl Bot Genet Sel [Bull Appl Bot & Genet Sel] 16(2):139–248. [in Russian]

Vavilov NI (1935) The phytogeographical basis of plant breeding. In: Love D (ed) (Transl), Origin and geography of cultivated plants vol 1 pp 330–336, University of Cambridge Press, Cambridge, 1992

Zeven AC, De Wet JMJ (1982) Dictionary of cultivated plants and their regions of diversity. Center for Agricultural Publishing and Documentation, Wageningen

Zeven AC, Zhukovsky PM (1975) Dictionary of cultivated plants and their centers of diversity. PUDOC, Wageningen

Part I

Wild Relatives

2 Defining a Wild Relative

While searching and gathering food along with other materials used from nature (bio-resources) to fulfill human needs, the men collected selectively economically important plants from the wild in the first step. During the beginning of agriculture, centralized growing promoted them. On domestication, selection based on reproductive potential and adaptation to climatic vagaries and for traits associated with the higher production of economic component(s) evolved them further. However, in this process, there was predominant concentration on selection for economic yield and yield-related traits, resulting in narrowing the genetic diversity to limited gene combinations, reducing the potential of selected progenies/populations to stand against natural vagaries, and making them more vulnerable to various biotic and abiotic stresses. This process got further intensified with the discovery of Mendelian genetics, concentrating increasingly on combining small number of gene combinations related with agronomy and productivity, further narrowing the genetic base through the process of funneling (Tanksley and McCouch 1997). Consequently, the cultigens of economic species got limited with or no genetic diversity for several traits, particularly those related to plant immunity system, i.e., resistance/tolerance against biotic and abiotic stresses, affecting the plant health and thereby reducing potential to contribute to economic factors/yield.

During this course of evolution of agriculture, many genetically close wild species and/or wild and weedy forms of these domesticated/cultivated species were left to themselves in the nature. They kept fighting; competing with other companions/associated bio-resources, both micro- and macro-organisms; and responding to climatic changes. In the process, they developed or evolved genes/gene combinations with greater resilience toward such negative factors/stresses. These stresses are common to both the wild species and their related cultivated species, because of the ancestral genetic affinities. Wild species could overcome them, because of their inbuilt genetic resistance, while cultigens through human protection. Many of these wild species are cross-compatible with cultivated species and still exchange genes on getting chance through physical proximity and cross-pollination. Whereas, others have been genetically isolated and distanced to different degrees. The recent

A.K. Singh, *Wild Relatives of Cultivated Plants in India*,
DOI 10.1007/978-981-10-5116-6_2

developments in biotechnological research have opened the possibilities of carrying the genes beyond taxonomic boundaries. Thus, the available genetic diversity of wild relatives of cultivated plant species now provides greater potential/opportunities to facilitate further genetic enhancement of cultivated species through conventional recombinant breeding and nonconventional parasexual biotechnological techniques. This is in addition to exploiting the genetic variability within the cultivated species through normal conventional recombinant breeding.

The wild-related species from whom the cultivated species were domesticated and brought into cultivation shall include the genetic variation within a population and among the populations of a species, generally referred to as the genetic diversity. This genetic variation allowed a species to adapt to diverse ecologies and over time to the environmental changes, by evolving for adaptation and resistance and tolerance to diseases and climatic stresses, mechanisms for predator avoidance, etc., and to diverse man-made agroclimatic conditions.

The floristic richness of Indian Subcontinent, including those of higher plants (angiosperms; see Chap. 1), has provided opportunities for economic exploitation of a large number of plant species and domestication of many. Regarding cultivated species, Indian Subcontinent has been recognized as one of the important centers of origin of crop plants (Vavilov 1935; Zeven and Zhukovsky 1975; Harlan 1975; Zeven and De Wet 1982). The archeological evidence and the scientifically authenticated Vedic literature from the northern part (Lallanji and Srivastava 2008; Fuller 2006; Murphy and Fuller 2014) and evidence from South India (Fuller et al. 2004) suggest domestication of many of the crops in the subcontinent. As per one of the recent overviews on agrobiodiversity in India, a total of around 215 plant species have been domesticated in India, falling in various crops groups from cereals to ornamental to industrial plants (Singh 2017).

Geography and biological components of the region influence the species distribution and so is the distribution of wild relatives of crop/cultivated plant species and vice versa. Therefore, diversity of the two together has evolved biogeographical or phytogeographical zones. This is also expressed in terms of association plant species diversity, including wild relatives of crops found in the Indian Subcontinent. The Indian Subcontinent is a confluence of at least two biogeographical realms, i.e., Palearctic and Oriental. The most commonly acknowledged classification of biogeographical zones was proposed by Udvardy (1975). It recognizes 12 biogeographical provinces in India falling in these two realms. Phytogeographical zonation, based on the type of vegetation, was attempted by Gadgil and Meher-Homji (1990) and classified the subcontinent into 16 phytogeographical zones. Rodgers and Panwar (1988) tried to distinguish biogeographic regions based on faunistic information. However, their classification is also predominantly dependent on plant community distribution, based on the assumption that plant species components provide key even to identify the probable habitat/presence of animals. Their classification includes 10 biogeographic zones and 26 biotic provinces. Rodgers et al. (2002) have further revised their maps by, using GIS techniques, again projecting 10 zones and 26 provinces. This classification was done including various physical factors such as altitude, moisture, topography, rainfall, etc. Considering that the natural distribution of the wild species related to crops/cultivated species is influenced by geography and associated biological and physical factors, their distribution is again found associated with commonly agreed ten biogeographic zones. However, Singh (2017) classified the Deccan Peninsula into Indian Peninsula and Eastern Ghats to accommodate the considerable physical variation found in Eastern Ghats in forest and vegetation types as per altitude.

The centers of origin of crop/cultivated plant species, most frequently, are also centers of diversity of the wild relatives from whom the cultivated species were domesticated and brought into cultivation. As related wild species can participate in further genetic expansion of cultivated species by providing gene(s) for introgression through either conventional methods or with assistance from modern biotechnological tool, they become one of the crucial and important components of agrobiodiversity and a reservoir of plant genetic resources (PGR). They are source of gene/allelic variation for use in genetic improvement of cultivated species (crop improvement) by the present and future generations. Therefore, they need to be valued and protected. They enlarge the spectrum of PGR encompassing genetic diversity of the available gene pool beyond the genetic diversity of cultivated species. The present context of wild relatives of cultivated species indigenous to India will include:

1. *Landraces*: Primitive or wild forms of the cultivated species evolved over times through natural selection, without the human intervention or the processes of modern breeding or selection. Local names may be known for them based on uniqueness, but they are genetically highly diverse. They will include wild, cultivated, semi-cultivated, marginally cultivated, and diverse forms of cultivated or economic species exploited/harvested from nature. As per the taxonomic classification, these are weedy races that are the close companions of cultivated plant species and together form a genetically compatible species complex. As per gene pool concept, they will be part of primary gene pool group B.
2. *Wild progenitors*: These will include the genetically distinct closest ancestral wild relatives and encompass freely cross-compatible wild species and subspecies or may be the botanical varieties of the cultivated species.
3. *Wild/weedy relatives*: It shall include genetically distant congeneric wild species belonging to the same genera. They may be reproductively isolated to different degrees. They constitute an important but scarcely exploited component of the gene pool of domesticated species available to plant breeders. They have extensive potential and have been used in resistance breeding. Based on gene pool concept of taxonomic/generic affinity, they may fall in the category of secondary or tertiary gene pool.
4. *Wild cross-incompatible species*: These consist of those congeneric wild species that are genetically distant from the cultivated species to the extent that they are nearly reproductively isolated and either do not cross or weakly cross directly with the cultivated species. However, they may cross with other congeneric species, providing a possibility/opportunity for genetic introgression through bridging technique or the parasexual biotechnological approaches. They are classified

as quaternary gene pool and are part of outermost limits of genera (Singh and Nigam 2016).

5. *Wild compatible species*: Cross-compatibility relationship is one of the criteria of gene pool concept and applied for tracing phylogenetic relationship and genetic distance between wild and cultivated species. The plant species belonging to other genera, but phylogenetically close and cross-compatible with cultivated species offering opportunity of gene transfer by conventional hybridization with or without assistance of nonconventional approaches, become integral part of available wild genetic resource, for example, in case of cereal species belonging to *Aegilops*, *Elymus, Elymordeum*, and *Eremopyrum* for wheat and barley; *Rhynchosia*, *Dunbaria*, and *Eriosema* in case of pigeonpea; and *Sclerostachya*, *Miscanthus*, and *Narenga* in case sugarcane.
6. *Wild useful species*: It includes wild relatives of those plant species that are not yet domesticated but extensively used by human beings under the discussed cultivated group as per their basic value. This is particularly true in case of medicinal and aromatic plants, ornamental plants, and forage plants and fruit and vegetable crops. Therefore, they may be potential genetic resources of economically exploited species.
7. *Wild species of potential uses*: It refers to those plant species which are neither domesticated nor extensively used but possess great economic potential for future domestication and cultivation. They are considered genetic resource, based on their economic value. They are more common in the plant groups of ethano-medicinal value, multipurpose tree species, etc. that can be used in agroforestry, silviculture, industrial crops, etc.
8. *Wild species naturalized to India*. These are wild relatives of cultivated species known to be of foreign origin (Africa, Americas) and do not justify their inclusion as per the scope of the present effort. However, many of them are now grow wild in India, either because of their extended distribution range, caused by the natural forces of dispersal, or getting escaped and naturalized to Indian Subcontinent to appear indigenous after introduction centuries or millenniums back (Singh 2016). They have thrown a great amount of genetic diversity in the wild to the extent of evolving new species, subspecies, and botanical varieties, thus further contributing toward the expansion of the plant genetic resources for a specific species. In this regard, wild species related to the temperate fruits, found in Himalayan region, present the best example.

A recent review on the use of crop wild relatives in plant breeding (Maxted and Kell 2009) reported that for 29 major crop species, only 183 wild taxa, including from allied genera, had been used in breeding. In addition, others have argued that in the future, plant breeding will require broadening of the genetic base, if we are to meet the challenges of climate change and production demands for the twenty-first century (Feldman and Sears 1981; Gepts 2004), thereby meaning greater use of wild species in crop improvement.

Thus, the wild relatives of crop/cultivated species will include taxa ranging from immediate wild ancestor of the domesticated plant to species from another closely related taxon. They can be defined as the wild plant taxon that has an indirect use, derived from its relatively close genetic relationship to a crop/cultivated plant. Therefore, they constitute an increasingly important alternative genetic resource to help improve genetic potential of crop/cultivated species. They have a role in maintaining sustainable environment, agroecosystems, and agricultural production. Therefore, crop wild relatives are the wild plants related to socioeconomically important plant species including food, fodder and forage crops, medicinal plants, spices and condiments, ornamental, and forestry species. They shall include plants used for industrial purposes, such as oils and fibers and those which can contribute beneficial traits.

In addition to using wild species in crop improvement, people also gather certain species from the wild on regular basis and eat them directly, as in the case of minor fruits, or cook them as food, as in the case of many of the leafy or tuberous vegetables. Besides, many plant species of ethano-medicinal value are harvested from the wild and used in the local traditional herbal medicinal systems of rural or tribal health. In Indian context, these are medicinal plants used under the Ayurveda, Siddha, Unani, and Tibetan medicinal systems. These can be sold with non-timber forest produce, providing livelihood support to rural households with an additional source of income. Some of these have also been included among wild relatives, because of their wild habitat. They need protection for sustainable economic use.

References

Feldman M, Sears ER (1981) The wild gene resources of wheat. Sci Am 244:98–109

Fuller DQ (2006) Agricultural origins and frontiers in South Asia: a working synthesis. J World Prehist 20:1–86. doi:10.1007/s10963-006-9006-8

Fuller DQ, Korisettar R, Venkatasubbaiah PC, Jones MK (2004) Early plant domestications in southern India: some preliminary archaeobotanical results. Veg Hist Archaeobotany 13:115–129

Gadgil M, Meher-Homji VM (1990) Ecological diversity. In: Daniel JC, Serrao JS (eds) Conservation in developing countries: problems and prospects, Bombay Natural History Society. Oxford University Press, Delhi

Gepts P (2004) Crop domestication as a long-term selection experiment. Plant Breed Rev 24:1–44

Harlan JR (1975) Crops and man, 2nd edn. American Society of Agronomy, Madison

Lallanji G, Srivastava VC (ed.) (2008) History of agriculture in India (up to C.1200 AD) Vol. V. Part I. In: History of Indian Science, Philosophy and Culture in Indian Civilization (Ed: Chattopadhyaya G) p 946 New Delhi, PHISPC (Center of Studies in Civilization)

Maxted N, Kell SP (2009) Establishment of a global network for the *In Situ* conservation of crop wild relatives: status and needs. FAO, Rome

Murphy C, Fuller DQ (2014) Plant domestication in India. In: Encyclopedia of the history of science, technology and medicine in non-western cultures. Springer, Dordrecht, pp 1–8. doi:10.1007/978-94-007-3934_10005-1

Rodgers WA, Panwar HS (1988) Planning a wildlife protected area network in India. Vol. 1 and 2 A report prepared for the Department of Environment, Forests and Wildlife. Government of India at the Wildlife Institute of India, Dehra Dun

Rodgers WA, Panwar HS, Mathur VB (2002) Executive summary. In: Wildlife Protected Area Network in India: a review, Wildlife Institute of India, Dehra Dun, India, p 44

Singh AK (2016) Exotic ancient plant introductions: part of Indian 'Ayurveda' medicinal system. Plant Genet Resour C 14(4):356–369

Singh AK (2017) Revisiting the status of cultivated plant species Agrobiodiversity in India: an overview. Proc Indian Natn Sci Acad 83(1):151–174. doi:10.16943/ptinsa/2016/v82/48406

Singh AK, Nigam SN (2016) *Arachis* gene pools and genetic improvement in groundnut. In: Rajpal VR, Rao SM, Raina SN (eds) Gene pool diversity and crop improvement. Springer, Cham, pp 17–77

Tanksley SD, McCouch SR (1997) Seed banks and molecular maps: unlocking genetic potential from the wild. Science 277:1063–1066

Udvardy MDF (1975) A Classification of the Biogeographical Provinces of the World. ICUN Occasional Paper No. 18, Morges, Switerzerland, p 49

Vavilov NI (1935) The phytogeographical basis of plant breeding. In: Love D (ed) (Transl), Origin and geography of cultivated plants vol 1 pp 330–336, University of Cambridge Press, Cambridge, 1992

Zeven AC, De Wet JMJ (1982) Dictionary of cultivated plants and their regions of diversity. Center for Agricultural Publishing and Documentation, Wageningen

Zeven AC, Zhukovsky PM (1975) Dictionary of cultivated plants and their centers of diversity. PUDOC, Wageningen

Importance of Wild Relatives as Genetic Resource and Otherwise 3

3.1 Introduction

Man has exploited more than 7000 plant species for food and agriculture; of these, 150 species have entered commerce. The main crops that produce food are wheat, rice, corn, barley, sorghum, millet, soybean, groundnut, potato, tomato, grape, sugarcane, banana, watermelon, etc. Wild relatives of crop species in crop improvement have been tried in most commercial crops through interspecific hybridization with varying success. Predominant importance of wild relatives revolves around their capacity to contribute many useful genes and the opportunities they offer for transfer of desirable genes into their congeneric crop/cultivated species modern varieties. Consequently, most major crop cultivars contain genes from their wild relatives. Crop wild relatives are the wild plants that are genetically close to crop/cultivated species. They continue to evolve in wild, with genes conferring traits to resist biotic and abiotic stresses, such as resistance to drought or pest and diseases, facilitating their continued survival even under adverse conditions. These genes can be and have been introgressed into the domesticated cultivated species, by the farmers and breeders through conventional recombinant breeding or modern biotechnological approaches to produce new cultivars with superior genetic potential. Further use shall help face the vagaries of the nature and provide greater economic yield. Thus, wild species have an important role to play in meeting the challenges for the twenty-first-century agriculture, in facing adverse effects of climatic changes, and in finding solution to food and nutritional security pressure of ever-growing population.

Farmers have often been planting such species alongside domesticated crops to promote natural crossing for beneficial traits. Genes from wild plants have provided cultivars with resistance against pests and diseases and improved tolerance to abiotic stresses. They have also been used to improve the yields and nutritional quality of crops since the beginning of agriculture by the traditional farming communities.

Breeders, in scientifically designed breeding programs, have also often concentrated on involving wild relatives of crops to handle problems of pest and diseases

A.K. Singh, *Wild Relatives of Cultivated Plants in India*,
DOI 10.1007/978-981-10-5116-6_3

and most recently the complex characteristics related with product quality and husbandry traits. It was estimated that about 30% of increased crop yields in the late twentieth century, since 1945, valued worldwide at around US$ 100 billion, can be attributed to the use of wild relatives of crop in the breeding programs (Pimentel et al. 1997). Around 39% of use of wild relatives is associated with improving disease resistance, 17% with pest resistance, 13% with abiotic stress, 10% with yield increase, 11% with quality improvement, 6% with husbandry improvement, and 4% with cytoplasmic male sterility (CMS) and fertility restoration to promote hybrid breeding technology (Maxted and Kell 2009). Hajjar and Hodgkin (2007) reviewed the use of crop wild relatives by Consultative Group on International Agricultural Research (CGIAR) in 16 of their mandate crops. They found that over 80% of usage has been for disease and pest resistance. However, in 13 of the 16 mandate crops, some traits besides resistance have been successfully transferred from crop wild relatives, representing a rise in the usage of wild relatives in breeding since an earlier review by Prescott-Allen and Prescott-Allen (1986). From 2011 onward, Springer has brought out a series of ten volumes on *Wild Crop Relatives: Genomic and Breeding Resources* edited by Chittaranjan Kole (2011), dedicated to wild relatives of cereals, millets and grasses, oilseeds, legume crops and forages, vegetables, temperate fruits, tropical and subtropical fruits, industrial crops, plantation and ornamental crops, and forest trees. The chapters of these volumes are devoted to descriptions of the geographical origin, distribution, genetic diversity, karyotype and genome size, morphology, and available genetic and genomic diversity of numerous wild relatives of crop species, as well as the information on their evolution and phylogenetic relationships. Each chapter has been authored by the pioneers of the exploration, description, and utilization of the wild species of those genera. Another review by Brumlop et al. (2013) of 104 molecular-assisted breeding papers published from 1995 to 2012 found that approximately 74% of these studies were focused on introgression of traits that confer disease resistance, whereas the rest focused on traits involved in abiotic stress tolerance, improved yield, and growth habit.

However, these reviews reveal that much of the examples of wide hybridization in crop breeding have been ad hoc in the usage of wild germplasm. None of these efforts has screened the total spectrum of variability of existing wild relatives to face the challenges of the twenty-first century. They are confined to few wild species and a few traits, and none has used crop wild relative collections that were systematically built to represent the range of adaptations found in natural populations, thereby demanding an urgent focus on systematic detailed study on wild relatives of crops to identify gaps and promote use in crop improvement programs.

3.2 Wild Species as Source of Genetic Enhancement to Broaden Crop Genetic Base

Distant hybridization is one of the key factors in plant evolution, including crop plants. However, in most crops, it has been observed that most cultivars are based on relatively narrow genetic diversity. An estimated 75% crop genetic diversity has been lost in the twentieth century. It is likely that wide hybridization with wild genetic diversity will play an increasing role in pre-breeding, with genetic enhancement and conversion of common crop breeding gene pool with greater variability, incorporating wild genetic diversity. This will help facilitate sustainable crop improvement and evolution of new cultivars. Major advances in breeding of crops will require a broadening of the genetic base, which will include crosses between distant species, including the ones from different genera. Therefore, there is a strong case for documenting and conserving a much wider plant diversity, including wild related species from all geographical regions of the world, than among those plants which are of the same species. The potential for genetic gains from use of crop wild relatives is well documented (Pimentel et al. 1997; Maxted and Kell 2009).

In sugarcane, utilization of wild relatives has been an integral part of sugarcane improvement since the late 1800s, especially for the enhancement of genetic variability incorporating disease resistances into cultivars. Nearly all modern sugarcane varieties are derived from interspecific hybrid clones and contain three to five species in their pedigree (Price 1963). Increases in chromosome number are a common feature when species of *Saccharum* are hybridized; consequently most modern varieties have very high aneuploid chromosome numbers ranging from 2n = 100 to 125 (Price 1963). Since vegetative propagation is practiced for commercial production, fertility is not an issue in hybrid sugarcane clones. The increase in chromosome number is important for adding buffering capacity and the desired characters from wild species because there is little gene exchange between *S. officinarum* and another species of the genus (Price 1963). Hence, in sugarcane most varietal products are of interspecific hybrid origin, where wild species clones were crossed with commercial canes with the objective of genetic enhancement leading to the production of a large number of hybrid clones. These clones were utilized in developing cultivars like Co7908 and Co8318 in the Indian program from 1979 onward (Sreenivasan and Amalraj 2004). Similarly, regarding species from other genera, such as *Erianthus* and *Sclerostachya*, promising hybrids were obtained, leading to the transfer of tolerance to drought and salinity and heavy tooling from *Erianthus* species and tolerance to water logging from *Sclerostachya fusca*. Since 1980, crosses in the Indian breeding program involving *Erianthus* have been made. For example, Co87003 was a hybrid involving Co6304 and *E. arundinaceus*. Encouraged with the results of such efforts, crosses involving Co7201 and several clones of *Erianthus* were made (Sreenivasan and Sreenivasan 2000). Most hybrids obtained were huge and tall with good stalk weight, stalk diameter, brix percentage, and sucrose content and were male fertile. These hybrids at different stages of nobalization are being evaluated. These results have led sugarcane breeders' world over to realize the problems of restricted genetic base and benefits of wide hybridization;

consequently, they are operating parallel basic hybridization programs to infuse new germplasm into commercial cane breeding programs. Pre-breeding with wild relatives in sugarcane for germplasm enhancement has become a regular feature in the development of commercial cane cultivars, with developing elite genetic stocks as the first step. These genetic stocks are expected to help in broadening the genetic base of commercial cultivars. Indian program maintains many such genetic stocks holdings at different research centers of Sugarcane Breeding Institute. The hybrid canes developed by breeders of different sugarcane research stations in India are being maintained in the National Hybridization Garden (Sreenivasan and Amalraj 2004).

The genus *Zea* (maize) has five species: *Z. mays* ssp. *mays* (cultivated maize) and *Z. mays* ssp. *mexicana* (annual teosinte), *Z. perennis*, *Z. diploperennis*, *Z. luxurians*, and *Z. nicaraguensis*. Maize and several of its sister wild subspecies, commonly called annual teosinte, hybridize naturally. However, because of some genetical problems, their use in maize improvement has been limited. *Tripsacum* species have been found good resource for improvement of maize. Hybridization involving both diploid and tetraploid *Tripsacum* with maize, followed by backcrossing, has shown elimination of chromosomes (Harlen et al. 1970). Stalker et al. (1977, 1978) concluded that the probability of gene transfers between maize and *Tripsacum* chromosomes was high during the early generations of backcrossing. Hybrid derivatives of tetraploid *Tripsacum* x diploid maize are being used in genetic enhancement leading to their use in commercial breeding. *Zea mays* L. is crossed with species of *Tripsacum* resulting in production of highly complex hybrids. Highly tripsacoid maize with $2n = 20$ were recovered. They were found with genetically dominant resistance to six common maize diseases, anthracnose, fusarium stalk rot, northern and southern corn leaf blight, rust, and Stewart's bacterial blight. These resistances were successfully transferred into commercial maize inbreds. Other useful traits transferred were gametophytic apomixes, weak perennialism, a tillering habit, an increase in inflorescence bearing clum nodes, and an increase number of fully developed ears at each node (de Wet 1979).

There have been efforts of genetic enhancement through interspecific breeding in several other cases, for example, interspecific hybridization efforts at ICRISAT in groundnut breeding program from 1978 to 1989, using many wild diploid *Arachis* species belonging to different gene pools. They produced 209 *A. hypogaea*-like interspecific derivatives, involving cross-compatible diploid species from secondary gene pool, incorporating genes conferring resistance to various groundnut diseases and pest, resulting in production of lines with higher haulm yield and resistance to foliar, viral, and insect pest and multiple diseases (Singh 1989; Singh and Nigam 2016). Subjection of these interspecific derivatives for evaluation against various stress factors and agronomic potential resulted in selection of lines with resistance to various diseases and cultivars such as ICGV 86775. Similarly, in *Vigna* spp., interspecific populations were raised, which were subjected to screening against diverse stress factors to select desired recombinants with resistance potential against target stress (Bisht and Singh 2013).

3.3 Wild Species as Source of Genetic Variability for Crop Improvement

3.3.1 Disease and Insect Pest Resistance

By far the most common reason for attempts of utilizing wild species is to transfer disease resistance. Most successes have been with single genes where adequate selection pressure could be applied. Knott and Dvorak (1976) presented an extensive bibliography of specific examples, where wild species have been used in resistance breeding. The cultivated hexaploid wheat, *Triticum aestivum*, contains three different genomes found in the ancestral wild related species and is a classic example of utilization of wild species in crop improvement through interspecific hybridization and genomic reconstitution. Sears (1969, 1975) reviewed the extensive work done in wheat cytogenetics and interspecific hybridization, which includes transfer of genes through direct hybridization, induction of homoeologous pairing, and use of irradiation to translocate alien chromosome segments. The induction of homoeologous pairing is by far the easiest method of transferring genes (Sears 1972; Riley et al. 1968), but induced translocations have also produced favorable results (Sears 1956; Knott 1961; Sharma and Knott 1966; Johnson 1966). Substitution lines have also been successfully utilized for transferring genes into the cultivated wheat genomes. Consequent to these efforts, stem rust resistance in wheat was introduced from *Triticum timopheevii* and *Agropyron* spp. and gene for eye-spot resistance from *Aegilops ventricosa* (Knott 1978). The rice bacterial blight disease-resistant gene Xa21 from *Oryza longistaminata* has been transferred to *O. sativa* (Khush et al. 1991), and the resulting progenies have been widely used for breeding new varieties resistant to *Xanthomonas* infection. A major dominant gene for resistance to the grassy stunt virus found in *Oryza nivara* was also transferred into cultivated rice (Khush et al. 1997). Blast resistance and insect resistance have been successfully transferred from wild species to cultivated rice (Amante-Bordeos et al. 2004). In maize, the corn blight caused by *Helminthosporium maydis* that destroyed the crop in the southern United States (Stolton et al. 2006) was overcome with introduction of blight resistance genes into the crop from wild Mexican maize plants (Shands 1993).

In grain legumes, breeders have successfully introgressed genes conferring resistance to cyst nematode from *Cicer reticulatum* and for cold tolerance from *C. reticulatum* and *C. echinospermum* in chickpea (Singh and Ocampo 1997). Genes for resistance to insects (e.g., bruchid beetle seed predators, *Apion* pod weevils) and pathogens (e.g., *Fusarium*) and higher nitrogen, iron, and calcium seed contents have been harnessed from existing collections of wild *Phaseolus* (Acosta-Gallegos et al. 2007). These efforts have contributed to both higher yields and improved nutritional quality and have lessened the environmental impact of crop production by facilitating reduced pesticide, herbicide, and fertilizer use. Reviewers have argued persuasively that the wild diversity can be better used if converted or incorporated into the common bean breeding pool.

Interspecific hybridization efforts at ICRISAT for transfer of foliar disease resistance from cross-compatible diploid wild *Arachis* species, such as *A. cardnesii*, into cultivated groundnut successfully produced *A. hypogaea*-like interspecific derivatives incorporating genes conferring resistance to foliar diseases (Singh and Gibbons 1985). Many of these have been registered as genetic resource and released as variety (Singh and Nigam 2016). They became the basis of resistance breeding worldwide. The resulting increase in harvest due to this work is estimated to be worth US$ 500 million (Sasson 1996).

Among vegetables, potato is the pioneer crops, which benefited most from the use of wild species to overcome biotic stresses and sustain productivity. The genus *Solanum* contains more than 2000 species. Several methods have been employed to incorporate wild and cultivated species germplasm into *Solanum tuberosum* overcoming the ploidy/genomic barriers. Direct hybridization and backcrossing with recurrent *S. tuberosum* parent have been used most commonly. However, bridge crosses, the use of mentor pollen (mixture of treated compatible and incompatible pollen), chemical treatments to promote growth of pollen tube or to support fertilized zygote, hybridization of induced autotetraploids with the cultigen (Livermore and Johnstone 1940), and haploid breeding techniques (Hougas et al. 1958) have also been used for gene introgression. Chase (1963) and Hanneman and Ruhde (1978) reviewed haploid breeding in potatoes in which haploids are produced from tetraploid potatoes, hybridized with selection at the diploid level, and polyploids are then resynthesized. Consequently, most potato varieties have two or more wild species in their pedigree (Hougas and Ross 1956). *Solanum demissum* carries genes conditioning resistance to late blight, leaf roll, and virus X (Ross 1966) and has been the most common germplasm resource for potato improvement. Other potato species have been used in variety development for resistances to viruses A, X, and Y, *Leptinotarsa decemlineata,* and nematodes (Ross 1966). Using the dihaploid of subspecies *tuberosum*, durable resistance to late blight has been transferred from diploid *S. verrucosum* and *S. microdontum* into *S. tuberosum* (Sharma et al. 1982; Birhman et al. 1991), whereas resistance to charcoal rot was transferred from *S. chacoense* to *S. tuberosum* (Upadhya et al. 1977) while to cyst nematode was derived from *S. vernei*. Jansky (2000) reviewed the interspecific breeding efforts for diseases resistance in potato. Resistance to cyst nematode was also introgressed in sugar beet by interpolid and autotetraploid hybridization (Jung et al. 1986).

In the tomato, *Lycopersicon esculentum* has been crossed with two wild relatives, *L. hirsutum* and *L. peruvianum*, for fungus resistance and with *L. peruvianum* for nematode worm resistance. Utilization of the rootstocks of wild species has also helped in elimination of many diseases and insect pests of plant species that are now commonly grafted in horticulture crops, such as grape, citrus, and rubber (Wellman 1972).

3.3.2 Abiotic Stress Tolerance

Comparatively, there are few examples of wild relatives contributing to abiotic stresses. The Indian Agricultural Research Institute developed BG1103 involving *Cicer reticulatum* with tolerance to drought and temperature. Cold tolerance has been successfully transferred from wild species of peppermint (Bugaenko et al. 1975), tomato (Robinson and Koualewski 1974), grape (Filippenko and Lebedev 1971), strawberry (Jones 1976), wheat (Kuvarin 1973), rye (Yakovlev 1972), onion (Meer 1975), and potato (Ross 1966) into their cultivated relatives. *Oryza rufipogon* Griff. genes have been exploited for tolerance to acidic sulfate (Nguyen et al. 2003) and *O. longistaminata* A. Chev. & Roehrich genes for drought tolerance (Brar 2005). *Aegilops speltoides*, a wild ancestor of bread wheat, *T. aestivum*, has provided wheat with tolerance to drought, heat, salinity, and water logging. Mickelbart et al. (2015) have recently reviewed the use of wild relatives in increase of tolerance to abiotic stresses such as resistance to water submergence, drought, and salt.

The use of wild relatives in crop improvement to increase tolerance to biotic and abiotic stress has contributed to improved productivity in most major crops. Common examples of crops/crop groups in this regard are wheat (Singh et al. 2012a), rice (Gothandam 2012), maize (Singh et al. 2012b), vegetables (Kunchge et al. 2012), grain legumes (Reddy et al. 2012), sugarcane (Shrivastava and Srivastava 2012), fruits (Nath et al. 2012), and oil crops (Gill et al. 2012; Lakhanpaul et al. 2012).

3.3.3 Yield Traits

Increase in economic yield is generally either due to an increase in the production of vegetative parts or an increase in seed yields. In addition, it can also be due to increased output of other economic parts, for example, fiber length in case of cotton (Schwendiman and Lefort 1974) and tuber production in case of potato (Tarn and Tai 1977), obtained through interspecific hybridization involving wild and cultivated species. In a recent effort, three independent yield-promoting genomic regions introduced from *Solanum pennellii* has been pyramided, leading production of hybrids with 50% yield increase over a leading variety (Gur and Zamir 2004).

The probability of increasing seed yield from utilization of wild species was seen to be more unlikely. Nevertheless, Frey (1976) reported that *Avena sativa* x *A. sterilis* resulted in a 25–30% seed yield increase over the recurrent parent. After this, there have been several examples of wild species contributing to increase in yield. Recently in wild relatives of tomato, a quantitative trait locus (QTL) has been identified contributing to increase in yield (Tanksley and McCouch 1997). The best example of obtaining yield increases utilizing wild species germplasm is in commercial production of sugarcane (Sreenivasan and Amalraj 2004).

3.3.4 Quality Traits

Wild species germplasm has had limited use in quality characteristics, because of the genetic complexity of most quality traits. However, wild grasses usually have a higher protein percentage than related cultivars (Harlan 1967). Protein quality and quantity has been altered in several species of the Poaceae with incorporation of quality traits. Varieties of durum wheat, *T. durum*, with improved protein content have been bred by crossing it with wild relative, *T. dicoccoides* (see review by Maxted and Kell (2009). Alleles from *Oryza rufipogon* increased grain weight in Hwaseongbyeo, a Korean rice cultivar (Xie et al. 2006). QTLs for yield and yield components and other agronomic characters were identified in *O. rufipogon* (Moncada et al. 2001; Septiningsih et al. 2003) and transferred to elite rice varieties. High seed protein content has been incorporated from cross-compatible wild *Cajanus* species, such *C. albicans*, *C. sericeus*, and *C. scarabaeoides*, into cultivated pigeonpea (Reddy et al. 1979) and in groundnut from *Arachis cardanesii* (Guok et al. 1986). In tomatoes, the soluble solid content of commercial varieties has been increased substantially by hybridizing cultivated varieties with a wild green-fruited species (Rick 1974). Crossing with *Lycopersicon chmielewskii* resulted in 2.4% increase in solid content of the fruit, worth an estimated US$ 250 million in California alone (Esquinas-Elcázar 1981). Many fruit quality parameters were incorporated from wild *L. cheesmanii* (Rick 1982). Starch content has been improved in potato tubers by utilizing wild species germplasm (Osipchuck 1970). Thus, wild relatives may provide important options for enhancing the nutritional profile of agricultural species.

3.3.5 Earliness and Adaptation

The variability found in wild species offers a valuable resource for genes conferring earliness and wider adaptations. In wheat, gene for adaptation to low fertilizer input conditions has been transferred from *Triticum polonicum* to durum wheat cultivated in rainfed areas of the Mediterranean (Doussinalt et al. 1983). Ram et al. (2007) indicated that there was a great scope to introgress such genes from wild relatives of rice into cultivars. In tomato, crossing of *Lycopersicon esculentum* with *L. cheesmaniae* has helped improved adaptation to drier conditions. Wild relatives found in a wide range of environments provide a source of genotypes adapted to new or changing environments in agricultural production regions (Henry 2014) and can be utilized for breeding crops which can grow in environments adversely affected by climate change (Redden 2015).

3.3.6 Modes of Reproduction

Male sterility, "deficiency in producing viable pollen," is one of the most common alterations in reproduction mechanism that results from interspecific or intergeneric crosses due to alloplasmic interaction. It is economically one of the most important derivatizations that helps exploit heterosis and facilitate hybrid seed production without cumbersome emasculation and hybridization, supporting cost-effective hybrid seed production technology with significant yield gains. Cytoplasmic male sterility using wild relatives has been obtained in many crops, for example, transfer of cytoplasmic male sterility from the wild *Helianthus petiolaris* into cultivated sunflower and from *Oryza rufipogon* and *O. perennis* into rice (Dalmacio et al. 1995; Xian-Hua et al. 2013). Breakthrough in hybrid pigeonpea production was achieved through the development of two CMS sources—A2 system (derived from *C. scarabaeoides*; Tikka et al. 1997) and A4 system (derived from *Cajanus cajanifolius*; Saxena et al. 2005). Dionne (1961) and Grun and Aubertin (1965) indicated that genes conditioning cytoplasmic male sterility could be extracted from several *Solanum* species for induction of male sterility in potato. Introduction of cytoplasmic male sterility to cultivars from wild species is one of the potentials offered by wild *Gossypium* germplasm. Other crops, where cytoplasmic male sterility was produced because of the cross between wild and cultivated species, include wheat, barley, *Brassica*, tobacco, ryegrass, potato, etc. Harvey et al. (1972) reviewed the uses and potentials of cytoplasmic male sterility for several crops.

3.3.7 Miscellaneous Traits

In addition to the above desirable features, the wild species have also contributed to incorporation of miscellaneous features, such as hard-seededness to cultivated bramble fruits, color and leaf texture to lettuce, and carotenoid synthesis gene and delayed ripening of fruits to cultivated tomato (Rick 1967). Plant stature and other morphological feature traits have also been introgressed from wild species. For example, Hurtado and Ramos (1970) reported interspecific hybrid in oil palm producing dwarf plants. Selection among *Triticum* x *Agropyron* hybrids resulted in semidwarf wheats (Vasilenko 1973). In rice, new varieties are being released based on the "new plant-type" concept, which has been supplemented with genes from wide crosses involving hybrids with wild *Oryza longistaminata* and *O. rufipogon*.

3.3.8 Quantitative Traits

Crop wild relatives have been established as a source of quantitative trait locus (QTL) that particularly have potential to improve the yields of many crop plants (Swamy and Sarla 2008). However, unlike direct introgression of qualitative characteristics, which is supported by backcrossing for incorporation of major gene, quantitative traits governed by several loci (polygenes) that are scattered over several chromosome segments present difficulties. Foremost in these is the difficulty in breaking their association with undesirable wild traits. In these cases, indirect selection for part components can help improve transfer methods, and there is reason to expect that the progress in this direction using biochemical or molecular markers should be the only answer. QTL in wild soybean has indicated potential for contributions to improved yields (Li et al. 2008). Yield QTL have been identified in inbred lines (Ils) originating from a cross between the green-fruited species *Lycopersicon pennellii* and the cultivated tomato (Eshed and Zamir 1995), in crosses between cultivated and wild pepper (Rao et al. 2003) and wild bean (wild *Phaseolus vulgaris*) (Blair et al. 2006). QTL for increased yield have been identified and transferred into cultivated rice (*O. sativa*) from *O. rufipogon* (Septiningsih et al. 2003). New sources in wild rice germplasm have recently been identified (Marta et al. 2015).

However, despite the above established credentials of wild relatives in the transfer of desirable traits, still questions are asked on the access to the genetic diversity, genotype interaction during introgression from wild germplasm, number of backcrosses, when does phenotypic screening be performed, how much wild germplasm be screened to find novel or useful genes, will analysis of the climate of origin help in selecting the most appropriate germplasm, etc. There are no generic answers to these questions and the approaches will vary. Genomics does provide the tools to evaluate all these questions (chapter 20 has answers to some of the questions).

3.4 Resource for Other Economic Uses

Besides being the genetic resource for crop improvement, the wild relatives of crop species can also be a source of several other by-products and economic and ecological gain.

3.4.1 Source of Alternative Food Crops

Over centuries Indian tribal and farming communities, in search of food, have developed vast amount of knowledge on food uses of wild species, particularly the sister wild species of crops/cultivated species. For example, in Bastar region of Chhattisgarh, India, only, the local people/tribes have identified many alternative sources of food. Around 88 plant species have been identified, many of which are used in their regular diet (Jain 1964). Of the 88 plants used by the local people for food, around 50 are vegetables, 29 fruits, 8 nuts, 6 beverage and drinks, 4 grains, 3 oilseeds, etc. Most belong to cultivated genera such as *Bauhinia* (3), *Dioscorea* (7), *Diospyros* (2), *Ficus* (3), *Grewia* (3), *Zizyphus* (3), etc., with significant genetic diversity offering a reservoir of alternative food resource both for direct consumption and as genetic resource (Jain 1964). Similarly, in Leh Ladakh and its adjacent areas in the Tibetan Plateau, which are characteristic high-altitude cold deserts under the rain shadow of the Himalayas, many species belonging to cultivated genera are harvested from nature and used as vegetable. Examples are *Amaranthus spinosus* Willd., *Smelowskia tibetica* (Thomson) Lipsky (syn. *Capsella thomsonii* T. Anderson), *Allium carolinianum* DC (syn. *A. thomsonii*), *Lactuca dolichophylla* Kitam, *Chenopodium foliosum* (Moench) Asch., *Lepidium latifolium* L., *Orobanche hansii* A. Kern., and *Polygonum aviculare* L. (Singh 2015). These can be further evaluated and domesticated based on their potential.

3.4.2 Sources of Ethano-medicinal and Floriculture Crops

The forest regions of Indian Subcontinent are known for medicinal plant diversity at the global level. Around 9000 species of ethano-medicinal value have been reported to be distributed in the different parts of the Indian Subcontinent. Some regions such as Himalayas, Northeast hills, Bastar region of Chhattisgarh, etc. are known for richness in ethano-medicinal plant diversity. Many of these have been documented in the local medicinal system like Ayurveda (1100–1270) from times immemorial (Singh 2016), while many more are being reported in different studies. For example, around 70 and 266 medicinal plant species have been listed by Uniyal (2006) and Chaurasia et al. (2008), respectively, from Ladakh and its adjacent areas in the Tibetan Plateau. This richness of diversity for medicinal properties in the plants of the region has led to the development of globally known Tibetans' medicine system called *Amchi*. Dr. HF Muni carried out a survey of the medicinal plants of Dantewada area in Bastar region between 1938 and 1943 and was awarded a PhD for the survey and documentation of the medicinal plants. It was recorded as the eighth richest area in the world. The region is full of diverse terrains, and much of the forest remains unexplored; it is highly probable that this area may be containing many undocumented species. Mr. Bharat, a highly decorated retired forest officer, listed a total of 166 plants with medicinal properties in the database created by him (www.dantewada.gov.in/medi.htm).

Floristic survey studies have reported many of the wild relatives of crops/cultivated species with potential source of ethnomedicine. For example, *Corchorus depressus* (L.) Stocks (syn. *Corchorus antichorus* Raeuch.), *Cynodon dactylon* (L.) Pers., *Desmodium triquetrum* (L.) DC., *Momordica balsamina* L., *Mimusops elengi* Linn., *Myrica esculenta* Buch.-Ham. ex D. Don, *Solanum anguivi* Lam. (syn. *S. indicum* L.), *S. erianthum* D. Don, *S. mammosum* L., *S. viarum* Dunal, and *Zingiber nimmonii* (J. Graham) Dalzell have been found to have medicinal properties. These need to be further evaluated from pharmaceutical point of view with identification of principal components. Based on the potential value, they can be either used directly in pharmaceutical industry or may be as genetic resource to improve the medicinal potential of related cultivated species.

Similarly, the Indian Subcontinent is known for richness of plant species diversity with attractive appearance (Khoshoo 1968). Many wild relatives have been reported to be potentially ornamental, such as *Allium sikkimense* Baker, *Alpinia malaccensis* (Burm. f.) Roscoe, *Areca triandra* Roxb. ex Buch.-Ham. (syn. *Areca nagensis* Griff), *Barleria grandiflora* Dalzell, *Begonia griffithiana* (A.DC.) Warb, *Colocasia affinis* Schott, *Crinum brachynema* Herb., *Curcuma kudagensis* Velay., V.S. Pillai & Amalraj, *Garcinia spicata* Hook., *Jasminum parkeri* Dunn, *Lactuca serriola* L. (syn. *L. scariola* L.), *Malva sylvestris* L., *Miscanthus nepalensis* (Trin.) Hack., *Musa velutina* Wendl. ex Drude, *Prunus arborea* (Blume) Kalkman,

Saccharum bengalense Retz., *Solanum erianthum* D. Don, *Zingiber rubens* Roxb., and *Zingiber spectabile* Griff. These can be further evaluated for their ornamental features and market value in comparison to existing ornamental species. Based on the results, they can either be brought into cultivation or used in genetic improvement of existing cultivated species to improve their attractiveness and/or productivity.

3.4.3 Sources of Renewable Bioenergy

The renewable energy can be produced from plant biomass, which has stored solar energy in the form of chemical energy/organic material. It may include wood, straw, manure, sugarcane, and by-products like seed oil. Traditionally, grain and oilseed crops and their crop residues, perennial herbaceous and woody crops, perennial oilseed plants, halophytes, etc. are used for raw material. However, now there are specific group of crop species grown for biofuel production, such as corn, wheat, barley, soybean, sugarcane, switchgrass, and willow in different parts of the world. The production process of sugar and ethanol from sugarcane has been fully exploited in Brazil. Bagasse is currently burned at the mill to provide heat for distillation and electricity to run the machinery. In India, non-edible oil-producing plants such as *Pongamia*, *Jatropha*, *Ricinus communis* (castor), *Sapindus mukorossi*, and *S. trifoliatus* (soap nut) are being used for biodiesel production. Hemp has also been proven to work as a biofuel.

The plant species that are found suitable for bioenergy production have fast growth; low requirements for biological, chemical, and physical input; tolerance to biotic and abiotic stresses; and the capacity to produce large volume of biomass/energy on marginal soils. Both biomass and biofuels can be derived from dedicated bioenergy crops, agricultural produce, or waste materials. Switchgrass (*Panicum virgatum* L.), elephant grass (*Pennisetum purpureum* Schum.), poplar (*Populus* L. spp.), willow (*Salix* L. spp.), mesquite (*Prosopis* L. spp.), etc. have been reported as the crops with the most widespread promise.

Based on biomass production and their use as bioenergy crop, these crops are classified under the following categories (Priyanka 2013):

1. Traditional bioenergy crops: Local natural vegetation.
2. First-generation bioenergy crops: Food and field crops, such as corn, sugarcane, oil palm and rapeseed.
3. Second-generation bioenergy crops: Providing fuel from cellulose and non-oxygenated, pure hydrocarbon of their biomass. They include perennial forage crops, switchgrass, reed canary grass, *Medicago sativa* L., elephant grass, and *Cynodon* L. spp. (Oliver et al. 2009). Non-edible plant oils and soap nut are new source of biodiesel.
4. Third-generation bioenergy crops: They include boreal plants, C4 plants, *Eucalyptus* spp., and microalgae. They provide feedstocks for direct cellulose fermentation. *Eucalyptus* produce bioenergy through thermo-conversion,

whereas algae are a potential source of biodiesel. Other crops being tested are African palm, coconut, and grain of castor bean, peanut, etc.

5. Dedicated bioenergy crops: This included cellulosic plants, including short-rotation trees and shrubs. They are more environmentally friendly and can also contribute to mitigation of global climatic change (Petersen 2008) by providing certain ecosystem services. They include (1) trees such as eucalyptus (*Eucalyptus* L Her spp.), poplar (*Populus* spp.), willow (*Salix* spp.), and birch (*Betula* L. spp.); (2) perennial grasses such as giant reed (*Arundo donax* L.), reed canary grass (*Phalaris arundinacea* L.), switchgrass (*Panicum virgatum*), elephant grass (*Miscanthus x giganteus* Greef et Deu.), Johnson grass [*Sorghum halepense* (L.) Pers], and sweet sorghum (*Sorghum bicolor* (L.) Moench.); and (3) non-edible oil crops such as castor bean (*Ricinus communis* L.), physic nut (*Jatropha curcas* L.), oil radish (*Raphanus sativus* L.), and *Pongamia* or *Karanj* (*Pongamia* L. spp.).

Thus, the wild species related to local high biomass production of crop plants, such as grasses, agroforestry tree or shrub species, and perennial non-edible oil-producing species, discussed under respective crop groups can be the genetic resources to improve the potential bioenergy-producing characteristics of these species. Wild species of the most promising bioenergy crops indigenous to India, are discussed in the chapters of respective crop groups associated with their primary use. After systematic evaluation, they can either be exploited directly as the bioenergy plant and/or for introgressing bioenergy characteristics into these crops for dual or diversified exploitation.

3.4.4 Source of Restricting Soil Erosion, Soil Reclamation, and Soil Fertility Improvement

There are many wild species related with forage and agroforestry crops, particularly from the extreme arid and semi-arid ecologies and from the family Fabaceae/Leguminosae. They have been reported to possess economic potential for soil binding with the help of their extensive root system and improve fertility with their natural nitrogen fixation capacity via their symbiotic association with *Agrobacterium*. They formulate important genetic resource for direct exploitation or indirect use in genetic improvement of related cultivated species to facilitate harnessing of the additional benefits from their cultivation. For example, wild *Bothriochloa intermedia* (R. Br.) A. Camus, *B. pertusa* (L.) A. Camus, *Chrysopogon aciculatus* (Retz.) Trin., *C. hamiltonii* (Hook. f.) Haines, *Eragrostis curvula* (Schrad.) Nees, *Populus ciliata* Wall. ex Royle, and *Salix tetrasperma* have been found to help restrict soil erosion, whereas *Medicago lupulina* L., *M. monantha* (C.A. Mey.) Trautv (syn. *Trigonella polycerata* L.), and *Alnus nepalensis* D. Don have been found to possess potential for improving soil fertility. They can be directly used in restricting soil degradation and/or reclamation of degraded soils, besides other benefits.

3.4.5 Securing Ecological Benefits Through Carbon Sequestration, Phytoremediation, Promoting Ecology, and Environment

The wild relatives of short-rotation trees and shrubs used in agroforestry, of many tree species of medicinal value and/or used in landscaping or as avenue tree, and the grasses used in pasture lands can contribute to mitigation of global climatic change besides their primary economic produce (Petersen 2008). They are beneficial in providing certain ecosystem services, including carbon sequestration (capturing of CO_2 from atmosphere through biological process); salinity, contaminants (metals, pesticides, solvents), and toxicity mitigation; enhancement of soil and water quality; and biodiversity enhancement. These plants are used in biological process of carbon sequestration through their cultivation in reforestation and urban forestry programs, and use as cover crop and green manure crops, providing protection and enrichment of the soil, and restoration of degraded lands. Their plantation would virtually mean creation of artificial grasslands and woodlands. Forests have been found typically more than 10 times, as effective as grasslands at storing of carbon on per hectare basis.

The characteristic features required in a tree species suitable for carbon sequestration and other ecological benefits are:

(i) Fast growth, as storage of the most carbon occurs during their first decades, often the most productive period of a tree.
(ii) Long life that can facilitate storage of carbon for generations without releasing it in decomposition.
(iii) Large leaves and wide crowns enable maximum photosynthesis.
(iv) Native and higher adaption to local conditions that will help them thrive better in local soil and would provide best support for local wildlife.
(V) Low at input, maintenance, and resistant to biotic and abiotic stresses. Such species will do better without greenhouse-gas-producing fertilizers and equipments.

Some of the common species used in biological carbon sequestration program are *Acacia* spp., *Eucalyptus* spp., *Prosopis* spp., *Populus* spp., *Salix* spp., *Betula* spp., *Pinus* spp., Bamboo, etc. among trees; *Bothriochloa*, *Cenchrus*, *Cynodon*, *Dichanthium*, etc. in grasses; and *Desmodium*, *Lablab*, *Mucuna*, etc. in forage legumes. The wild relatives of these species have been documented under forage crops, agroforestry, and industrial crops, respectively. These can be evaluated for the above characteristic features and used directly or in genetic enhancement or introgression of genes required to improve carbon sequestration capacity in the respective crop/cultivated species for diversification or additional use.

Phytoremediation is the use of living green plants for in situ, or in place, removal, degradation, or containment of contaminants in soils, sludges, sediments, surface water, and groundwater. A number plant species has been found to accumulate various toxic substances; for example, *Brassica juncea* L. (Indian mustard) and

B. carinata A. Braun, *Hordeum vulgare* L., and Zea *mays* L. have been found to take up incredibly high levels of lead; *Helianthus annuus* L., arsenic; and *Salix viminalis* L., cadmium. Other species reported to have been used in phytoremediation are *Cannabis sativa* L., *Festuca arundinacea* Schreb., *Panicum virgatum*, *Pennisetum purpureum*, *Phalaris arundinacea*, etc. The wild relative of these species discussed under their basic group can be evaluated for phytoremediation properties and can either be used directly for greater effectivity or to improve the genetic potential of cultivated species used for phytoremediation.

The locally available economically potential wild relatives, particularly tree, shrub, and grass species, can also be used in biodiversity conservation through revegetation. This refers to strategies of planting native plants in remnant habitat areas and/or forest margins (Hobbs 1993). This would help restricting the degradation of lands. Revegetation can be done in buffer strips, corridors, patches, and planting of natural habitat strips within farming systems. Economically important wild species can provide additional produce/income to local communities/farmers, as well as biodiversity and ecological conservation. Some of the wild species reported suitable for revegetation are *Bothriochloa intermedia*, *B. pertusa*, and *Eragrostis curvula*. The last one is commonly known as love grass or cane grass.

3.5 Perspective

It is quite evident from the above discussion that there are many examples to demonstrate that the wild species in general and wild relatives of cultivated species hold great potential value. However, they need to be properly characterized and bioprospected, both for direct economic exploitation, ecological benefits, and indirect use as genetic resource in genetic improvement of related cultivated species with improved resilience against adverse environment/stresses and productivity. These splendored resources, if not taken care with required management and protection practices in situ, may be lost forever with the loss of innumerable opportunities. If taken care, they may serve the mankind in facing the challenges of climatic change and food and nutritional security of the present generation and future generations. Therefore, they need greater research attention from identification to collection, characterization, conservation, and documentation of generated information to facilitate their conservation and profitable use.

References

Acosta-Gallegos JA, Kelly JD, Gepts P (2007) Prebreeding in common bean and use of genetic diversity from wild germplasm. Crop Sci 47:S44–S59

Amante-Bordeos A, Sitch LA, Nelson R, Damacio RD, Oliva NP, Aswidinnoor H, Leung H (2004) Transfer of bacterial blight and blast resistance from the tetraploid wild rice *Oryza minuta* to cultivated rice, *Oryza sativa*. Theor Appl Genet 84:345–354

Birhman RK, Singh BP, Khushu CL, Bhattacharjje SK (1991) Nature of horizontal resistance to late blight in *Solanum microdontom*. In: Golden jubilee symposium on genetic research and education: current trends and the next fifty years. Indian Society of Genetics and Plant Breeding/Indian Agricultural Research Institute, New Delhi, pp 231–232

Bisht IS, Singh M (2013) Asian *Vigna*. In: Singh M, Upadhyaya HD, Bisht IS (eds) Genetic and genomic resources of grain legume improvement. Elsevier, Heidelberg, pp 237–268

Blair MW, Iriarte G, Beebe S (2006) QTL analysis of yield traits in an advanced backcross population derived from a cultivated Andean × wild common bean (*Phaseolus vulgaris* L.) cross. Theor Appl Genet 112:1149–1163

Brar DS (2005) Broadening the gene pool of rice through introgression from wild species. In: Toriyama K, Heong KL, Hardy B (eds) Rice is life: scientific perspective for 21 century. Proceeding of the world rice research conference, Tokyo and Tsukuba, Japan, 4–7 Nov 2004. IRRI, Phillipines, pp 157–160

Brumlop S, Reichenbecher W, Tappeser B, Finckh MR (2013) What is the SMARTest way to breed plants and increase agrobiodiversity? Euphytica 194:53–66

Bugaenko LA, Reznikova SA, Popovich AL (1975) Experiments of continuous cropping of peppermint. Tr Prikl Bot Genet Scl 54:267–274

Chase SS (1963) Analytic breeding in *Solanum tuberosum* L.-a scheme utilizing parthenotes and other diploid stocks. Can J Genet Cytol 5:359–363

Chaurasia OP, Khatoon N, Singh SB (2008) Field guide. Floral Diversity of Ladakh/Field Research Laboratory/Defence Research and Development Organization, Leh, p 198

Dalmacio R, Brar DS, Ishii T, Sitch LA, Virmani SS, Khush GS (1995) Identification and transfer of a new cytoplasmic male sterility source from *Oryza perennis* into indica rice (*O. sativa*). Euphytica 82:221–225

De Wet JMJ (1979) *Tripsacum* introgression and agronomic fitness in maize (*Zea mays* L.). In: Proc. Conf Broadening Genetic Base of Crops, Wageningen, 1978, Pudoc. Wageningen, pp 203–210

Dionne LA (1961) Cytoplasmic sterility in derivates of *Solanum demissum*. Am Potato J 38:117–120

Doussinalt G, Delibes A, Sanchez-Monge, Garica-Olmedo F (1983) Transfer of dominant gene for resistance to eye spot disease from a wild grass to hexaploid wheat. Nature 303:968–700

Eshed Y, Zamir D (1995) An introgression line population of *Lycopersicon pennellii* in the cultivated tomato enables the identification and fine mapping of yield-associated QTL. Genetics 141:1147–1162

Esquinas-Elcazar JT (1981) Genetic resources of tomatoes and wild relatives. International Board for Plant Genetic Resources (IBPGR), Rome

Filippenko IM, Lebedev AV (1971) Grapes. Tr Tsentr Genet Lab IV Michurina 12: 46-56

Frey KJ (1976) Plant breeding in the seventies: useful genes from wild plant species. Egypt J Genet Cytol 5:460–482

Gill SS, Gill R, Kumar G, Pareek A, Sharma PC, Anjum NA, Tuteja N (2012) Mustard: approaches for crop improvement and abiotic stress tolerance. In: Tuteja N, Singh Gill S, Tiburcio AF, Tuteja R (eds) Improving crop resistance to abiotic stress. Wiley-VCH Verlag GmbH & Co. KGaA, Weinheim, pp 1351–1368

Gothandam KM (2012) Rice: improving cold stress tolerance. In: Tuteja N, Singh Gill S, Tiburcio AF, Tuteja R (eds) Improving crop resistance to abiotic stress. Wiley-VCH Verlag GmbH & Co. KgaA, Weinheim, pp 733–750

Grun P, Aubertin M (1965) Evolutionary pathways of cytoplasmic male sterility in *Solanum*. Genetics 51:399–409

Guok HP, Wynne JC, Stalker HT (1986) Recurrent selection within a population from an interspecific peanut cross. Crop Sci 26:249–253

Gur A, Zamir D (2004) Unused natural variation can lift yield barriers in plant breeding. PLoS Biol 2:1610–1615

Hajjar R, Hodgkin T (2007) The use of wild relatives in crop improvement: a survey of developments over the last 20 years. Euphytica 156:1–13

Hanneman RE, Ruhde RW (1978) Haploid extraction in *Solanum tuberosum* group Andigena. Am Potato J 55:259–263

Harlan JR (1967) A wild wheat harvest in Turkey. Archaeology 20:197–201
Harlen JR, De Wet JMJ, Naik SM, Lambert RJ (1970) Chromosome pairing within the genomes in maize – *Tripsacum* hybrids. Science 167:1247–1248
Harvey PH, Levings CS III, Wernsman EA (1972) The role of extrachromosomal inheritance in plant breeding. Adv Agron 24:1–27
Henry RJ (2014) Genomics strategies for germplasm characterization and the development of climate resilient crops. Front Plant Sci 5:68
Hobbs R (1993) Can Revegetation assist in the conservation of biodiversity in agricultural areas? Pac Conserv Biol 1:29–38
Hougas RW, Ross RW (1956) The use of foreign introductions in breeding American potato varieties. American Potato Journal 33(11):328–339
Hougas RW, Peloquin SJ, Ross RW (1958) Haploids of the common potato. J Heredity 49:103–106
Hurtado JR, Ramos NG (1970) Study of noli palm (Elaeis melanococca Gaert.) and its preliminary breeding in Colombia. Acta Agron 20:9–23
Jain SK (1964) Wild plant-foods of the tribals of Bastar (Madhya Pradesh). Proc Nat Inst Sci India 30B(2):56–80
Jansky S (2000) Breeding for disease resistance in potato. In: Janick J (ed) Plant breeding reviews, vol 19. Wiley, Oxford, pp 86–155. isbn:0-471-38787-8 © 2000
Johnson R (1966) The substitution of a chromosome from *Agropyron elongatum* for chromosomes of hexaploid wheat. Can J Genet Cytol 8:279–292
Jones JK (1976) Strawberry, *Fragaria ananassa* (Rosaceae). In: Simmonds NW (ed) Evolution of crop plants. Longman Press, London, pp 237–242
Jung C, Wehling P, Loptien H (1986) Electrophoretic investigations on nematode resistant sugar beets. Plant Breed 97(1):39–45
Khoshoo TN (1968) Genetic improvement of ornamentals in India-prospects, experimental approaches, and suggestions. Indian J Genet 28a:87–98
Khush GS, Bacalangco E, Ogawa T (1991) A new gene for resistance to bacterial blight from *Oryza longistaminata*. Rice Genet Newsl 7:121–122
Khush GS, Ling KC, Aquino RC, Aquiero VM (1997) Breeding for resistance to grassy stunt in rice. In: Proceeding of the 3rd International congress SABRAO. Plant Breeding Papers, Canberra, pp 3–9
Knott DR (1961) The inheritance of rust resistance VI. The transfer of stem rust from *Agropyron elongatum* to common wheat. Can J Plant Sci 41:109–123
Knott DR (1978) The transfer of genes for rust resistance to wheat from related species. In: Ramanujam S (ed) Proceeding of the Fifth International wheat genetics Symposium. Indian Soc Genet Plant Breed, New Delhi, pp 354–357
Knott DR, Dvorak J (1976) Alien germplasm as a source of resistance to disease. Ann Rev Phytopathol 14:211–235
Kole C (2011) Wild crop relatives: genomics and breeding resources, vol 1-10. Springer, Berlin/Heidelberg
Kunchge N, Kumar K, Firke P (2012) Vegetable crops (chili pepper and onion): approaches to improve crop productivity and abiotic stress tolerance. In: Tuteja N, Singh Gill S, Tiburcio AF, Tuteja R (eds) Improving crop resistance to abiotic stress. Wiley-VCH Verlag GmbH & Co. KGaA, Weinheim, pp 951–978
Kuvarin VV (1973) Methods of increasing winter hardiness in cereals. *Sel Semenovod* 6:69–71
Lakhanpaul S, Singh V, Kumar S, Bhardwaj D, Bhat KV (2012) Sesame: overcoming the abiotic stresses in the queen of oilseed crops. In: Tuteja N, Singh Gill S, Tiburcio AF, Tuteja R (eds) Improving crop resistance to abiotic stress. Wiley-VCH Verlag GmbH & Co. KGaA, Weinheim, pp 1251–1283
Li D, Pfeiffer TW, Cornelius PL (2008) Soybean QTL for yield and yield components associated with *Glycine soja* alleles. Crop Sci 48:571–581
Livermore JR, Johnstone FE Jr (1940) The effect of chromosome doubling on the crossability of *Solanum chacoense, Solanum jamesii* and *S. bulbocastanum* with *S. tuberosum*. Am Potato J 17:170–173

Marta B, Agnelo F, Henry Robert J (2015) Genomics of crop wild relatives: expanding the gene pool for crop improvement. Plant Biotechnol J 14:1070–1085

Maxted N, Kell SP (2009) Establishment of a global network for the *In Situ* conservation of crop wild relatives: status and needs. FAO, Rome

Mickelbart MV, Hasegawa PM, Bailey-Serres J (2015) Genetic mechanisms of abiotic stress tolerance that translate to crop yield stability. Nat Rev Genet 16:237–251

Moncada P, Martinez CP, Borreo J, Chatel M, Gauch H Jr, Guimaraes E, Tohme J, McCouch SR (2001) Quantitative trait loci for yield and yield components in an *Oryza sativa* x *Oryza rufipogon* BC2F2 population evaluated in an upland environment. Theo Appl Genet 102:41–52

Nath P, Sane VA, Asif MH, Sane AP, Trivedi PK (2012) Fruit crops: omic approaches toward elucidation of abiotic stress tolerance. In: Tuteja N, Singh Gill S, Tiburcio AF, Tuteja R (eds) Improving crop resistance to abiotic stress. Wiley-VCH Verlag GmbH & Co. KGaA, Weinheim, pp 1033–1048

Nguyen B, Brar D, Bui B, Nguyen T, Pham L, Nguyen H (2003) Identification and mapping of the QTL for aluminum tolerance introgressed from the new source, *Oryza rufipogon* Griff. into indica rice (*Oryza sativa* L.) Theor Appl Genet 106:583–593

Oliver RJ, Finch JW, Taylor G (2009) Second generation bioenergy crops and climate change: a review of the effects of elevated atmospheric CO2 and drought on water use and the implications for yield. GCB Bioenegy 1:97–114

Osipchuck AA (1970) The use of Solanum gibberulosum Juz. et Buk in potato breeding. Kartoplyarstvo. Resp. mizhvid. temat. nauk. zb, 1:32–37

Petersen JE (2008) Energy production with agricultural biomass: environmental implications and analytical challenges. Eur Rev Agric Econ:1–24. doi:10.1093/erae/jbn016

Pimentel D, Wilson C, McCullum C, Huang R, Dwen P, Flack J, Tran Q, Saltman T, Cliff B (1997) Economic and environmental benefits of biodiversity. Bioscience 47:747–757

Prescott-Allen C, Prescott-Allen R (1986) The first resource: wild species in the North American economy. Yale University Press, New Haven

Price S (1963) Cytogenetics of modern sugar canes. Econ Bot 17:97–105

Priyanka D (2013) Bioenergy crops an alternative energy. Int J Environ Eng Manag 4:265–272

Ram T, Majumder ND, Mishra B, Ansari MM, Padmavathi G (2007) Introgression of broad spectrum blast resistance gene(s) into cultivated rice (*Oryza sativa* ssp *indica*) from wild rice, *O. rufipogon*. Curr Sci 92:225–230

Rao GU, Ben Chaim A, Borovsky Y, Paran I (2003) Mapping of yield-related QTLs in pepper in an interspecific cross of *Capsicum annuum* and *C. frutescens*. Theor Appl Genet 106:1457–1466

Redden R (2015) Wild relatives for the crop improvement challenges of climate change: the adaptation range of crops. In: Redden R, Yadav SS, Maxted N, Dulloo ME, Guarino L, Smith P (eds) Crop wild relatives and climate change. Wiley, New Jersey, pp 61–76

Reddy LJ, Green JM, Singh U, Bisen SS, Jambunathan R (1979) Seed protein studies on *Cajanus cajan*, *Atylosia* spp., and some hybrid derivatives. In: Seed protein improvement in cereals and grain legumes Vol. II. International Atomic Energy Agency, Vienna, pp 105–117

Reddy DS, Bhatnagar-Mathur P, Vadez V, Sharma KK (2012) Grain legumes (soybean, chickpea, and peanut): omics approaches to enhance abiotic stress tolerance. In: Tuteja N, Singh Gill S, Tiburcio AF, Tuteja R (eds) Improving crop resistance to abiotic stress. Wiley-VCH Verlag GmbH & Co. KGaA, Weinheim, pp 995–1032

Rick CM (1967) Fruit and pedicel characters derived from Galapagos tomatoes. Econ Bot 21:171–184

Rick CM (1974) High soluble-solids content in large-fruited tomato lines derived from a wild green-fruited species. Hilgardia 42:493–510

Rick CM (1982) The potential of exotic germplasm for tomato improvement. In: Vasil IK, Scowcroft WR, Frey KJ (eds) Plant improvement and somatic cell genetics. Academic, New York, pp 1–28

Riley R, Chapman V, Johnson R (1968) Introduction of yellow rust resistance of *Aegilops comosa* into wheat by genetically induced homoeologous recombination. Nature (London) 217:383–384

Robinson W, Koualewski E (1974) Tomato. Proc Int Hortic Congr XIX I Sect VII:76
Ross H (1966) The use of wild Solanum species in German **potato** breeding of the past and today. Am Potato J 43:63–80
Sasson A (1996) Biotechnologies and the use of plant genetic resources for industrial purpose: benefits and constraints for developing countries. In: Castri F di, Younes J (eds) Biodiversity, Science and Development: Towards a new Partnership. CAB International, pp 469–487
Saxena KB, Kumar RV, Srivastava N, Shiying B (2005) A cytoplasmic-nuclear male-sterility system derived from a cross between *Cajanus cajanifolius* and *C. cajan*. Euphytica 145:291–229
Schwendiman J Lefort PL (1974) hybrid lines derived from a cross between Gossypium hirsutum L. and G. barbadense L. -IV. Correlations between characters, heterosis effects of inbreeding and combining ability in diallel cross. Cotton Fibers Too 29: 383-394
Sears ER (1956) The transfer of leaf rust resistance from *Aegilops umbellulata* to wheat. Brookhaven Symp Biol 9:1–22
Sears ER (1969) Wheat cytogenetics. Ann Rev Genet 3:451–468
Sears ER (1972) Agropyron-wheat transfers through induced by homoeologous pairing. Can J Genet Cytol 14:736
Sears ER (1975) The wheats and their relatives. In: King RC (ed) Handbook of genetics Vol.2. Plenum Press, New York, pp 59–91
Septiningsih EM, Prasetiyono J, Lubis E, Tai T, Tjubaryat T, Moeljopwiro S, McCouch SR (2003) Identification of quantitative trait loci for yield and yield components in an advanced backcross population derived from *Oryza sativa* variety IR64 and the wild relative *O. rufipogon*. Theo Appl Genet 107:1419–1432
Shands H (1993) Harvesting nature's diversity. FAO, Rome
Sharma D, Knott DR (1966) The transfer of leaf rust resistance from *Agropyron to Triticum* by irradiation. *Can J Genet Cytol***8**:137–143
Sharma KP, Pandey SK, Tyagi BR (1982) Breeding for fileld resistance to late blight of potato. In: Nagaich BB, Shekhawat GS, Gaur PC, Verma SC (eds) Potato in developing countries. Indian Potato Association, Central Potato Research Institute, Shimla, pp 92–95
Shrivastava AK, Srivastava S (2012) Sugarcane: physiological and molecular approaches for improving abiotic stress tolerance and sustaining crop productivity. In: Tuteja N, Singh Gill S, Tiburcio AF, Tuteja R (eds) Improving crop resistance to abiotic stress, vol 2. Wiley-VCH Verlag GmbH & Co. KGaA, Weinheim, pp 885–922
Singh AK (1989) Exploitation of *Arachis* species for improvement of the cultivated groundnut, Part I: Transfer of genes from compatible species of section *Arachis* and incompatible species of section *Ericoides* into *A. hypogaea*. Report of work done in 1988. Legumes Program, ICRISAT, Patancheru AP 502324, India, p 72
Singh AK (2015) Agricultural biodiversity heritage sites and Systems in India. Asian Agri-History Research Foundation, Secunderabad, India, p 467
Singh AK (2016) Exotic ancient plant introductions: part of Indian 'Ayurveda' medicinal system. Plant Genet Resour C 14(4):356–369
Singh AK, Gibbons RW (1985) Wild species in crop improvement: groundnut – a case study. In: Gupta PK, Bahl JR (eds) Advances in genetics and crop improvement. Rastogi Publi., Meerut, pp 297–308
Singh AK, Nigam SN (2016) Arachis gene pools and genetic improvement in groundnut. In: Rajpal VR, Rao SM, Raina SN (eds) Gene pool diversity and crop improvement. Springer International Publishing, Switzerland, pp 17–77
Singh KB, Ocampo B (1997) Exploitation of wild species for yield improvement in chickpea. Theo Appl Genet 95:418–423
Singh K, Chugh V, Sahi GK, Chhuneja P (2012a) Wheat: mechanisms and genetic means for improving heat tolerance. In: Tuteja N, Singh Gill S, Tiburcio AF, Tuteja R (eds) Improving crop resistance to abiotic stress. Wiley-VCH Verlag GmbH & Co. KGaA, Weinheim, pp 657–694
Singh I, Nepolean T, Rajendran RA, Shono M (2012b) Maize: physiological and molecular approaches for improving drought tolerance. In: Tuteja N, Singh Gill S, Tiburcio AF, Tuteja

R (eds) Improving crop resistance to abiotic stress. Wiley-VCH Verlag GmbH & Co. KGaA, Weinheim, pp 751–778

Sreenivasan TV, Amalraj VA (2004) Sugarcane. In: Dhillon BS, Tyagi RK, Saxena S, Agrawal A (eds) Plant genetic resources: oilseeds and cash crops. Narosa Publishing House, New Delhi, pp 199–213

Sreenivasan TV, Sreenivasan J (2000) Intergenric hybrids for sugarcane improvement. Sugarcane Breed Inst Newsl 19:1–2

Stalker HT, Harlan JR, de Wet JMJ (1977) Observation on introgression of *Tripsacum* into maize. Am J Bot 64:1162–1169

Stalker HT, Harlan JR, de Wet JMJ (1978) Genetics of maize *Tripsacum* introgression. Caryologia 31:271–282

Stolton S, Maxted N, Ford-Lloyd B, Kell S, Dudley N (2006) Food stores: using protected areas to secure crop genetic diversity. WWF, Gland

Swamy BPM, Sarla N (2008) Yield-enhancing quantitative trait loci (QTLs) from wild species. Biotechnol Adv 26:106–120

Tanksley SD, McCouch SR (1997) Seed banks and molecular maps: unlocking genetic potential from the wild. Science 277(5329):1063–1066

Tarn TR, Tai GCC (1977) Heterosis and variation of yield components in F1 hybrids between Group of Tuberosum and Group Andigena potatos. Crop Sci 17:517–521

Tikka SBS, Parmar LD, Chauhan RM (1997) First record of cytoplasmic-genic male-sterility system in pigeonpea and its related wild species. J Plant Physiol 137:64–71

Uniyal MR (2006) Potential of medicinal plants from Ladakh. S & T Entrepreneurs Park, IIT Kharagpur

Upadhya MD, Khushu CL, Bhattacharjje SK, Ray S (1977) Breeding potato varieties for charcoal rot (PL 480 Project). In: Final technical report. Central Potato Research Institute, Shimla

Van der Meer QP (1975) Onion. Zaadbelangen 29:178

Vasilenko II (1973) *Triticum* x *Agropyron* hybrids. Sel Semenovod 6:71–73

Wellman FL (1972) Tropical American plant disease. Scarecrow Press, Metuchen, p 989

Xian-Hua SHEN, Song YAN, Ren-Liang HUANG, Shan ZHU, Hong-Liang XIONG, Lin-On Jun SHEN (, 2013) Development of Novel cytoplasmic MALE Sterile Source from Dongxiang Wild Rice (*Oryza rufipogon*). Rice Sci 20 is (. 5) : 379-382

Xie X, Song M, Jin F, Ahn S, Suh J, Hwang H, McCouch SR (2006) Fine mapping of a grain weight quantitative trait locus on rice chromosome 8 using near–isogenic lines derived from a cross between *Oryza sativa* and *Oryza rufipogon*. Theo Appl Genet 113:885–894

Yakovlev GV (1972) Tr Prikl Bot Genet Sel 50:229–238

Part II

Wild Relatives Distribution and Diversity in Different Crop Groups

4 Cereals

4.1 Introduction

Cereals occupy maximum hectares of land and hold world's largest grain production. Rice, wheat, and maize account for 80% of the grain produced in the world. The major distribution areas of wild species of cereals in India are the Western Himalayas, the northeastern region, and eastern peninsular tracts. Archeological and historical evidence reveal the region extending from Northeast Hills in India to mountain ranges of Southeast Asia and South-West China to be the primary center of origin of *Oryza sativa* L. India has abundant wild species genetic resources of rice spread all over the subcontinent except for the western arid and semi-arid plains (Fig. 4.1). It is particularly rich for *Oryza nivara* Sharma & Shastry, *O. rufipogon* Griff., *O. officinalis* Wall. ex Watt, *O. malamphuzhensis* and *O. granulata* Nees & Arn. ex Watt. (Fig. 4.1). Most species are found in ditches, ponds, and paddy fields; *O. nivara* is largely found in the Deccan Plateau (Fig. 4.2a). The natural hybrid populations (*O. sativa* ssp. *spontanea*) of *O. sativa* with wild relatives, *O. nivara*, and *O. rufipogon* in rice field are found all over South and Southeast Asia. Maximum diversity in *Oryza* species occurs in the peninsular tract of eastern part of the country with *O. nivara* (annual), *O. rufipogon* (perennial), and weedy *O. spontanea* types in *indica* cultivars. In addition, *Porteresia coarctata* (Roxb.) Tateoka (= *O. coarctata*), a tetraploid species, also exists in India and is found in the tidal swamps of Sunderbans and the Godavari delta. The other members of Oryzeae of potential use are *Hygroryza aristata* (Retz.) Nees, *Leersia hexandra* Swartz, and *L. hackelii* Keng, a cold-tolerant species.

In case of wheat, species belonging to the tribe Triticeae of family Gramineae/Poaceae are mostly distributed wild in the Himalayan region. In relation to bread wheat, *Triticum aestivum* L., the Northwest part of the Indian Subcontinent adjoining the Hindukush Mountains is regarded as the secondary center of diversity; *T. sphaerococcum* Perc. (dwarf) and *T. compactum* Host are supposed to have originated in this region (Pal 1966). The Western/Northwestern Himalaya also harbors wild species of comparatively distant but cross-compatible genera, such as *Aegilops*

A.K. Singh, *Wild Relatives of Cultivated Plants in India*,
DOI 10.1007/978-981-10-5116-6_4

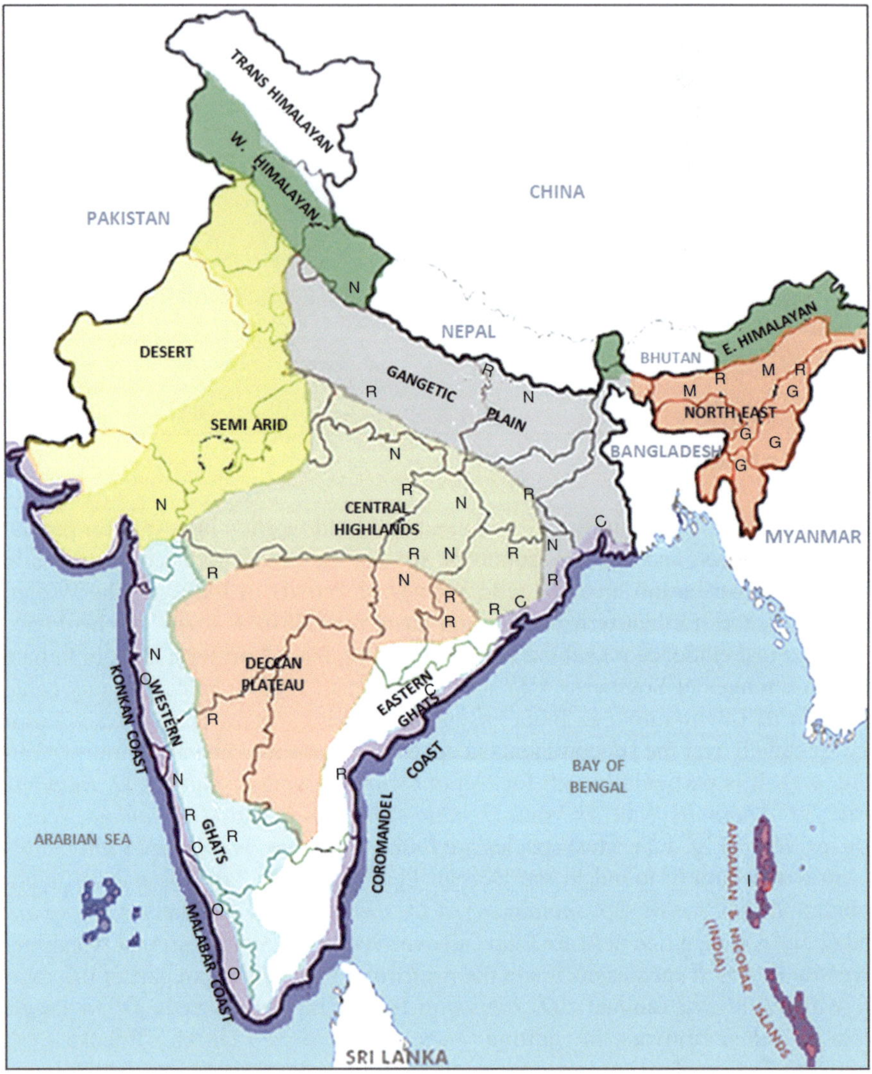

Fig. 4.1 Biogeographical regions associated with the distribution of wild *Oryza* species C = *O. coarctata*; G = *O. granulata*; I = *O. indandamanica* (*meyeriana*); M = *O. minuta*; N = *O. nivara*; O = *O. officinalis*; R = *O. rufipogon*

and *Elymus,* occurring in the cultivated fields. Types related to wheat occurring here are *Elymus dahurcus* Turcz., *E. dasystachys* Trin., and *E. nutans* Griseb. Wild species related to barley, *Hordeum spontaneum* L. and others, mainly occur in Western Himalayan region with possibility of occurrence of *Hordeum agriocrithon* A. E. Aberg., in Sikkim. Additionally, distant but compatible relatives present are *Elymordeum* and *Eremopyrum*. A natural sterile intergeneric hybrid *Elymordeum schmidii* Meld. (*Elymus* x *Hordeum*), a product of crossing between *Elymus nutans*

Fig. 4.2 Representative wild relatives of economically important cereal crops: (**a**) *Oryza nivara*, (**b**) *Aegilops tauschii*, (**c**) *Echinochloa colona*, and (**d**) *Panicum psilopodium*

and *Hordeum turkestanicum* Nevski (habit of *Hordeum* with long-awned spike and two-rowed arrangement like *Elymus*), also occurs in this region (bordering Pakistan) (Bor 1960), while in the case of oat, wild *Avena ludoviciana* Durieu is found.

Millets together represent 19% of the world's cereal production. The minor millets are important in terms of world food production and essential as food crops in their respective agroecosystems. They are mainly grown in the marginal areas, under agricultural conditions, where major cereals fail to produce consistently. In many parts of India, sorghum [*Sorghum bicolor* (L.) Moench.], pearl millet [*Pennisetum glaucum* (L.) R. Br.], and six small millet crops are cultivated for grain and fodder. They are finger millet or Ragi [*Eleusine coracana* (L.) Gaertn.], Italian or foxtail millet, [*Setaria italica* (L.) P. Beauv.], common or proso millet (*Panicum milliaceum* L.), kodo millet (*Paspalum scrobiculatum* L.), little millet [*Panicum sumatrense* Roth. ex Roem. & Schult.; syn. *P. miliare* auct. Non Lam. ssp. *psilopodium* (Trin.) Wet.], and barnyard millet, *Echinochloa frumentacea* Link. Two of these, kodo millet, *Paspalum scrobiculatum* and *Panicum sumatrense*, are indigenous to India. Wild-related types of millet *Panicum sumatrense* are found in *tarai* range and of the allied, ssp. *psilopodium* in hills of Tamil Nadu. In addition, *P. hippothrix* K. Schum. ex Engl. and *P. trypheron* Schult. are also found in India. Much variability occurs for taxa related to the foxtail millet, *Setaria italica*, viz.,

S. verticellata (L.) P. Beauv. and *S. viridis* (L.) P. Beauv., both in plains and hills. Among the wild relatives of finger millets, wild types of *Eleusine coracana*, the annual *E. indica* (L.) Gaertn., have a wider distribution compared to the perennial, *E. compressa* (*Ochthochloa compressa*), of the northwestern plains. The allied species of barnyard millet, *Echinochloa frumentacea*, *E. crus-galli* (L.) P. Beauv., is widely distributed and has several variants occurring in rice fields and elsewhere, mainly in waterlogged habitats. In addition, several members of Maydeae, *Polytoca digitata* (L.f.) Druce, *P. wallichiana* (Nees ex Steud.) Benth., and *Trilobachne cookie* (Stapf) Schenck, as also of *Coix*, *C. aquatica* Roxb., and *C. gigantea* Koenig ex Roxb., are distributed, chiefly in the eastern or western peninsular humid/subhumid tracts, extending to northeastern or to southern region. Among other cereals, important wild types are the forms of *Digitaria cruciata* (Nees) A. Camus and *Coix lacryma-jobi* L., domesticated in Indo-Myanmar region (Arora 1977). *Digitaria sanguinalis* (L.) Scop. is cultivated as a cereal in the Caucasus and Kashmir. Kodo millet, *Paspalum scrobiculatum*, is another indigenous millet cultivated in India with several weedy types. Two minor millets, *Setaria pumila* (Poir.) Roem. and Schult. and *Brachiaria ramosa* (L.) Stapf., are indigenous millets of hills of Central India.

The following paragraphs list and discuss the distribution of wild and weedy relatives of cereals found in India with characteristic features and information on their potential value, if available. For simplicity, they are being summarized in alphabetical order.

4.2 Wild Species Found in India

1. ***Aegilops tauschii* Coss. (Poaceae/Gramineae).** It is an annual grass, native of temperate Asia (Fig. 4.2b), distributed in the Northwestern Himalayas in the state of Jammu and Kashmir in stony alpine habitats and as weed in the fields. It is a diploid ($2n = 2\times = 14$), which has contributed D genome to common wheat. Thus, part of the secondary gene pool of wheat and tertiary gene pool of rye. Source of resistance to disease and pest of wheat and for transfer of multiple diversities to wheat.
2. ***Amaranthus atropurpureus* Roxb. (Amaranthaceae).** A common pot herb, native of Americas, naturalized to Gangetic delta of West Bengal as a weed.
3. ***Amaranthus blitum* L.** Purple amaranth, an annual growing 1 m tall, native range is obscure. In Indian Subcontinent, it occurs as weed and cultivated as vegetable. Frost tolerant. A variant, var. *oleracea* (L.) Hook. f., reported from Kerela.
4. ***Amaranthus dubius* Mart. Ex Thell.** An annual growing to 1 m, native of Americas, naturalized to Indian Subcontinent (Maharashtra, Tamil Nadu), called "Araikeerai." Cultivated as vegetable and used medicinally. Frost sensitive. Genetically close to other *Amaranthus* spp.
5. ***Amaranthus hybridus* L.** A summer annual herb growing to 2 m, native to North America as weed. Commonly called smooth amaranth. Naturalized to

warmer regions of the world, including India. Frost sensitive. Cultivated in Peninsular India.

6. ***Amaranthus spinosus* L.** Prickly amaranth, an annual erect herb, 0.6 m tall, native of Americas, naturalized in India. Abundant near village wastelands, as weed, colonizer of new areas in plains. Frost tolerant. Used as vegetable and folklore medicine.
7. ***Amaranthus tenuifolius* Willd.** A careless weed, prostrate or ascending annual, much branched from the base, 0.35 m tall. Only found in India (Punjab).
8. ***Amaranthus tricolor* L.; syn. *A. gangeticus* L., *A. mangostanus* L., *A. polygamus* L.** It is a rapidly growing annual shrub, up to 1 m tall. Native of South Central America. Naturalized worldwide, including India. Cultivated as ornamental, vegetable, and medicine. Possess significant variability, drought, and heat tolerant.
9. ***Amaranthus viridis* L.** Cosmopolitan annual erect herb, densely branched, 0.80 m tall, native of South America. Widely naturalized as tropical weed, including India. Traditionally cultivated as vegetable and used in traditional Ayurveda medicine too.
10. ***Avena barbata* Pott. ex Link. (Poaceae/Gramineae).** A winter annual grass with thin tillers growing up to 0.80 m. It is a diploidized autotetraploid ($2n = 4x = 28$). Native to Central Asia with distribution extending to Northwest Himalayas (Kumaon). Probably originated in Spain; it is a member of tertiary gene pool of oat. Source of disease resistance and quality of oat, also used as forage.
11. ***Avena fatua* L.; syn. *A. fatua* ssp. *meridionalis* Malzev; *A. fatua* var. *pilosissima* Gray; *A. meridionalis* (Malzev) Roshev.** A green grass with hollow, erect stems 30–120 cm tall, bearing nodding panicles of spikelets, known as the common wild oat. Native of Eurasia, Mediterranean, North Africa, and Ethiopia. Naturalized throughout temperate regions, as weed, including Northwest Himalayas. The genetic variation found in the subcontinent is classified into *A. fatua* var. *pilosa* Syme., confined to Northwest Himalayas, and *A. fatua* var. *glabrata* Peterm., extending to Sikkim Himalayas. Resistant to drought and diseases and have potential for high yield. It belongs to primary gene pool of oat (Loskutov and Rines 2011).
12. ***Avena sterilis* subsp. *ludoviciana* (Durieu) Gillet & Magne; syn. *A. sterilis, A. ludoviciana* Durieu.** Native of Africa, Asia, and Europe, found as dominant tufted annual weed grass, in winter season crops of West Himalayas up to Kumaon Hills. Resistant to diseases and herbicides and potential for high yield. It belongs to primary gene pool (Loskutov and Rines 2011). Used for livestock feeding.
13. ***Chionachne koenigii* (Spreng.) Thwaites; syn. *C. gigantea* (J. Koenig) Veldkamp.** Robust perennial with culms to 2.5 m high; nodes bearded. Grows along stream sides in moist deciduous forests of Indo-Malesia region, including warmer parts of India (Punjab and Kashmir). Genetically close to *Coix* and maize.

14. ***Chionachne semiteres* (Benth.) Henrard.** Culms to 2 m tall. Localized distribution in Peninsular India and Tamil Nadu in deciduous forest openings. Cultivated.
15. ***Coix aquatica* Roxb.** A perennial, aquatic herb, both a diploid (2n = 10) and an autogamous apomict, used as fodder. Distributed in central peninsular tract, extending to Western Ghats in marshy habits. Genetic resource for job's tears.
16. ***Coix lacryma-jobi* L.** It is a tall grain-bearing perennial, native to Southeast Asia. Hard-shelled wild form, which appears to be the immediate ancestors of cultivated thin-shelled type, is distributed over the Northeast Hill region. It presents significant variability. It is widely distributed in peninsular and northeastern tracts in lowlands waterlogged and marshy habitat bordering rice fields. Medicinal, closely related to cultivated var. *ma-yuen*. Variety *gigantea* (Koenig ex Roxb.) Stapf; syn. *C. gigantea* Koenig ex Roxb., is a perennial robust grass with up to 3 m tall culms. Sporadically distributed in peninsular region and in marshy and hygrophytic habitats.
17. ***Digitaria cruciata* (Nees) A. Camus; syn. *Panicum cruciatum Nees*.** An annual with decumbent culms 0.5–1.3 m long. Native to Asia, with much variability in Eastern Himalayas and Northeast Hills of India. Used as fodder. Wild relative of *D. cruciata* var. *esculenta* Bor. cultivated in Khasi Hills of Northeast India.
18. ***Digitaria sanguinalis* (L.) Scop.** It is an annual grass with a branched inflorescence, native to North Africa, Asia, and Europe. Wild type related to ssp. *agyptica* var. *frumentacea* are distributed in the Western Himalayas, particularly Kashmir, where it is cultivated for animal fodder and seeds are also eaten.
19. ***Echinochloa colona* (L.) Link; syn. *E. colonum* var. *zonalis* (Guss.) Wooton & Standl.** Jungle rice, a wild annual grass 10–80 cm high (Fig. 4.2c), native to tropical Asia and Africa, and widespread in Indian Subcontinent in wastelands of wet soils, like rice field. It is the ancestor of the cultivated *Echinochloa frumentacea* Link., sawa millet.
20. ***Echinochloa crus-galli* (L.) P. Beauv.** Barnyard grass, 1.5 m tall, of tropical Asian origin. Widely distributed in waterlogged rice fields with significant variation in eastern peninsular tract, Bengal plains, and the northeastern region, particularly in Brahmaputra Valley.
21. ***Echinochloa stagnina* Beauv.; syn. *Panicum stanninum* Retz.** It is a perennial or annual semiaquatic grass, widespread in tropical Africa and Asia, with an invasive nature. Distributed over the peninsular tract and northeastern parts of the Indian Subcontinent. It has potential as sugar and fodder yielding plant.
22. ***Eleusine indica* (L.) Gaertn.** A small annual C_4 grass naturally distributed throughout the warmer areas of the world, including India, both in hills and plains. Hilly forms are more robust, but occurrence is wider in plains. Seeds are edible and used as famine food. It is an ancestral diploid species, belonging to secondary gene pool of allotetraploid, *E. coracana*, the finger millet, donor of A genome (Hiremath and Salimath 1992; Bisht and Mukai 2001).
23. ***Elymus dahuricus* Turcz. ex Griseb. ssp. *excelsus* (Turcz. ex Griseb.) Tzvelev; syn. *E. tangutorum* (Nevski) Hand. -Mazz.** A tufted perennial,

native of Asia, including cold dry tracts of Northwestern Himalayas in barley and wheat fields. Two varieties are found, var. *dahuricus* and var. *micranthus* Melderis. A source of salt tolerance. The genus *Elymus* belongs to tertiary gene pool of wheat and barley.

24. ***Elymus dasystachys* Trin.; syn. *Leymus secalinus*.** Distributed at higher reaches of Northwestern Himalayas among grasses and forbs in open exposed slopes.
25. ***Elymus nutans* Griseb.** A tufted perennial forage grass 25–110 cm tall. Widely distributed in higher reaches of Northwestern Himalayas in grass-mixed alpine desert vegetation. Extend up to Russia and China. Several variants found. It belongs to tertiary gene pool of wheat and can also be used in improving other related cereals.
26. ***Eremopyrum buonapartis* (Spreng.) Nevski.** A tufted annual up to 30 cm tall (2n = 28). Confined to Central Asia and higher elevations of Western Himalayas in alpine meadows or stony habitat.
27. ***Eremopyrum distans* (C. Koch) Nevski.** A tufted annual up to 25 cm tall with 2n = 14. Native to Asia, including alpine stony habitat of Western Himalayas. Relative to wheat.
28. ***Eremopyrum orientale* (Linn.) Jaub. Et. Spach.** A tufted annual forage plant up to 25 cm tall (2n = 28). Native to North Africa, Asia, and Europe, including grass-mixed alpine vegetation of Western Himalayas. A close relative of cultivated wheat (Liu and Ding 1996).
29. ***Fagopyrum acutatum* (Lehm.) Mansf. ex K. Hammer; syn. *F. cymosum* (Trevir.) Meisn.** Perennial tall buckwheat up to 1 m, native to China, Himalayan India, and Far East. Domesticated and used as vegetable, fodder, and medicinal and ornamental plant. Rhizome has been widely used to treat various diseases. Hybridize with other *Fagopyrum* spp., including *F. esculentum* (Woo et al. 1999) revealing possibility of genetic introgression.
30. ***Fagopyrum tataricum* ssp. *potanini* Batalin.** A weed with 2n = 16, distributed from mountains in southern China to Western Himalayas (Kashmir/Pakistan). It is the immediate ancestor and a wild form of cultivated tatary buckwheat, *F. tataricum*.
31. ***Hordeum aegiceras* Nees ex Royle.** An annual with erect 40–80 cm long culms, found in Asia, including India, in exposed alpine grass-mixed vegetation of Northwestern Himalayas. A six-rowed barley with three-lobed appendage at the tip of the lemma. Cultivated in Kashmir.
32. ***Hordeum brevisubulatum* (Trin.) Link; syn. *H. turkestanicum* Nevski.** A perennial with erect or geniculately ascending 15–100 cm-long culm. Native to Asia, including Northwestern Himalaya (Kashmir), in higher elevations (3000 m), midst grasses. It is a member of tertiary gene pool of barley (von Bothmer and Jacobsen 1986).
33. ***Hordeum murinum* L.** False barley is an annual grass growing to 30 cm tall with unbranched spikes. It has subsp. *leporinum* (Link) Arcang.; syn. *H. leporinum* Link., referred as mouse barley, grows in tufts, from 10 to 40 cm in height. It is native to Africa, Asia, and Europe, including Northwestern Himalayas,

where it is found in alpine grass-mixed vegetation. It is a member of tertiary gene pool of barley. Another subspecies, *glaucum* (Steud.) Tzvelev; syn. *H. glaucum* Steud, called smooth barley, is distributed in semi-arid regions where rainfall is less. The two differ primarily in their chromosome numbers, spikelet morphology, and geographical distribution.

34. ***Hordeum spontaneum* K. Koch; syn. *H. vulgare* L. ssp. *spontaneum* (K. Koch) Thell., *H. agriocrithon* Aberg.** An annual wild grass like cultivated barley (*H. vulgare* L.) with small grains. Native to North Africa, Middle East, Northwestern Himalaya, and Southwest China. Found in grass-mixed vegetation mainly in high elevation zone. Progenitor and member of primary gene pool of barley. Gene source for crop quality, yield, disease resistance, cold, salt, and waterlogging tolerance.
35. ***Leersia hackelii* Keng.; syn. *L. oryzoides* (L.) Sw.** A tufted perennial with ascending 30–70 cm tall culm, distributed in Northwest India, including grassy meadows of Kashmir. The genus *Leersia* is closely allied to cultivated rice.
36. ***Leersia hexandra* Sw.** A perennial grass with hollow decumbent and creeping stems rooting at nodes, maintained as pasture. Pantropic in wetlands, including the Northeastern hills of India, Manipur; amidst grasses near lakes.
37. ***Ochthochloa compressa* (Forssk.) Hilu.; syn. *Eleusine compressa* (Forssk.) Asch. & Schweinf. ex C.Chr.** A hardy drought-tolerant genetic resource for finger millet. Distributed mainly in Northwestern arid tracts of the subcontinent. It is tetraploid species with 2n = 40, excluded from the genus *Eleusine* and included into genus *Ochthochloa* (Hilu 1981).
38. ***Oryza coarctata* Roxb; syn. *Porteresia coarctata* (Roxb.) Tateoka; *Sclerophyllum coarctatum* (Roxb.) Griff.** Perennial with extensive creeping rhizomes. Commonly found on the banks of coastal rivers in India, including Lower Gangetic Plains in Sundarbans, estuarine deltas along river margins, and saline water of tidal swamps in Orissa and Andhra Pradesh. Highly salt resistant. Tetraploid belongs to tertiary gene pool of rice (Tang et al. 2010). Crosses with cultivated rice, with difficulty.
39. ***Oryza granulata* Nees et Arn. ex Watt.** An upland short perennial wild rice with lanceolate dark green leaves and compact panicles. Occurs in the foothills and plains of Eastern India. Tolerant to shade and drought, immune to bacterial blight, and resistant to brown plant hopper.
40. ***Oryza indandamanica* J.L. Ellis; syn. *O. meyeriana* (Zoll. & Moritzi) Baill. var. *indandamanica* J.L. Ellis.** A perennial herb growing on a rocky slope over light loam soil, on bank of streams. A new wild diploid species (2n = 24) discovered in Rutland and Andamans Islands (Ellis 1985).
41. ***Oryza minuta* J. Presl & C. Presl.** Perennial with geniculately ascending or decumbent culms, 50–100 cm long. Native to tropical Asia, including Assam Valley. Source of disease resistance, like bacterial blight. Close relative of cultivated rice. Probable progenitor and member of secondary gene pool (Ge et al. 1999).
42. ***Oryza nivara* SD Sharma & Shastry; syn. *O. rufipogon* Griff.; *O. sativa* f. *spontanea* Roshev.; *O. sativa* L.** Biosystemically same as above, but wild

annual, found near waterlogged fields, ponds, etc. in Peninsular India and eastern plains. Progenitor of cultivated rice, carrying AA genome (Ge et al. 1999), part of primary gene pool. Source of disease resistance. Genome is sequenced.

43. ***Oryza officinalis* Wall. ex Watt; syn. *O. officinalis* ssp. *malampuzhaenensis* (Krishnasw. & Chandras.) Tateoka.** It is a perennial rhizomatous erect herb of variable height and open panicles, diploid 2n = 24 with CC genome. Native to Asia including India. Source of disease resistance. Member of secondary gene pool (Ge et al. 1999). Subspecies *malampuzhaenensis* is a tetraploid *O. officinalis* with 2n = 48. It is endemic to the Western Ghats at less than 1000 m.
44. ***Oryza rufipogon* Griff.; syn. *O. jeyporensis* Govindasw. & K.H. Krishnam.** Wild perennial red rice. Native to Asia, including India with much variability in peninsular tract, extending to northeast, Bengal plains, and Assam Valley and bordering areas, as a weed near rice field borders and ponds. Probable progenitor (carrier of AA genome) of cultivated *O. sativa*, belonging to primary gene pool (Ge et al. 1999). Natural weedy hybrids with cultivated rice do occur. Source of quality, yield, metal and salt tolerance, and disease resistance.
45. ***Panicum atrosanguineum* Hochst. ex A. Rich.** A tufted annual grass with 10–40 cm-long culm. Distributed in central to Northwest India. Excellent fodder.
46. ***Panicum curviflorum* Hornem.; syn. *P. trypheron* Schult.** An annual loosely tufted grass, found around paddy fields in Deccan Peninsula. Grains cooked like rice.
47. ***Panicum garadei* Sunderaragh. & Karth.** Conspecific with *P. plenum* Hitchc. & Chase, which is a predominantly American species, recorded from Coimbatore Tamil Nadu, India. Both are treated as synonym (Kabeer and Nair 2009).
48. ***Panicum hippothrix* K. Schum. ex Engl.** An annual grass with 30–100 cm-long robust culms. Found in western India and Deccan Peninsula. Grain cooked like rice.
49. ***Panicum notatum* Retz.** Perennial with 2 m tall erect culms having glabrous nodes. Distributed in moist deciduous forests of South and Southeast Asia, including India.
50. ***Panicum psilopodium* Trin.** A tufted annual with 14–120 cm-high culms (Fig. 4.2d). Based on panicle color, there are two varieties, *psilopodium* with green panicle, distributed in peninsular tract (Tamil Nadu), and var. *coloratum* Hook.f., with purple. Largely found in *tarai* range of Western Himalaya. A close relative of little millet, *P. sumatrense*.
51. ***Panicum sumatrense* Roth.** An annual or perennial, with straight or folded blades, 20–60 cm tall. Pantropical in distribution, including *tarai* range of northern India, in moderate elevations (1500 m) of Himalayas. Wild form of cultivated little millet.
52. ***Panicum turgidum* Forssk.** It is a perennial bunchgrass/desert grass, native to hotter region of Africa and Asia, including western plains of Indian Subcontinent, growing in dense bushes up to 1 m. It is drought- and salt-tolerant and used for flour, fodder, thatch, and erosion control.

53. ***Paspalum scrobiculatum* L.** Wild-type Kodo millet, annual or perennial 1 m tall, distributed in damp habitats across the Indian Subcontinent, but much variability occurs in drier, rainfed pockets of upper Gangetic Plains and peninsular region on slopes, bordering forests. Wild types are like cultivated. Wild are extremely drought and salt resistant. Also, attributed with several medicinal properties.
54. ***Pennisetum orientale* Rich.** Ornamental grass with large clumps, sporadically distributed in Western Ghats, eastern peninsular hills, Bihar, and in the Western Himalaya. A potential forage, drought resistant, remains green throughout the year.
55. ***Pennisetum pedicellatum* Trin**. Herbaceous perennial grass 1 m tall, called, *Deenanath* grass in India. Native of Ethiopia, naturalized to tropics of Indian Subcontinent, particularly Bihar, West Bengal, Haryana, Punjab, M.P., and U.P. Used as fodder.
56. ***Pennisetum polystachion* (L.) Schult.** A vigorous annual or perennial grass growing to over 1 m tall, native to tropical Africa. It is an invasive species introduced to Indian Subcontinent. Cultivated as an ornamental throughout the tropics of the world.
57. ***Pennisetum purpureum* Schum.** A C4 perennial grass, native to tropical Africa. Introduced to Indian Subcontinent as forage grass in natural pastures. It has escaped and naturalized. Genetic resource for Napier/elephant grass and pearl millet.
58. ***Setaria glauca* (Linn.) P. Beavu.** Yellow foxtail, a bright green clumping annual or perennial weed grass. Wild type occurs in northern parts of Maharashtra often along field borders; cultivated types also occur. Related to *S. viridis*, the green foxtail.
59. ***Setaria intermedia* Roem. & Schult.; syn. *S. tomentosa* (Roxb.) Kunth.** Trailing or erect annuals, 40–50 cm tall, sporadically distributed in plains and hills of Indian Subcontinent, Myanmar, and Sri Lanka, in damp, shady habitats bordering forests.
60. ***Setaria pallide-fusca* (Sch.) Stapf. & CE Hubb.; syn. *S. pumila* (Poir.) Roem. & Schult. ssp. *subtesselata* (Büse) B. K. Simon.** Native of Europe. Occurs as common weed in northern/northeastern foothills and peninsular hills of India, sporadic inflorescence variability in Maharashtra. Part of primary gene pool of foxtail millet (*S. italica*).
61. ***Setaria sphacelata* (Sch.) Stapf. et C.E. Hubb. Ex M.B. Moss.** A tall C4 grass, native to tropical Africa. Introduced to India in 1950 for pastures, naturalized all over plains and hills in upper Gangetic Plains with significant variability.
62. ***Setaria verticillata* (L.) P. Beauv.** Bristly foxtail, a loosely tufted annual growing to 1 m tall, native to Europe. Widely neutralized to plains and hills of Indian Subcontinent, more in humid areas, and used as famine food.
63. ***Setaria viridis* (L.) P Beauv.; syn. *S. italica* ssp. *viridis* (L.) Thell.** A hardy C4 annual grass with decumbent or erect 1 m long stem, native to Eurasia, including Northwestern Himalaya. Adapted as weed to colder tracts extending up to

3300 m. Probable progenitor of foxtail millet (*S. italica*) (Benabdelmouna et al. 2001).

64. ***Sorghum arundinaceum*** **(Desv.) Stapf; syn.** ***S. stapfii*** **(Hook.f.) C.E.C. Fisch.;** ***S. bicolor*** **(L.) Moench ssp.** ***verticilliflorum*** **(Steud.) de Wet ex Wiersema & J. Dahlb,** ***S. bicolor*** **(L.) Moench ssp.** ***drummondii*** **(Steud.) de Wet ex Davidse.** Conspecific annual or short-lived perennial weedy forms of sorghum. Native of Africa, introduced and naturalized in tropical India and sporadically in Tamil Nadu with much variability. Part of primary gene pool of sorghum, evolved independently.
65. ***Sorghum controversum*** **(Steud.) Snowden; syn.** ***S. halepense*** **(L.) Pers.,** ***S. miliaceum*** **(Roxb.) Snowden.** Perennial with erect 50–200 cm long culm. Part of tetraploid rhizomatous species complex. Distributed in tropical Indian Subcontinent, including Pakistan. Member of secondary gene pool of sorghum. Weedy hybrids with cultivated sorghum found. Used as fodder.
66. ***Sorghum deccanense*** **Stapf ex Raizada.** Annual with erect 80–150 cm long culm. Confined to Deccan Plateau and Maharashtra. Used as fodder.
67. ***Sorghum nitidum*** **(Vahl) Pers.** Perennial with 80–200 cm-high culms; nodes bearded. Found margins of shola forests and grasslands in India, Sri Lanka, Myanmar, and Indochina to Australia. Used as forage.

4.3 Perspective

There are several wild taxa among various cereal crops with multiple synonyms, such as *Oryza nivara*, *O. rufipogon*, and *Sorghum arundinaceum*, reflecting taxonomic confusion. Further biosystematic investigations are needed in such cases for ascertaining the essential features of the species, identity, and the taxonomic entity of variants, if present. Collection, characterization, and evaluation of wild species, for example, in case of rice with A genome have not received desired attention in this regard for conservation and use. They may contain genes capable of improving both yield and grain quality. Considering this, identification of yield quantitative traits loci (QTLs) in species like *O. rufipogon* (Xiao et al. 1996) is desired. In addition, in underutilized cereals, like minor millets, there is a need for further biosystematic studies to facilitate classification of wild species and weedy relatives into primary, secondary, and tertiary gene pool, based on cross-compatibility relationships. This shall help promote wider use of wild relatives using appropriate breeding approaches, to harness the benefits of efforts made on their collection and conservation.

The weakest link in promoting utilization of wild species is the lack of evaluation of wild species for desired traits to assess the genetic potential and usefulness for their genetic diversity (traits) from economic point of view. Evaluation of wild relatives in their site of origin and over environments and proper documentation of information generated on their genetic potential in a computerized database is another area to emphasize up on, making use of information technology and promoting their use in crop improvement.

In situ conservation of genetic diversity of wild species of crops such as rice, barley, and minor millet in Himalayan region and other areas of species and genetic diversity needs greater emphasis. This will ascertain their maintenance and evolution of rich genetic diversity responding to changing climate, on one hand, and would make them available on the other in improvement of these crops.

References

Arora RK (1977) Job's tears (*Coix lacryma-jobi*) – a minor food and fodder crop of northeastern India. Econ Bot 31:358–366

Benabdelmouna A, Darmency MA, Darmency H (2001) Phylogenetic and genomic relationships in *Setaria italica* and its close relatives based on the molecular diversity and chromosomal organization of 5S and 18S-5.8S-25S rDNA genes. Theor Appl Genet 103:668–677

Bisht MS, Mukai Y (2001) Genomic *in situ* hybridization identifies genome donor of finger millet (*Eleusine coracana*). Theor Appl Genet 102:825–832

Bor NL (1960) The grasses of Burma, Ceylon, India and Pakistan. Pergamon Press, London

Ellis JL (1985) *Oryza inandamanica* Ellis, a new rice plant from islands of Andamans. Bull Bot Surv India 27:225–227

Ge S, Sang T, Lu BR, Hong DY (1999) Phylogeny of rice genome with emphasis on origins of allotetraploid species. Proc Natl Acad Sci U S A 96:14400–14405

Hilu KW (1981) Taxonomic status of the disputable *Eleusine compressa* (Gramineae). Kew Bull 36:559–562

Hiremath SC, Salimath SS (1992) The 'A' genome donor of *Eleusine coracana* (L.) Gaertn. (Gramineae). Theor Appl Genet 84:747–754

Kabeer KAA, Nair VJ (2009) *Panicum plenum* Hitchc. & Chase (Poaceae), a species: a new record for India. Indian J For 32:473–476

Liu JW, Ding M (1996) Morphological and cytogenetical studies on intergeneric hybrids between *Triticum aestivum* and *Eremopyrum orientale*. Acta Genet Sin 23:117–123

Loskutov IG, Rines HW (2011) Chapter 3. Avena. In: Kole C (ed) Wild crop relatives: genomic and breeding resources cereals. Springer, Berlin/Heidelberg, pp 109–183

Pal BP (1966) Wheat. ICAR, New Delhi

Tang L, Zou XH, Achoundong G, Potgieter C, Second G et al (2010) Phylogeny and biogeography of the rice tribe (Oryzeae): evidence from combined analysis of 20 chloroplast fragments. Mol Phylogenet Evol 54:266–277

von Bothmer R, Jacobsen N (1986) Interspecific crosses in *Hordeum* (Poaceae). Plant Syst Evol 153:49–64

Woo SH, Wang YJ, Cambell CG (1999) Interspecific hybrids with *Fagopyrum cymosum* in the genus. Fagopyrum 16:13–18

Xiao J, Grandillo S, Ahn SN, McCouch SR, Tanksley SD, Li J, Yuan L (1996) Genes from wild rice improve yield. Nature 384:223–224

Grain Legumes 5

5.1 Introduction

India is the world's largest producer and consumer of grain legumes, because of the vegetarian dietary habits, in which they meet the protein requirement. Pigeonpea, chickpea, and dry beans, including several *Vigna* species, are the major pulses. For several important ones, India is the center of origin and primary center of diversity, particularly the pigeonpea and Asiatic *Vigna* species. For these reasons diversity in grain legumes and their wild relatives is high in India. Pigeonpea, *Cajanus cajan* (L.) Millsp., originated in India and spread to Africa quite early, where a secondary center of diversity developed. Seventeen species of *Cajanus* including the closest species, *C. cajanifolius* (Haines) Maesen, occur in India. *C. cajanifolius*, the most probable progenitor of pigeonpea, is found only in India. The only distinguishable morphological feature by which it differs from cultivated pigeonpea is the presence of seed strophiole in *C. cajanifolius*. The species readily crosses with pigeonpea. It is extremely rare and probably at the verge of extinction, because its natural habitat consists of open areas, which are vulnerable to grazing. It still occurs in rare patches in southern part of Orissa and Chhattisgarh. Of the 17 species reported native to India, 10, namely, *C. albicans* (Wight & Arn.) Maesen, *C. candollei* (Wight & Arn.) Maesen, *C. barbata* (nov.sp.), *C. cajanifolius, C. grandiflorus* (Benth. ex Bak.) Maesen, *C. kulnensis* (= *heynei*), C. *mollis* (Benth.) Maesen, *C. rostrata*, *C. rugosus* (Wight & Arn.) Maesen, and *C. scarabaeoides* (L.) Thours, are found in Eastern Peninsular India (Fig. 5.1). *Cajanus barbata, C. scarabaeoides, and C. villosus* (Benth. ex Baker) Maesen are found in the northeastern region. *Cajanus mollis, C. platycarpus* (Benth.) Maesen, and *C. villosus* are found in the upper Gangetic Plains. *Cajanus albicans, C. barbata, C. candollei, C. elongatus* (Benth.) Maesen, *C. grandiflorus, C. lineatus* (Wight & Arn.) Maesen, *C. mollis, C. nivea* Graham, *C. platycarpus, C. rugosus, C. scarabaeoides, C. sericeus* (Baker) Maesen, and *C. villosus* occur in the western peninsular tract and C. *scarabaeoides* in the Indus plains. Thus, a wide array of diversity in wild *Cajanus* species occurs in India with maximum concentration in the Western Ghats and in the northeast region (Fig. 5.1). The

A.K. Singh, *Wild Relatives of Cultivated Plants in India*,
DOI 10.1007/978-981-10-5116-6_5

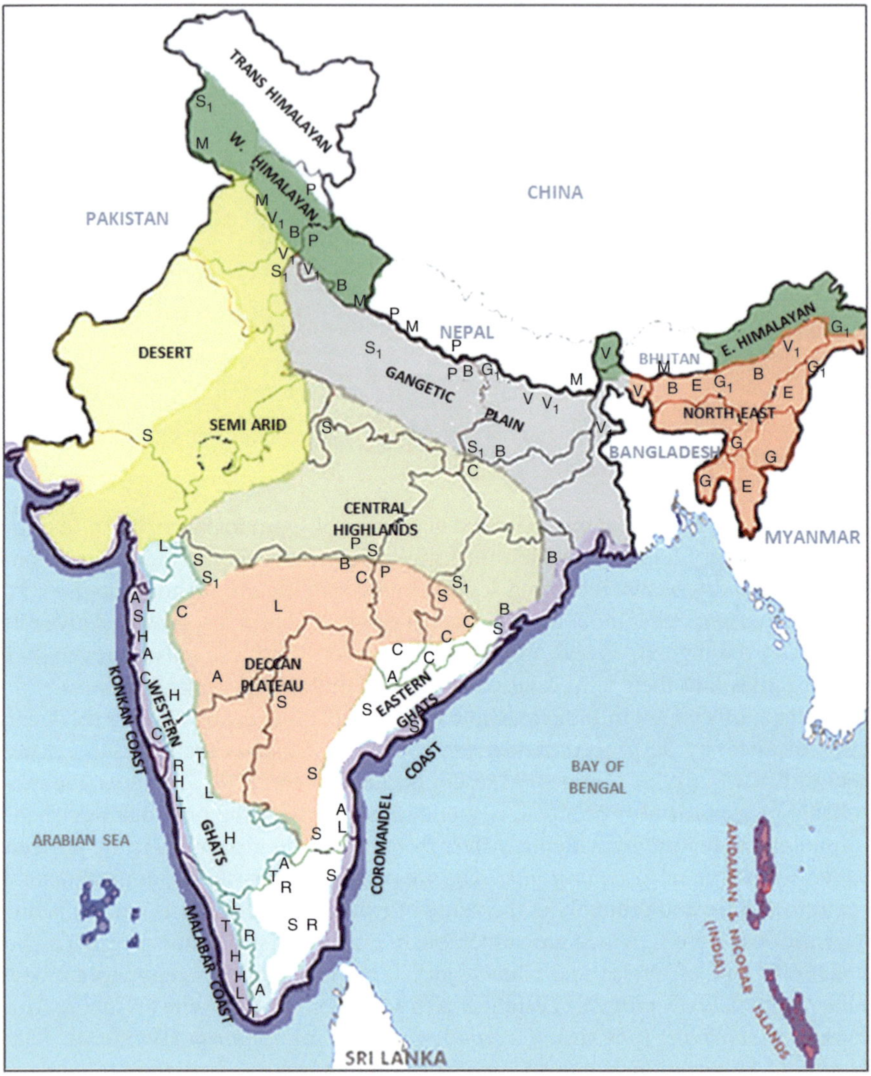

Fig. 5.1 Biogeographical regions associated with the distribution of wild *Cajanus* species A = *C. albicans*; C = *C. cajanifolius*; B = *C. crassus*; E = *C. elongatus*; G = *C. goensis*; G1 = *C. grandiflorus*; H = *C. heynei*; L = *C. lineatus*; M = *C. mollis*; P = *C. platycarpus*; R = *C. rugosus*; S = *C. sericeus*; S1 = *C. scarabaeoides*; T = *C. trinervius*; V = *C. villosus*; V1 = *C. volubilis*

area of variability of cultivated pigeonpea, *C. cajan*, is more in the eastern peninsular tract, along with closest wild relative counterpart, *C. cajanifolius*, endemic to Belladilla range in Orissa.

For Asiatic *Vigna* species, India is the center of diversity. These taxa include *Vigna mungo* var. *silvestris* Lukoki, Marechal & Otoul., related to black gram, *V. radiata* (L.) R. Wilczek, distributed in Khandala Ghats and Konkan belt of Western

Ghats; *V. radiata* var. *sublobata* in *tarai* range of sub-Himalayan tract and sporadically in western and eastern peninsular tracts; *Vigna grandis* akin to *V. radiata*, confined to hills of Khandala Ghats and Konkan belt; *Vigna dalzelliana* (Kuntze) Verdc., akin to the cultivated *V. angularis* (Willd) Ohwi & H. Ohashi and *V. umbellata* (Thunb.) Ohwi & H. Ohashi, occur in the Western Ghats; and wild *V. umbellata* types, akin to their cultivated forms, occur in peninsular hills, in the sub-Himalayan tract, and in the northeastern hills (Fig. 5.2). Among others, the wild forms of moth bean, *V. aconitifolia* (Jacq.) Marechal, are reported to occur sporadically, mainly in the northern or northwestern plains and in Deccan Plateau, while those of *V. trilobata*, besides occurring in these areas, extend to the sub-Himalayan tract also (Fig. 5.2). Yet another species, *V. capensis* (Thunb.) Burtt Davy, is reported to occur in the subtropical Himalayas and in peninsular hills, extending to the northeastern region.

Cicer microphyllum Royle, a semiprostrate-type Indian wild relative of chickpea, occurs in the high elevation zones (2700–3500 m) of the northwestern Himalayas.

The major botanical/phytogeographical areas of distribution of wild-related taxa of cultivated legume crops correspond to the areas with maximum species concentration in the eastern and western peninsular tracts, the northeastern parts, and the Western Himalayas (Figs. 5.1 and 5.2).

5.2 Wild Species Found in India

1. ***Cajanus albicans* (Wight & Arn.) Maesen (Leguminosae/Fabaceae)**. A climber, distributed in scrub vegetation and dry deciduous forests of Peninsular India and Sri Lanka. It is cross-compatible with pigeonpea, *C. cajan*, and part of secondary gene pool. Source of high seed protein (HSP) and sterility mosaic virus (SMV).
2. ***Cajanus cajanifolius* (Haines) Maesen**. A perennial shrub, endemic to dry deciduous forests and open grassland of Eastern Peninsular India, Orissa, and Bastar. Nearest wild relative of pigeonpea and probable ancestor. Member of secondary gene pool. Source of nuclear male sterility (NMS) and cytoplasmic male sterility (CMS).
3. ***Cajanus crassus* (Prain ex King) Maesen**. A tall climber with short pods, found in dry forests of northwest Himalayan foothills, Central India, Assam, and Eastern Ghats. Variability has been classified into two botanical varieties, var. *crassus* and var. *burmanicus*. The latter is a robust climber, climbing on trees, edges of dry forests with long pods, in Bihar and Orissa. Part of tertiary gene pool. Resistant to SMV.
4. ***Cajanus elongatus* (Benth.) Maesen**. A slender climber with flattish-oblong pods. Distributed in grasslands and open hillsides of Northeast India and western peninsular tract of the Indian Subcontinent.

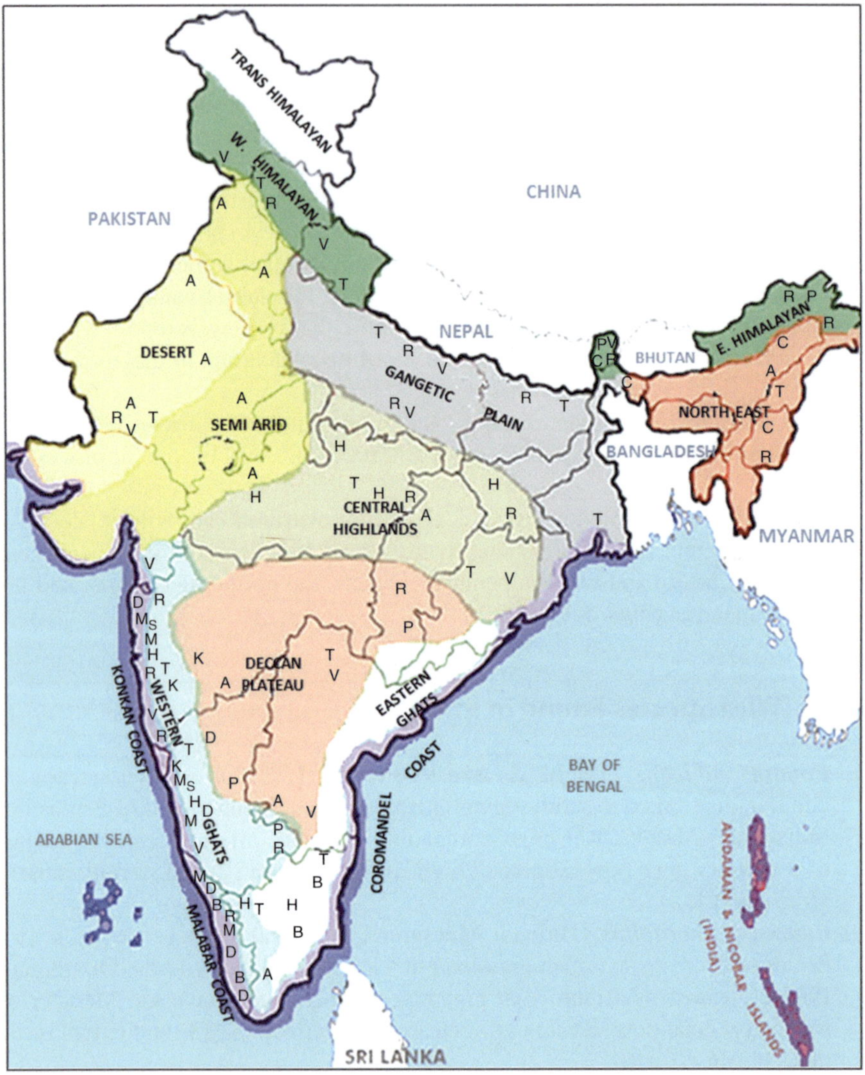

Fig. 5.2 Biogeographical regions associated with distribution of wild *Vigna* species
A = *V. aconitifolia*; B = *V. bournaea*; C = *V. clarkei*; D = *V. dalzelliana*; H = *V. hainiana*; K = *V. khandalensis*; M = *V. minima*; Ms. = *V. mungo var. silvestris*; P = *V. pilosa*; R = *V. radiata* var. *setulosa*; T = *V. trilobata*; V = *V. vexillata*

5. ***Cajanus goensis* Dalz**. A hairy climber with linear pods. Distributed in tropical dry deciduous/slight wet forests of Eastern Himalayas of Arunachal Pradesh and Assam.
6. ***Cajanus grandiflorus* (Benth. ex Bak.) Maesen**. A tall climber with sturdy pods. Distributed in Himalayas and eastern peninsular tract of India.

Fig. 5.3 Representative wild relatives of economically important grain legume crops: (**a**) *Cajanus scarabaeoides*, (**b**) *Cicer microphyllum*, (**c**) *Lathyrus aphaca*, and (**d**) *Vigna minima*

7. ***Cajanus heynei* (W. & A.) Maesen**. A climber with dark green leaves and sticky pods. Described by Van der Maesen. Distributed in forests of South Western Ghats of India and Sri Lanka.
8. ***Cajanus lineatus* (Wight & Arn.) Maesen**. An erect shrub 2.5 m tall with small pods. Distributed in tropical forests of western peninsular tract of the Indian Subcontinent, Western Ghats, and Nilgiri Hills. Cross-compatible with pigeonpea, *C. cajan*. Part of secondary gene pool. Source of cleistogamy and CMS.
9. ***Cajanus mollis* (Benth.) Maesen**. A sturdy climber with oblong pods. Distributed in pine or broadleaf forests of Himalayan foothills, upper Gangetic Plains, and eastern and western peninsular tract of India. Part of tertiary gene pool. Source of highest seed protein (33.4%) content.
10. ***Cajanus platycarpus* (Benth.) Maesen**. Creeper or climber flowering in 48 days and have flattish-oblong pods. Distributed in grasslands of upper Gangetic Plains, Central India, and western peninsular region. Crosses with pigeonpea, *C. cajan* with difficulty. Part of tertiary gene pool. Resistant to *Phytophthora* blight.
11. ***Cajanus rugosus* (Wight & Arn.) Maesen**. Climber-creeper with oblong pods 14–23 mm. Distributed in eastern and western peninsular tract of South India and Sri Lanka. Part of tertiary gene pool.

12. ***Cajanus scarabaeoides*** **(L.) Thours**. Gray-green climber-creeper with oblong pods (Fig. 5.3a). Distributed in Indus plains, northeastern region, and eastern and western peninsular tract of India. Cross-compatible with pigeonpea, *C. cajan*. Part of secondary gene pool. Used as source of CMS, high seed protein, and dwarfism. Also for resistance to pod borer (PB), SMV. Cultivated as a forage crop.
13. ***Cajanus sericeus*** **(Baker) Maesen**. Erect shrub to 1.5 m tall. Distributed in dry deciduous monsoon forests of western peninsular tract of the Indian Subcontinent and Mount Abu. Cross-compatible with pigeonpea, *C. cajan*, and part of secondary gene pool. Source of HSP, CMS, and resistance to both *Phytophthora* blight and SMV.
14. ***Cajanus trinervius*** **(DC.) Maesen**. Erect shrub to 2 m tall with oblong pods 2–4 cm long. Distributed in peninsular tract of South India, including Nilgiri Hills, and Sri Lanka. Cross-compatible with pigeonpea, *C. cajan*. Part of secondary gene pool,
15. ***Cajanus villosus*** **(Benth. ex Baker) Maesen**. Creeper-climber, 1 m tall. Distributed in Eastern Himalayas, Sikkim, Darjeeling, *terai* region, upper Gangetic Plains, and western peninsular tract of the Indian Subcontinent. Part of tertiary gene pool.
16. ***Cajanus volubilis*** **(Blanco) Blanco**. A twining vine with robust stem. Extended from southern China to Siwalik range, northern Indian plains, northeast West Bengal, and Assam. Also, recorded in Tamil Nadu and Kerala. Part of tertiary gene pool.

Khoury et al. (2015) have reviewed the distribution and potential of wild relatives of pigeonpea as genetic resources, particularly for abiotic tolerance.

17. ***Cicer macracanthum*** **Popov**. Wild perennial species confined to cold arid zone of Jammu and Kashmir, Ladakh, and adjacent areas. Possess medicinal property.
18. ***Cicer microphyllum*** **Royle**. A perennial species (Fig. 5.3b). Distributed in Asia temperate and Western Himalayas, between 2700 and 3550 m in Himachal Pradesh, Jammu and Kashmir, and Uttarakhand. Wild relative of cultivated chickpea *C. arietinum*, potential source for drought resistance (Toker et al. 2007).
19. ***Glycine javanica*** **Linn.; *G. max* subsp. *soja* (Siebold & Zucc.) H. Ohashi; syn. *G. soja* Siebold & Zucc**. A perennial species, distributed in Northeast Hill region of India, bordering with China and in hills of Western Ghats (Khandala, Mysore, Nilgiri and Pulneys). Natural hybridization between *G. max* ssp. *soja* and *G. max* ssp. *max* the cultivated soybean, by pollen flow has contributed to designing of genetic introgression from ssp. *soja* in cultivated soybean breeding program. Member of primary gene pool and source for resistance to yellow mosaic virus.

20. ***Glycine pentaphylla*** **Dalz.; syn.** ***Ophrestia pentaphylla*** **(Dalzell) Verdc**. Twining herb, distributed in dry deciduous forests Western Ghats, largely in Konkan region, including northern Karnataka, Wayanad, etc.
21. ***Lathyrus aphaca*** **L**. A trailing or scrambling annual plant producing yellow pea flowers growing as weed (Fig. 5.3c). Native to Africa, Europe, and Asia, including temperate zone and northern/upper Gangetic Plains. It is related to cultivated *L. sativus* Linn. (*Khesari* dal), whose weedy form also occur in northern, central, and western India up to 1200 m.
22. ***Macrotyloma ciliatum*** **(*Willd.*) Verdc**. A deciduous climber with smooth stem. Habitat to disturbed areas in Orissa, Andhra Pradesh, and Tamil Nadu to Western Ghats.
23. ***Macrotyloma sar-garhwalensis*** **R.D. Gaur & L.R. Dangwal**. It is a noon-twining annual herb found in Central Himalayas of India. Wild relative of horse gram, *M. uniflorum* (Lam.) Verdc., with high protein content of 38.35%, which can be introgressed into cultigens.
24. ***Rhynchosia*** **spp. Lour**. *Rhynchosia* species are perennial herbs with twining or trailing stems, considered genetic resources belonging to tertiary gene pool of pigeonpea, *Cajanus cajan*. They are represented by *R. acutifolia* F.v. Muell. Ex Benth., distributed in many parts of Indian Subcontinent; *R. aurea* DC., found throughout the plains and hills up to 1500 m; *R. bracteata* Baker in Bihar and Deccan Peninsula; and *R. rufescens* (Willd.) DC., in peninsular region and northeast and common *R. minima*. Other related genera are *Dunbaria* and *Eriosema*.
25. ***Vigna aconitifolia*** **(Jacq.) Marechal**. Wild-type moth bean, conspecific with cultigen types, is a creeping annual herb, native of arid and semi-arid regions of Rajasthan and Madhya Pradesh. Hot-season fodder, yielding palatable pasture and hay, drought resistant.
26. ***Vigna bournaea*** **Gamble**. A climber, distributed in Tamil Nadu and Malabar region of South India. Genetically distant from the *V. radiata*.
27. ***Vigna clarkei*** **Prain**. A slender woody climber 12–20 cm tall, distributed in northeastern region of the Indian Subcontinent. Only specimen recorded. It needs collection and further investigation.
28. ***Vigna dalzelliana*** **(Kuntze) Verdc**. Twining or creeping herb with glabrous stem. Native to tropical Asia, including southward to Khandala in Western Ghats. Belong to secondary gene pool of azuki bean, *V. angularis*, and tertiary gene pool of black gram, *V. mungo*; cowpea, *V. unguiculata*; mung bean *V. radiata*; and rice bean *V. umbellata* (Doi et al. 2002; Tomooka et al. 2011). Listed in the IUCN Red List of Threatened Species (Chadburn 2012a).
29. ***Vigna hainiana*** **Babu, Gopin. & S.K. Sharma**. A climber, distributed in Chota Nagpur Plateau, Central Highlands, Malwa Plateau, and Western Ghats. Closely related to *Vigna radiata* with 11% crossability and highest hybrid germination of 80%.
30. ***Vigna khandalensis*** **(Santapau) Raghavan & Wadhwa; syn.** ***V. grandis*** **(Dalz. & Gibbs.) Verdc**. A stout erect herb 50–100 cm tall. Endemic to Western Ghats (Khandala Ghats), sporadic in distribution. Closely related to *V. radiata*

with alike pods. Listed in the IUCN Red List of Threatened Species (Chadburn 2012b).

31. ***Vigna minima* (Roxb.) Ohwi & H. Ohashi**. Annual twining herbs, very variable (Fig. 5.3d). Native of Asia, including the parts of Western Ghats. Part of secondary gene pool of azuki bean *V. angularis* and rice bean *V. umbellata* and tertiary gene pool of black gram *V. mungo*, cowpea *V. unguiculata*, and mung bean *V. radiata* (Doi et al. 2002).
32. ***Vigna mungo* (Linn.) Hepper var. *silvestris* Lukoki, Marechal & Otoul**. Annual herb with small hardy seed coat. Native to Western Ghats. Progenitor and member of primary gene pool of black gram *V. mungo*, secondary gene pool of *V. radiata*, and tertiary gene pool of azuki bean *V. angularis*, cowpea *V. unguiculata*, and rice bean *V. umbellata* (Doi et al. 2002; Tomooka et al. 2011).
33. ***Vigna pilosa* (Willd.) Baker**. An herb, native to Asia, including the semi-evergreen forests of Peninsular India and Western Ghats. Contains medicinal properties. Needs further investigation on genetic potential.
34. ***Vigna radiata* var. *setulosa* (Dalzell) Ohwi & Ohashi; syn. var. *sublobata* (Roxb.) Verdec., *V. sublobata* (Roxb.) Bairig. et al**. An annual climber, 1 m tall. Found in *tarai* range of Himalayas, Siwalik Hills, and Kumaon up to 1200 m., sporadically in peninsular tract of Chota Nagpur and Western Ghats. Plants/pods/seeds are hairier in Konkan. Probable progenitor of cultigen mung bean *V. radiata* and member of primary gene pool (Bisht et al. 2005). Member of tertiary gene pool of azuki bean *Vigna angularis*, black gram *Vigna mungo*, cowpea *Vigna unguiculata*, and rice bean *Vigna unguiculata* (Doi et al. 2002). Source of drought and bruchid resistance.
35. ***Vigna trilobata* (L.) Verdc**. A regenerating annual twiner herb. Native to Indo-Malesia, including India in rocky areas of dry and moist deciduous forests. Belong to tertiary gene pool of all other cultivated *Vigna* species (Doi et al. 2002; Javadi et al. 2011). Sown as forage and green manure crop in semi-arid India.
36. ***Vigna vexillata* (L.) A. Rich.; syn. *Vigna capensis* (Thunb.) Burtt Davy**. A long strong trailing vine with stem clothed with spreading silky hairs, narrow pointed leaves, and fusiform tuberous rhizome/roots. It has a pantropical distribution in southern and eastern Africa, the Indian Subcontinent, and Southeast Asia. It is being considered as "linking" species between the African *Vignas* (subgenus *Plectrotropis*) and Asiatic *Vignas* (subgenus *Ceratotropis*). It is closest to the cowpea *V. unguiculata* and distant to the *V. radiata*. The genetic diversity is classified into botanical varieties, such as var. *vexillata* and var. *wightii* (Benth. ex Baker) Babu & Sharma, etc. Resistant to cowpea weevil. The seeds are boiled and consumed by the tribal living in the hilly region of Pune, India. Possess medicinal properties.

In addition, some more species such as *Vigna prainiana* Babu & S.K Sharma are described from Sikkim, Uttar Pradesh, and Southeast Asia but need collection and further investigations.

5.3 Perspective

The wild relatives of arid legumes such as *Vigna* spp., *Macrotyloma* spp., etc., are at risk, since they are relatively less known and collected less; therefore, there is a need for more collection and study of their relationships with cultivated species to facilitate use. Pigeonpea has been well worked out from taxonomic and phylogenetic point of view; however, there is a need for greater effort on the evaluation of wild genetic resources against biotic and abiotic stresses and nutritional traits, identification of associated genes or gene complexes, nomenclature of them, and dissemination of generated information to facilitate use. Further emphasis needs to be given on the exploitation of secondary and tertiary gene pool species employing biotechnological techniques for improved nutritional quality.

Asiatic *Vigna* species are represented by several economically important cultivated species and the wild-related species, presenting rich genetic diversity from Indian Subcontinent. There have been biosystematic studies, and it has been discovered that the cultivated species are phylogenetically closely related and most have conspecific wild forms or related species. They need to be collected and conserved along with cultivated species diversity. For these reasons, the wild species have great potential for widening the genetic base of cultivated *Vigna* species through interspecific hybridization, besides the targeted efforts for introgression of specific traits.

Several wild species have also revealed their own economic potential, such as *Vigna trilobata*, which can prove to be most useful forage crop in the semi-arid regions of the country, providing desired support to the livestock of the region.

References

Bisht IS, Bhat KV, Lakhanpaul S, Latha M, Jayan PK, Biswas BK, Singh AK (2005) Diversity and genetic resources of wild *Vigna* species in India. Genet Resour Crop Evol 52:53–68

Chadburn H (2012a) *Vigna dalzelliana*. The IUCN red list of threatened species 2012: e.T19892280A20163346

Chadburn H (2012b) *Vigna khandalensis*. The IUCN Red List of Threatened Species 2012: e.T19892969A20163711

Doi K, Kaga A, Tomooka N, Vaughan DA (2002) Molecular phylogeny of genus *Vigna* subgenus *Ceratotropis* based on rDNA ITS and atpB-rbcL intergenic spacer of cpDNA sequences. Genetica 114:129–145

Javadi F, Tun YT, Kawase M, Guan K, Yamaguchi H (2011) Molecular phylogeny of the subgenus *Ceratotropis* (genus *Vigna*, Leguminosae) reveals three eco-geographical groups and Late Pliocene-Pleisticene diversification: evidence from four plastid DNA region sequences. Ann Bot (Oxford) 108:367–380

Khoury CK, Castaneda-Alvarez NP, Achicanoy HA, Sosa CC, Bernau V, Kassa MT, Norton SL, van der Maesen LJG, Upadhyaya HD, Villegas AR, Jarvis A, Struik PC (2015) Crop wild relatives of pigeonpea [*Cajanus cajan* (L.) Millsp.]: distributions, *ex situ* conservation status, and potential genetic resources for abiotic stress tolerance. Biol Conserv 184:259–270

Toker C, Canci H, Yildirim T (2007) Evaluation of perennial wild *Cicer* species for drought resistance. Genet Resour Crop Evol 54:1781–1786

Tomooka N, Kaga A, Isemura T, Vaughan D (2011) Chapter 15. Vigna. In: Kole C (ed) Wild crop relatives: genomic and breeding resources, legume crops and forages. Springer, Berlin/Heidelberg, pp 291–311

Oilseeds

6

6.1 Introduction

Oilseeds constitute a very important group of commercial crops in India. Of the nine oilseed crops cultivated, five, rapeseed mustard, groundnut, sesame, linseed, and castor, are major. Of these Indian mustards, sesame, safflower, and linseeds are partially indigenous in nature and are represented by their wild types/species in genera such as *Brassica* and *Lepidium*, *Carthamus*, *Linum*, and *Sesamum*, distributed in the different parts of the continent (Fig. 6.1). In Brassiceae, two wild species, *Brassica trilocularis* and *B. quadrivalvis*, also considered varieties of *B. campestris* var. *sarson*, occur in northern plains. The former is distributed largely in Central and Eastern Himalaya and in the sub-hilly tracts of Assam, while the *quadrivalvis* types occur in the fields of *B. campestris* in the upper Gangetic Plains (Fig. 6.1). Another wild or spontaneous type is *B. tournefortii* occurring in northern India. In the Himalayas, much species diversity occurs in another Brassiceae, genera, *Lepidium*, represented by *L. capitatum*, *L. latifolium*, *L. ruderale*, and *L. draba*. *Lepidium sativum* botanically related to mustard is cultivated as culinary vegetable, whose leaves are eaten raw as salad and cooked, while seeds are used as condiments, medicine, and edible seed oil with potential for biofuel. *Carthamus oxyacantha*, related to the cultivated safflower, C. *tinctoria* (of which it may be a wild form), is a weed in the northern plains. While the wild species of *Sesamum*, such as *S. laciniatum*, are distributed in the southern peninsular tract in Andhra Pradesh and Tamil Nadu, *S. mulayanum* is distributed in diverse parts of the subcontinent, and naturalized populations of S. *indicum* occur sporadically in northern plains and in the Aravalli ranges.

The wild relatives of dual crops like linseed (*Linum usitatissimum*), where seed is the source oil and stem of fiber. Though the type of cultigens of the two are morphologically different, where seed producing are shorter compared to fiber producing, which are tall to facilitate greater fiber yield, the wild *Linum* species can be genetic resource for both. *Linum* is not native of India; however, wild species *L. mysorense* found in Western Ghats is related to cultivated *L. usitatissimum*, and

A.K. Singh, *Wild Relatives of Cultivated Plants in India*,
DOI 10.1007/978-981-10-5116-6_6

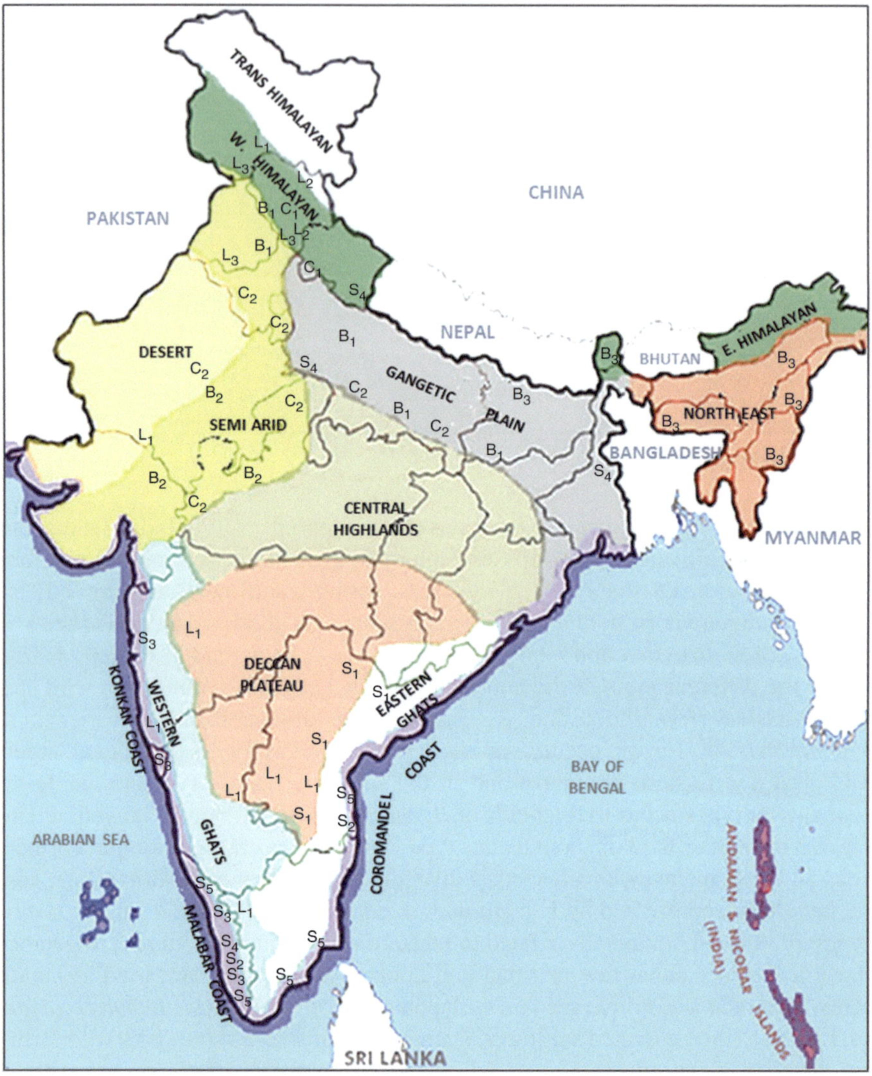

Fig. 6.1 Biogeographical regions associated with distribution of wild species of oilseed crops, *Brassica*, *Carthamus*, *Linum*, and *Sesamum*
B1 = *Brassica quadrivalvis*; B2 = *B. tournefortii*; B3 = *B. trilocularis*; C1 = *Carthamus lanatus*; C2 = *C. oxyacantha*; L1 = *Linum mysorense*; L2 = *L. perenne*; L3 = *L. strictum*; S1 = *Sesamum alatum*; S2 = *S. laciniatum*; S3 = *S. malabaricum*; S4 = *S. mulayanum*; S5 = *S. radiatum*

its distribution extends to Western Himalayas and in western India to Mount Abu in Rajasthan and the Aravallis. This is however different from the cold hardy *L. perenne* types available in Western Himalayas, in Lahaul and Spiti and *L. strictum* (Fig. 6.1). Thus, overall major distribution areas of the wild relatives of oilseed crops are in the Indo-Gangetic Plains and the Western Himalayas and to a lesser extent South India.

6.2 Wild Species Found in India

1. ***Brassica napus* ssp. *napus* var. *quadrivalvis* (Hook. f. & Thom.) O. Schulz; syn. *B. quadrivalvis* Hook.f. & Thom**. Annual or biennial, 30–150 cm tall, erect plant. Found in fields of *B. campestris* in northern parts of Punjab and upper and middle Gangetic Plains. It has four-valved-*quadrivalvis*-type capsule.
2. ***Brassica tournefortii* Gouan**. A spreading annual herb 1.25 m tall with dull yellow flower (Fig. 6.2a), referred as Asian mustard. Native to North Africa and Middle East, invading into semidry plains of western India. Source of cytoplasmic male sterility with potential for gene transfer in rapeseed. Member of secondary gene pool for cabbage and tertiary gene pool for other mustard and Brassiceae vegetable crops (Warwick et al. 2006; FitzJohn et al. 2007).
3. ***Brassica rapa* var. *trilocularis* (Roxb.) O. Schulz.; syn. *B. trilocularis* Hook.f. & Thomson; syn. *B. rapa* L. ssp. *trilocularis* (Roxb.) Hanelt; var. *trilocularis* (Roxb.) Kitam**. A biennial herb 0.9 m tall. It is distributed throughout the temperate regions of the world, including Central and Eastern Himalayas, Sikkim, and sub-hilly tracts of Assam. Cultivated. These are three-valved types, *trilocularis*. Contain AA genome, 2n = 20. Crossed successfully with *B. tournefortii*.
4. ***Carthamus lanatus* Linn**. It is an annual species growing to 0.6 m, of thistle, known as wooly distaff or saffron thistle. Native of Mediterranean basin with an extended distribution into Northwestern Himalayas. Closely related to safflower, *C. tinctorius*.
5. ***Carthamus oxyacantha* M. Bieb**. It is a spiny-leaved annual herb; many-branched, florets orange-yellow, fruit achene obovate or elliptic; noxious pernicious weed (Fig. 6.2b) in winter crops. Native of Asia, including arid and semi-arid northern plains of India. It is a wild form of C. *tinctorius* and member of primary gene pool (Mayerhofer et al. 2011). Highly drought tolerant and have medicinal properties.
6. ***Lepidium capitatum* Hook.f. & Thomson**. Annual or biennial herbs, glabrous or sparsely pubescent with simple or branched capitate hairs, 10–30 cm tall, not scapose. Distributed on damp grassy slopes of temperate Himalayas up to 4000 m, from Kumaon to Sikkim and West Bengal. Seeds are oily, used in traditional medicine. Genetic resource for *L. sativum*.
7. ***Lepidium draba* L**. Whitetop or hoary cress, a perennial mustard herb, glabrous or sparsely hirsute with simple or branched hairs, about 20–100 cm tall, not scapose. Native of West Asia and Eastern Europe, extended to Himalayas. Occur as weed in Punjab and northern and Northeast Hills. Genetic resource for *L. sativum*.
8. ***Lepidium latifolium* L**. A perennial pepperweed, 30 cm to 1 meter tall. Native to Southern Europe, Mediterranean, and Northwestern Himalayas. Used as phytofoods among cold arid region of Ladakh. Its leaves contribute significantly to people's diet and contain nutritionally significant levels of fatty acids and glucosinolates. Close ally of *L. sativum* and source of cold tolerance.

Fig. 6.2 Representative wild relatives of economically important oilseed crops: (**a**) *Brassica tournefortii*, (**b**) *Carthamus oxyacantha*, (**c**) *Linum strictum*, and (**d**) *Sesamum alatum*

9. ***Lepidium ruderale*** **L.** Narrow-leaf pepperwort, an annual/biennial weed, 0.3 m tall. Native to Asia. Grows on roadsides, near cultivated fields, and in pastures in Western Himalayas (Kashmir to Kumaon). Also, recorded from Tamil Nadu. The plant contains diterpene bitter principles, iridoid monoterpenes, and flavonoids including rutin and quercitrin and leonurin, hence medicinal. Ally of *L. sativum*.
10. ***Linum mysorense*** **Heyne ex Wall. (Linaceae)**. Mysore flax, *Undri*, is an endemic annual erect herb with slender stem branched at the top. Common in grasslands of Western Ghats, Ratnagiri, Deccan Hills, and Mahabaleshwar/Khandala Ghats down south to Nilgiri on open slopes up to 1800 m. Also, sporadically distributed in Western Himalayas and in western India in Mount Abu.
11. ***Linum perenne*** **Linn. (Linaceae)**. Perennial, native to Europe, probably extended or escaped and naturalized in the northwestern Himalayas up to 4000 m with cold hardy type growing in Lahaul and Spiti region of Himachal Pradesh and other parts of Western Himalayas. Belong to tertiary gene pool of *L. usitatissimum*, based on incompatibilities between flax and *L. perenne* (Jhala et al. 2008).

12. ***Linum strictum* Linn. (Linaceae)**. It is an annual plant (Fig. 6.2c), widespread in Southern Europe, escaped and naturalized to Western Himalayas and north-western plains. It has robust stems, 10–45 cm high, and therefore is known for fiber strictness, drawn close together very upright. *L. strictum* appear important in improvement of the cultivated linseeds regarding number of tillers (Seetharam 1972). Also, known for medicinal properties.
13. ***Sesamum alatum* Thonn**. A tall erect herb (Fig. 6.2d). Native to Africa and tropical India, habitat to disturbed moist places in Andhra Pradesh. Whole plant used as vegetable, fodder, medicine, and seeds for oil. Interspecific hybrid between *S. alatum* and *S. indicum* (both 2n = 26) has been produced through ovule culture to facilitate gene transfer (Rajeswari et al. 2010). Resistant to phyllody disease.
14. ***Sesamum laciniatum* Klein ex Willd**. Prostrate, much-branched herbs, to 1 m tall; stem and branches grooved, sparsely hairy, with 2n = 32. Endemic to Peninsular India, from coastal Andhra Pradesh to Tamil Nadu, extending up to Cochin. Companion of strong smelling species, *S. prostratum* Retz., found in similar habitat. Based on morphological similarity and similar 2n = 32, the two have been combined (Bedigian 2015). It has resistance potential against the diseases and pests of sesame.
15. ***Sesamum malabaricum* L.; syn. *S. indicum* var. *malabaricum* (Burm.) Christenh**. An annual herb, native to Malabar region of southern Western Ghats, extending up to Goa and Maharashtra. Most closely related to cultivated *S. indicum* with same chromosome number (2n = 26). A probable progenitor and member of primary gene pool. Source of cytoplasmic sterility and powdery mildew disease of sesame.
16. ***Sesamum mulayanum* Nair**. *Ran Til*, Nair, described this species in 1964 from North India, now recorded in southern Western Ghats, Garhwal Himalaya, and West Bengal. Morphologically similar with same chromosome number (2n = 26), cross-compatible with *S. indicum*, producing fertile hybrid, suggest that two are component of the same biological species (Kawase 2000). On archeological evidences, Fuller (2003) suggested *S. malabaricum* and *S. mulayanum* are synonyms. Therefore, *S. mulayanum* and *S. malabaricum* are wild form of cultivated *S. indicum* and part of primary gene pool. It is source of resistance to phyllody, powdery mildew and wilt.
17. ***Sesamum radiatum* Schumach**. An erect annual plant growing up to 1.5 m tall with 2n = 64. Native of West Africa, naturalized to sandy hill areas of Coastal Andhra Pradesh, Tamil Nadu, and Kerala. Used as vegetable and oilseed. Also, has medicinal properties.

6.3 Perspective

Despite significant systematic and biosystematic studies, confusion prevails in relation to subspecific or varietal status of certain taxa belonging *Brassica rapa*, which need further attention and resolution. Further collection of wild species and their evaluation against biotic and abiotic stress may promote use and thereby much needed progress in widening of genetic base of oilseed *Brassica* species.

In case of *Sesamum*, though recent biosystematic studies have helped in the generation of desired information regarding the phylogenetic relationships of the wild species with cultivated *S. indicum*, further genome analysis using molecular techniques is needed to facilitate better understanding and promote the use of wild species in genetic improvement of cultigens. Also, evaluation of wild accessions against specific traits, such as biotic and abiotic stresses and nutritional quality, is needed to understand their potential value as genetic resource. In other crops like, *Carthamus*, national program needs to be energized with collection of wild germplasm to facilitate use.

References

Bedigian D (2015) Systematics and evolution in *Sesamum* L. (Pedaliaceae), part 1: evidences regarding origin of sesame and its closest relatives. Webbia: J Plant Taxon Geogr 70(1):1–42. doi:10.1080/00837792.2014.968457

FitzJohn RG, Armstrong TT, Newstrom-Lloyd LE, Wilton AD, Cochrane M (2007) Hybridisation within *Brassica* and allied genera: evaluation of potential for transgene escape. Euphytica 158:209–230

Fuller DQ (2003) Further evidences on the prehistory of Seseame. Asian Agri-History 7(2):127–137

Jhala AJ, Hall LM, Hall JC (2008) Potential hybridization of flax with weedy and wild relatives: an avenue for movement of engineered genes. Crop Sci 48:825–840

Kawase M (2000) Genetic relationships of the ruderal weed type and the associated weed type of *Sesamum mulayanum* Nair distributed in the Indian subcontinent to cultivated sesame *S. indicum* L. Jpn J Trop Agric 44(2):115–122

Mayerhofer M, Mayerhofer R, Topinka D, Christianson J, Good A (2011) Introgression potential between safflower (*Carthamus tinctorius*) and wild relatives of the genus *Carthamus*. BMC Plant Biol 11:47. ttp://dx.doi.org/10.1186/14712229

Rajeswari S, Thiruvengadam V, Ramaswamy NM (2010) Production of interspecific hybrids between *Sesamum alatum* Thonn and *Sesamum indicum* L. through ovule culture and screening for phyllody disease resistance. S Af J Bot 26:252–258

Seetharam A (1972) Interspecific hybridization in *Linum*. Euphytica 21(3):489–495

Warwick SI, Francis A, Al-Shehbaz IA (2006) Brassicaceae: species checklist and database on CD-ROM. Plant Syst Evol 259:249–258

7 Fiber Crops

7.1 Introduction

Cotton, *Gossypium* spp., is the major fiber crop of global importance with high commercial value. It has diploid species (2n = 26) distributed in India. The great Indo-Gangetic alluvium of North India is the center of origin for the diploid species *G. arboreum* representing "A" genome, which is considered part of the secondary gene pool of the dominant New World tetraploid cotton and is cultivated and represented by several landraces and wild types. *Gossypium herbaceum* is the sister species found wild in Gujarat and Karnataka. In addition to wild species, weedy races, uncultivated geographical races, and perennial tree cotton species of genera related to *Gossypium*, such as *Hibiscus*, *Abutilon*, etc., with desired traits not found in *Gossypium* gene pool can be considered distant wild relatives for gene transfer.

Jute and allied fibers are other important commercial fiber crops. Eight different species constitute this group. These are tossa jute (*Corchorus olitorius* L.), white jute (*C. capsularis*), roselle/mesta (*Hibiscus sabdariffa* L.), kenaf/mesta (*H. cannabinus* L.), sunn hemp (*Crotalaria juncea* L.) (Fig. 7.1), flax (*Linum usitatissimum* L.), ramie [*Boehmeria nivea* (L.) Gaud.], and sisal (*Agave sisalana* Perr.). Nine species of *Corchorus* are indigenous to India (Mahapatra et al. 1998), including cultivated *C. capsularis* and *C. olitorius* (Fig. 7.1). *Corchorus olitorius*, introduced from Africa, has naturalized in most tropical regions as an escape from cultivation generating variability and creating complexity regarding its origin. *Corchorus aestuans* seems to be the most dominating species followed by *C. tridens*, *C. trilocularis*, and *C. fascicularis*. The other three species, *C. depressus* Stocks, *C. urticifolius* Lam., and *C. velutinus* Her., are restricted in their distribution and lower than the other species. In *Hibiscus* 36 species have been reported from India with taxonomic complexity. *Crotalaria* is represented by many species, while the cultivated species, *C. juncea* L., is believed to be the native of India. *Boehmeria* is represented by 40 species from India (Hooker 1885). About 19 species are reported from Himalayan and Northeast Hills (Kanjilal et al. 1940), 2 from Bengal (Prain 1903), 4 from the Western Ghats, and 5 from the Western Peninsula (Cooke 1967). Further Kundu

A.K. Singh, *Wild Relatives of Cultivated Plants in India*,
DOI 10.1007/978-981-10-5116-6_7

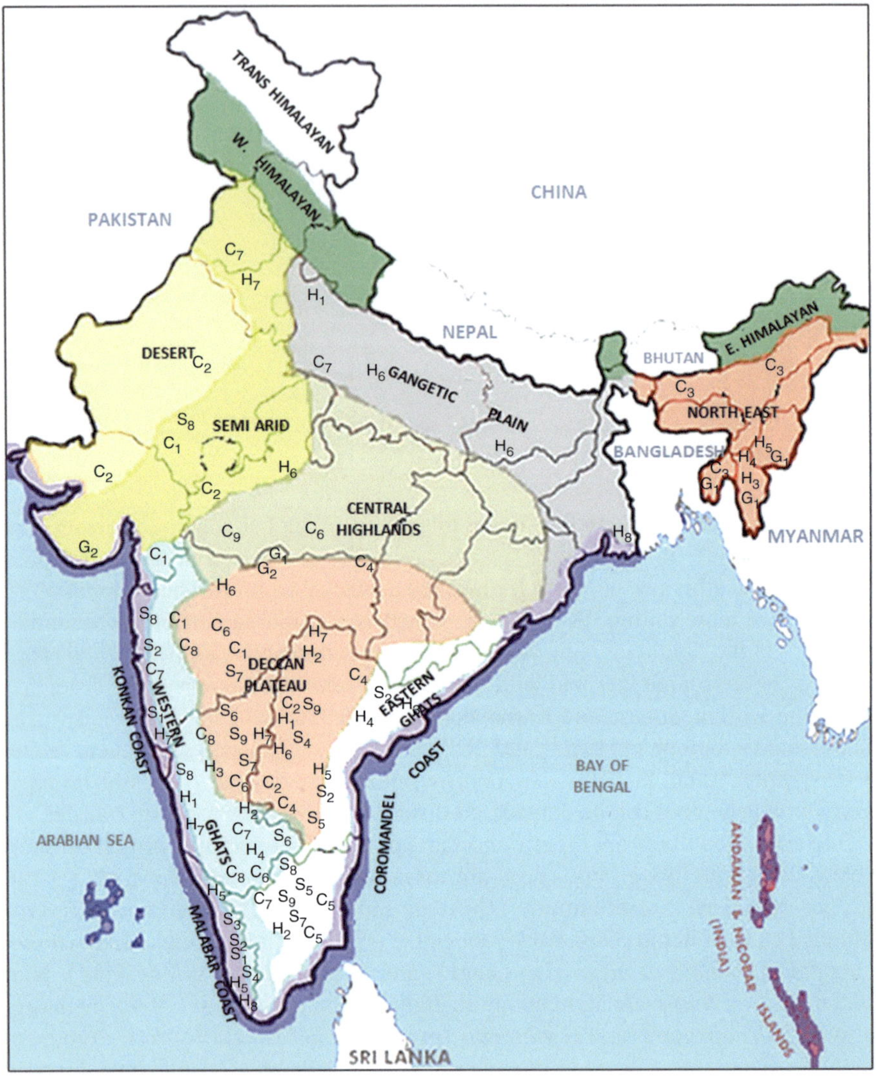

Fig. 7.1 Biogeographical regions associated with the distribution of wild species of fiber crops, *Corchorus*, *Gossypium*, *Hibiscus*, and *Crotalaria*

C1 = *Corchorus acutangulus*; C2 = *C. antichorus*; C3 = *C. capsularis*; C4 = *C. fascicularis*; C5 = *C. pseudo-olitorius*; C6 = *C. tridens*; C7 = *C. trilocularis*; C8 = *C. urticifolia*; G1 = *Gossypium arboreum*; G2 = *G. herbaceum*; H1 = *Hibiscus cannabinus*; H2 = *H. eriocarpus*; H3 = *H. ficulneus*; H4 = *H. furcatus*; H5 = *H. hispidissimus*; H6 = *H. radiatus*; H7 = *H. surattensis*; H8 = *H. tiliaceus*; S1 = *Crotalaria clarkei*; S2 = *C. digitata*; S3 = *C. grahamiana*; S4 = *C. madurensis*; S5 = *C. medicaginea*; S6 = *C. pallida*; S7 = *C. paniculata*; S8 = *C. retusa*; S9 = *C. willdenowiana*

(1958) reported natural occurrence of ramie (*Boehmeria*) in the parts of Arunachal Pradesh, Assam, Himachal Pradesh (Kangra Valley), and Tamil Nadu (Nilgiri Hills). In case of flax, where the seed produces oil and the stem fiber (flax), wild *Linum* species, *L. mysorense* and *L. perenne*, have already been discussed under oilseeds. *Urena lobata*, a marginal fiber crop traditionally used for making ropes and coarse textiles, though of African origin, occurs in India wild as forest undergrowth throughout the tropics and subtropics. *Urena sinuata* may be a form of this species, whereas *U. repanda,* yielding better quality fiber, is confined to the Gangetic Plains, with sporadic distribution in the *tarai* range and in Western and South India.

7.2 Wild Species Found in India

1. ***Boehmeria malabarica* Wall. ex Wedd. (*Urticaceae*)**. An erect shrub or small tree with spreading branches. Distributed in Eastern Himalayas and hills, up to 1500 m, and Western Ghats. A variety *leioclada* of Sect. Boehmeria, clustered into a clade with Sect. Tilocnide, provide solid support for the expansion of wild ramie core germplasm resources (Run-Qing et al. 2015).
2. ***Boehmeria macrophylla* Hornem**. A perennial herb or large shrub with brown bark. Native to Asia, including Assam, Meghalaya, Bihar, and Uttar Pradesh (Kumaun Hills). Produce excellent quality fiber from bark, which is shiny, white, and strong. Possess medicinal potential too.
3. ***Boehmeria platyphylla* D. Don**. A variable large spreading shrub with considerable variability, splitting into many varieties (Hooker 1885). Found in tropical and subtropical Himalayas and Deccan Peninsula. *Boehmeria platyphylla* var. *scabrella* or *B. scabrella* Gaud. is a variant distributed in Koina valley below Mahabaleswar, Orissa, Chota Nagpur, North Bengal, and Uttar Pradesh. Close to *B. macrophylla*, even considered synonym. Yields strong fiber, used for making ropes.
4. ***Boehmeria* rugulosa Wedd**. A medium-sized tree, 10 m tall called *Dar* in Nepalese. Distributed in Sikkim, Naga Hills, Someswar Hill of Orissa, and Uttar Pradesh.
5. ***Boehmeria sidaefolia* Wedd**. A slender diffused undershrub. Distributed in North Cachar Hills (Assam), Hantia Hills, Meghalaya, Arunachal Pradesh, Eastern Nepal, and Sikkim.

Many more species, such as *Boehmeria polystachya* Wedd., a robust shrub with stout branches, distributed in temperate Arunachal Pradesh, Meghalaya, Kumaun Hills, and Bhutan, have been reported. However, most require further investigations to ascertain their species entity and genetic potential as genetic resource.

6. ***Corchorus aestuans* L.; syn *C. acutangulus* Lam. (Tiliaceae)**. Annual, much-branched, spreading herb to undershrub. Native to Americas, naturalized as weed all over India in degraded deciduous forests and wastelands of the hotter

parts and western peninsular tract, indicating wide adaptability. Source of resistance to stem rot caused by *Macrophomina phaseolina*.

7. ***Corchorus depressus* (L.) Stocks; syn. *C. antichorus* Raeuch**. A mat-forming, prostrate, much-branched woody perennial. It is restricted to Central and Northwest India, particularly in hotter/drier parts of Rajasthan and Gujarat with drought-/heat-tolerant type. Also, a promising ethnomedicinal plant.
8. ***Corchorus capsularis* L.** Bushy/branched shrubby types found in wild/semiwild state in northeastern Assam plains and bordering hills up to Tripura. Exact native location is obscure due to extensive cultivation, but appears to be native of Indo-Myanmar region. Cross-compatible with *C. olitorius*. Fiber quality is finer than *C. olitorius*.
9. ***Corchorus fascicularis* Lam. (Tiliaceae)**. Semi-woody, erect, or prostrate annual plant widespread in hotter, drying moist places of Central and southern Peninsular India; drought-/heat-tolerant/adaptable. The extracted fiber is used in rope making. Medicinal too.
10. ***Corchorus pseudo-olitorius* Islam & Zaid**. Plant with narrow lanceolate leaves and glabrous stem. A new addition to Indian flora from Tamil Nadu (Mahapatra et al. 2003). Genetic resource, showed immune reaction to fungal diseases, stem rot, root rot, black band, soft rot, and anthracnose. Medicinal plant too.
11. ***Corchorus tridens* L.** A weedy annual herb. Pantropical, including dry deciduous forests/hotter parts of Central, Western, and southern parts of India. Used as vegetable (leaves) and fodder. It has high protein (19–25% dry weight), with essential amino acid composition meeting World Health Organization's standard.
12. ***Corchorus trilocularis* L.** An erect or suberect annual or biennial weedy herb. Paleotropic, in Indian Subcontinent adapted to dry deciduous forests and hotter lands of upper Gangetic Plains to central, western, and southern parts. Young tender leaves are cooked as vegetable.
13. ***Corchorus urticifolia* Wight & Arn**. An annual herb to subshrub 1 m tall, pubescent. Distributed in East Africa and Indo-Malesia region. In India, it is confined to western and southern parts. Yet to be collected and need characterization and evaluation.

Another species, *Corchorus velutinus* Her., with doubtful entity has been documented based on herbarium. It is expected to occur in southern and western parts of the country (Mahapatra et al. 1998).

14. ***Crotalaria clarkei* Gamble (Fabaceae/Leguminosae)**. A suberect herb with quadrangular, pilose stem. Endemic to evergreen forests and grasslands of Western Ghats (Malabar region), including Maharashtra. Medicine for neurological disorders. Listed indeterminate in the 1997 IUCN Red List of Threatened Plants (Walter and Gillett 1998).
15. ***Crotalaria digitata* Hook**. A woody shrub, native to evergreen forests of Western Ghats and Eastern Ghats. Shepherd community uses it in ethnoveteri-

nary formulations. Listed rare in the 1997 IUCN Red List of Threatened Plants (Walter and Gillett 1998).

16. ***Crotalaria grahamiana* Wight & Arn.** Bushy rattlepod is a dense stout subshrub, 1 m tall. Endemic to grasslands and montane scrub jungles of Southern Western Ghats. Soil improver, cultivated as green manure/cover crop. Poisonous too.
17. ***Crotalaria madurensis* Wight & Arn.** Much-branched, silky tomentose shrub. Endemic to Shola forests and grasslands of Southern Western Ghats and Eastern Ghats (Srisailam) and Andaman and Nicobar Islands. Antimicrobial and poisonous too.
18. ***Crotalaria medicaginea* var. *rigida* (Heyne ex Roth) Kuntze; syn. *C. rigida* Heyne ex Roth.** Endemic to dry districts of Andhra Pradesh and Tamil Nadu. Listed rare in 1997 IUCN Red List of Threatened Plants (Walter and Gillett 1998).
19. ***Crotalaria pallida* Aiton; syn. C. *mucronata* Desv.; *C. striata* DC.** Smooth rattlepod, perennial erect herb to subshrub, often well branched, found in wasteland and roadsides. Native to Africa, introduced and naturalized to India in western peninsular tract, extending up to Andaman and Nicobar. Yields fiber like sunn hemp. Soil improver, cultivated as green manure/cover/forage crop. Seed is poisonous.
20. ***Crotalaria paniculata* Willd.** A 1 m tall perennial undershrub. Native of India and endemic to Leeward semi-arid Deccan plateau (Andhra Pradesh, Karnataka, and Tamil Nadu). Two varieties are recorded, var. *nagarjunakondensis* Thoth. and var. *paniculata*, with distinct cordate bracts. Variety *nagarjunakondensis* is endemic to the Nagarjunakonda Valley. It is listed as threatened endemic plants from Eastern Ghats, India, by Sudhakar et al. (2003) and Chadburn (2012), respectively. Listed medicinal.
21. ***Crotalaria retusa* Linn.** Rattleweed, an annual herb or undershrub 1.75 m tall. Exact native land not known, found throughout tropics, including all vegetation types, in hotter parts of India. Common in coastal areas. Much variability in Western Ghats. Cultivated as fiber and cover crop for erosion control, poisonous.
22. ***Crotalaria verrucosa* L.** *Shanapushpi* (blue-flower rattlepod), erect branched shrub, 0.75 m tall. Pantropical, including deciduous forests and plains of India. Cultivated as soil improver cover crop. Used for treating indigestion, laryngitis, Kapha, etc.
23. ***Crotalaria willdenowiana* DC.** Erect perennial herbs/undershrubs, 20–75 cm tall. Endemic to Peninsular India. Variability is classified into subspecies *glabrifoliolata* and recently described ssp. *varmae* from Andaman (Reddy et al. 2007). Medicinal too.
24. ***Eulaliopsis binata* (Retz.) Hubb. (Poaceae/Gramineae).** *Babui*, a tufted perennial grass distributed in eastern Asia, including India, on dry base slopes and forest banks of sub-Himalayan areas. It is common in Bihar Orissa, Bengal, Central India, and Punjab. Collected or cultivated for papermaking, strings, ropes, and mats.

25. ***Gossypium arboreum* L. (Malvaceae)**. Tree cotton is a shrub attaining heights of 1–2 m, native to India and other parts of tropical old world. Wild types reported from Chittagong hills with possibility in hills of Tripura and Mizoram and Melghat Forest, Maharashtra (Singh et al. 2003). Evidence of its cultivation is found in Harappa; however, cultivated type appears to have originated in Gangetic Plains; several cultigens are still cultivated evolving landraces. Contributor of AA genome (2n = 26) to tetraploid cotton. Genome sequencing done. Interspecific hybrids have been obtained both with diploid tetraploid *Gossypium* species. Central Institute for Cotton Research (CICR), Nagpur and Coimbatore, conserves around 1900 accessions. Source for resistance pest and diseases and tolerance to drought (Liu et al. 2006; Ahmad et al. 2011).
26. ***Gossypium herbaceum* L**. *Rakta Karpasa*, an annual or perennial shrub with hairy leaves, diploid (2n = 26, AA). Native of Africa, but wild types do occur in Melghat forests, Maharashtra, and Gujarat. Cultivated in Maharashtra, Tamil Nadu, Andhra Pradesh, and Gujarat. It has wide adaptability and resistance to biotic (insects and diseases) and abiotic (drought) stresses. Exploitable genetic variability for yield, yield components, fiber length, and seed oil content is available (Singh et al. 1993). Member of primary gene pool of *G. arboretum* (Beasley 1942) and tertiary gene pool *G. hirsutum* and *G. barbadense* (Wendel et al. 2009).
27. ***Hibiscus cannabinus* L. (Malvaceae)**. Kenaf, *Deccan hemp*, semi-wild perennial erect shrub, 1.8 m tall, with many branches. Found mainly in Deccan Peninsula, Western Ghats, and all over the country up to an elevation of 1000 m, including lower Himalayas. Endemic to tropical areas, probably native of south Asia (India). Cultivated for its bast fiber, fodder and reported to be used in Ayurveda.
28. ***Hibiscus eriocarpus* DC.; syn. *H. platanifolius* (Willd.) Sweet**. Large shrub to tall tree. Found in deciduous to evergreen forests and exposed slopes of Peninsular India and Sri Lanka. Genetic resource for both kenaf and roselle.
29. ***Hibiscus ficulneus* L.; syn. *Abelmoschus ficulneus* (L.) Wight & Arn**. Native rosella, fibrous perennial erect shrub, 1 to 2 m tall, with woody stem. Native to Africa and tropical Asia, including India as weed throughout the plains, Northeast, Eastern, and Western Ghats. As genetic resource related to both rosella and okra. Medicinal too.
30. ***Hibiscus furcatus* Wall**. A prickly trailing shrub, up to 1 m, habitat to the river banks and stream areas throughout tropical Africa and Asia, including Indian plains. Antioxidant, genetic resource for kenaf and roselle.
31. ***Hibiscus hispidissimus* Griff**. Rambling or climbing shrub with lemon yellow flowers. Paleotropic, habitat to dry and moist deciduous forests, and plains of Northeast, Eastern and Western Ghats. Used in tribal medicine of Kerala.
32. ***Hibiscus radiatus* Cav**. It is a small shrub growing about 1 m tall with small prickles on the undersides of leaves and along stems and red velvet flowers. Native to tropical Asia, including India throughout the plains. Cultivated as vegetable, ornamental and folklore medicine. Genetic resource for kenaf and roselle.

33. ***Hibiscus surattensis*** **L.** A weak-stemmed, prostrate, or climbing plant covered with soft hairs and scattered prickles. Native to tropical Africa and Asia, including moist deciduous forests and plains of upper Gangetic Plains, Deccan Peninsula, and Western Ghats. Medicinal too.
34. ***Hibiscus tiliaceus*** **L.** *Bola*, coastal cottonwood tree 4–10 m tall. Native to old world, habitat to banks of tidal streams and mangroves. Common in eastern and western coasts of India near the backwaters in Sundarbans and Andaman Islands. Wood is used for diverse applications from firewood to seacraft construction. Cultivated.
35. ***Urena lobata*** **Linn. (Malvaceae)**. *Bachita*, Caesar weed, or Congo jute is an annual, variable, erect shrub with stem covered with minute starlike hairs, 0.5 to 2.5 m tall. Invasive, pantropical weed distributed in open areas, roadside, and plantations and as forest undergrowth in hotter parts of India. A hardy type var. *sinuata* is recorded in Narsapur, Medak, Telangana. Cultivated as fiber crop, as fine and supple-type jute, but not as strong as jute. Medicinal too.
36. ***Urena repanda*** **Roxb. ex Sm.** Medium-sized, perennial, much-branched, covered with star-shaped hairs. More confined to scrub forests of Gangetic Plains (Punjab, Uttar Pradesh, Bihar, West Bengal) with sporadic occurrence in South India, western peninsula, and the foothills of Himalayas. Possess tolerably fine fiber.

7.3 Perspective

In cotton the information generated by conventional genomic analysis has facilitated interspecific hybridization among the cultivated varieties and the wild species. Subsequent backcrossing has resulted in significantly successful introgression of desirable gene conferring required features, such as cytoplasmic sterility, fertility restoration, fiber length, fiber strength, high ginning outturn, drought resistance, blackarm resistance, rust resistance, hairiness, caducous bract, etc. (CICR 2000).

In the case of jute and allied fibers, there are several wild relatives, which have been documented from herbarium without ensuring physical availability of accessions in collections. They need to be collected or notified extinct. In both major and minor fiber crops, wild species and wild types of primary gene pool present high degree of variability. However, in most cases of the species, phylogenetic relationships between wild and cultivated species are not known, restricting their use, consequent to which the genetic base in these crops is very narrow. More basic biosystematic studies are needed in jute and allied fiber crops to alleviate this situation, promoting conservation and use. The major reason reported for non-utilization of wild relatives is cross-incompatibility, suggesting application of biotechnological approaches to circumvent the situation.

References

Ahmad S, Mahmood K, Hanif M, Nazeer W, Malik W, Qayyum A, Hanif K, Mahmood A, Islam N (2011) Introgression of cotton leaf curl virus-resistant genes from Asiatic cotton (*Gossypium arboreum*) into upland cotton (*G. hirsutum*). Genet Mol Res 10:2404–2414

Beasley JO (1942) Meiotic chromosome behavior in species, species hybrids, haploids, and induced polyploids of *Gossypium*. Genetics 27:25–54

Chadburn H (2012) *Crotalaria paniculata*. The IUCN Red List of Threatened Species 2012: e.T19893123A20049586

CICR (Central Institute of Cotton Research) (2000) Wild and cultivated species of cotton. Technical bulletin No 5. Central Institute of Cotton Research, Nagpur, India

Cooke T (1967) The flora of the presidency of bombay, vol 3. Botanical Survey of India, Calcutta, pp 135–136

Hooker JD (1885) Flora of India. VL Reve and Co Ltd, Ashford, p 910

Kanjilal UN, Kanjilal PC, Das RN, Das AN (1940) Flora of Assam, vol 4. Government Press, Shillong, pp 285–289

Kundu BC (1958) Ramie fibre and its development in India. Sci Cult 23:461–470

Liu D, Guo X, Lin Z, Nie Y, Zhang X (2006) Genetic diversity of Asian cotton (*Gossypium arboreum* L.) in China evaluated by microsatellite analysis. Genet Resour Crop Evol 53:1145–1152

Mahapatra AK, Saha A, Basak SL (1998) Origin, taxonomy and distribution of *Corchorus* species in India. Green J 1:64–82

Mahapatra AK, Abdul Nizar M, Saha D, Saha A, Gupta D (2003) *Corchorus pseudo-olitorius* Islam and Zaid – a new addition to Indian flora. Curr Sci 84(8):983–984

Prain D (1903) Bengal plants Vol I & II. Botabical Survey of India, Calcutta, p 1319

Reddy CS, Murthy EN, Prasad PRC, Raju VS (2007) *Crotalaria willdenowiana* DC. Subsp. *varmae*, a new taxon from Andaman, India. Int J Bot 3(4):398–400

ReddyCh S, Reddy KN, Prasad PRC, Raju VS (2003) Threatened endemic plants from eastern Ghats, India. The Eastern Ghats EPTRI-ENVI Newsletter 9(2):2–6

Run-Qing Y, Baloch SU, Li-Jun L, Ding-Xiang P (2015) The phylogenetic relationships among germplasm resources of wild ramie (*Boehmeria nivea* l. gaud) in China based on trnl-f and its sequences. Pak J Bot 47(4):1451–1457

Singh P, Narayanan SS, Singh M (1993) Breeding of *Herbaceum* cotton in India. J Cotton Res Dev 7(1):1–8

Singh VV, Mohan P, Kulkarni VN, Baitule SJ, Pathak BR (2003) Explorations within India for collection of cotton species germplasm. Plant Genetic Resources Newsletter 136:40–46

Walter KS, Gillett HJ (eds) (1998) 1997 IUCN Red List of Threatened Plants. Compiled by the World Conservation Monitoring Centre. IUCN – The World Conservation Union, Gland, Switzerland and Cambridge, UK. lxiv + p 862

Wendel JF, Brubaker CL, Alvarez JP, Cronn RC, Stewart JM (2009) Evolution and natural history of the cotton genus. In: Paterson AH (ed) Genetics and genomics of cotton. Springer, New York, pp 3–22

Forage Crops 8

8.1 Introduction

Forage crops are grasses and legume plant species that are domesticated/protected or propagated or grown for livestock feed. It is the vegetative portion of the plant, mainly leaves and stems, which are consumed by livestock. The Indian gene center possesses a rich genetic diversity in native grasses and legumes. For grasses, there are reports of 263 genera and 1291 species of Gramineae/Poaceae of which about 21 genera and 139 species are endemic. One-third of Indian grasses are considered to have fodder value. Most of the grasses belong to the tribes Andropogoneae (30%), Paniceae (15%), and Eragrosteae (9%) (www.icar.org.in/files/forage-and-grasses.pdf). Similarly, out of about 1192 species of 173 genera of Leguminosae/Fabaceae, 21 genera are reported to be useful as forage. The main centers of genetic diversity are peninsular India (for tropical types) and northeastern region (for subtropical types) besides some micro-centers for certain species spread over the subcontinent (Fig. 8.1).

Five coarse cereals, viz., barley (*Hordeum vulgare*), maize (*Zea mays*), pearl millet (*Pennisetum glaucum*), oat (*Avena sativa*), and sorghum (*Sorghum bicolor*), provide major feed for animal. Minor millets, such as finger millet/ragi (*Eleusine coracana*), little millet (*Panicum miliare*), kodo millet (*Paspalum scrobiculatum*), foxtail millet (*Setaria italica*), barnyard millet (*Echinochloa frumentacea*), proso millet (*Panicum miliaceum*), and savan millet (*Echinochloa colona*), are also important source of fodder. Similarly, several species belonging to grain legumes or leguminous vegetables, like *Lablab* and *Vigna*, are also important source of fodder. The wild relatives of these dual-purpose, food, and forage crops based on primary value have been documented and discussed under respective crop group and therefore are referred for consultation to respective chapters for the use of traits related with vegetative biomass production to genetically improve the forage yielding capacity.

Remaining species that are integral part of feed and fodder production are either part of permanent pasture or vast grasslands and rangelands. The major genera of forage value exhibiting genetic diversity include grasses like *Bothriochloa*,

A.K. Singh, *Wild Relatives of Cultivated Plants in India*,
DOI 10.1007/978-981-10-5116-6_8

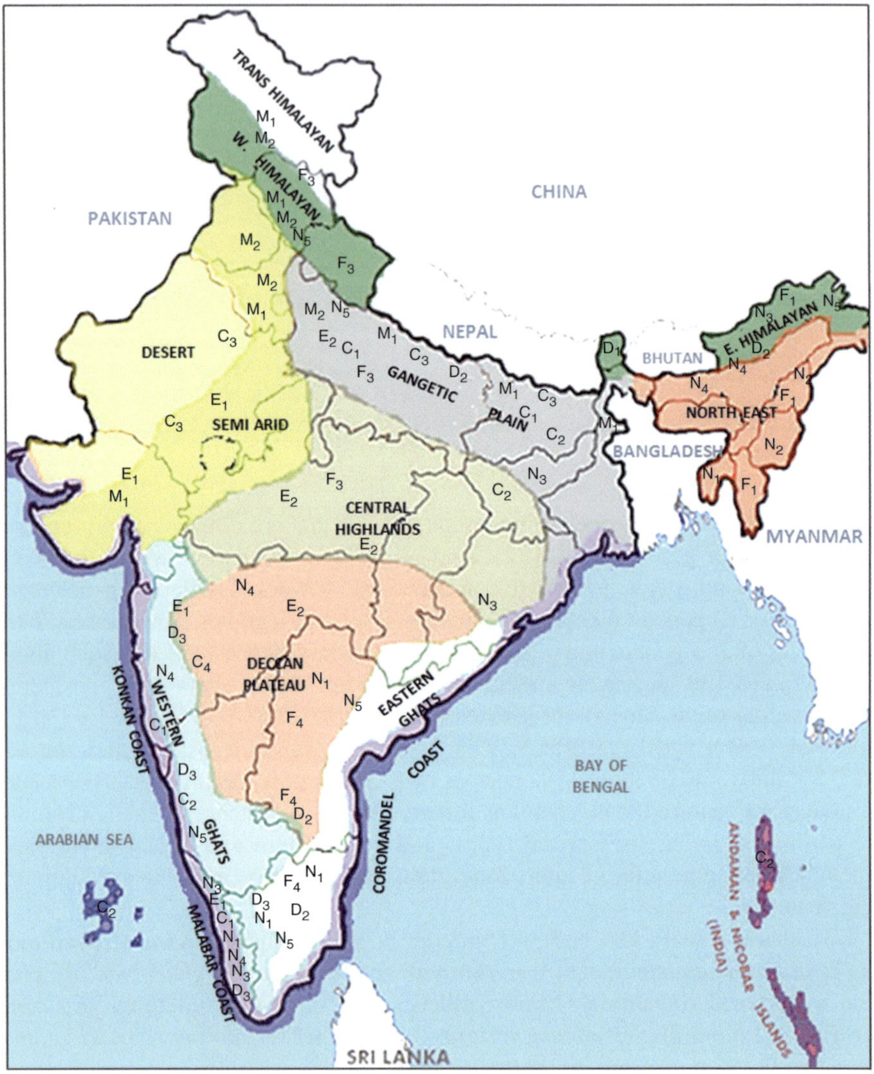

Fig. 8.1 Biogeographical regions associated with distribution wild species of forage crops, *Chrysopogon*, *Desmodium*, *Dichanthium*, *Digitaria*, *Medicago*, *Mucuna*

C1 = *Chrysopogon aciculatus*; C2 = *C. hamiltonii*; C3 = *C. fulvus*; D1 = *Desmodium kulhaitense*; D2 = *D. triquetrum*; D3 = *D. heterophyllum*; E1 = *Dichanthium annulatum*; E2 = *D. tuberculatum*; F1 = *Digitaria compacta*; F2 = *D. longiflora*; F3 = *D. sanguinalis*; F4 = *D. tomentosa*; M1 = *Medicago lupulina*; M2 = *M. monantha*; N1 = *Mucuna atropurpurea*; N2 = *M. bracteata*; N3 = *M. gigantea*; N4 = *M. monosperma*; N5 = *M. pruriens*

Cenchrus, *Chloris*, *Chrysopogon*, *Cynodon*, *Dichanthium*, *Digitaria*, *Lasiurus*, etc.; legumes like *Centrosema*, *Desmodium*, *Lablab*, *Mucuna*, etc. (Fig. 8.1); and graze plants such as *Albizia* (medicinal too), *Bauhinia* (ornamental too), *Sesbania*, etc. Seven species of *Bothriochloa* have been reported; out of these, *B. pertusa* and *B. intermedia* are widely spread in different parts of the country and are of excellent fodder value. Eight species of *Dichanthium* are distributed in various agroecological zones of India (Arora et al. 1975), but only two species, viz., *D. annulatum* (Forsk.) Stapf. and *D. caricosum* (L.) A. Camus, are important for their forage attributes (Fig. 8.1). In case of *Mucuna*, maximum diversity in primary gene pool of cultivated types reported to occurs in the eastern peninsular region, with the wild types exhibiting wider range of distribution (Fig. 8.1). *Mucuna pruriens* (L.) DC., syn. *M. cochinchinensis*, the cultivated form occurs wild in the western and eastern peninsular belt, extending to the northeastern region, where *M. bracteata* DC also occurs. The species belonging to these genera would be documented and discussed in desired details.

8.2 Wild Species Found in India

1. ***Bothriochloa intermedia* (R.Br.) A. Cam; syn. *B. bladhii* (Retz.) S.T. Blake. (Poaceae/Gramineae)**. Perennial, tufted, erect grass, native to tropical and temperate Africa, Asia (India), and Australia. In India, it is distributed throughout tropical to subtropical parts, from Punjab to West Bengal of Gangetic Plains. Used as forage, fodder, and revegetator.
2. ***Bothriochloa pertusa* (L.) A. Camus**. A perennial bunchy grass weed that spreads via stolons. Native to Asia, including India. Widely distributed in UP and other northern and southern states up to 2000 m altitude. Mainly known for forage and fodder for livestock and lawn. Occasionally seeded for erosion control, mine reclamations, pasture grass, etc.
3. ***Cenchrus setigerus* Vahl. (Poaceae/Gramineae)**. Birdwood or *Anjan*/*Dhaman* grass, a perennial tufted grass, 60 cm tall. Native to Africa and Asia, including northwest India with significant variability for tolerance to drought and frost and flooding. Cultivated. Source of drought resistance, for improving low rainfall grazing lands with *C. ciliaris* (buffelgrass).
4. ***Cenchrus rajasthanensis* Kanodia & Nanda; syn *C. biflorus* Roxb**. A new species collected and described from western Rajasthan (Kanodia and Nanda 1978). Used as forage, may be revegetator. Source of drought resistance for cultivated *C. setigerus*. Listed indeterminate in 1997 IUCN Red List of Threatened Plants (Walter and Gillett 1998).
5. ***Chloris bournei* Rang. & Tad. (Poaceae/Gramineae)**. Perennial tufted herb. Endemic to South India. Genetic resource for *C. gayana* Kunth (Rhodes grass).
6. ***Chloris dolichostachya* Lag**. Perennial tufted erect grass. Distributed in China, Indo-Malesia region to Australia, including semi-evergreen, dry, and moist deciduous forests, plains, hill slopes, and foothills of India. Cultivation limited. Genetic resource for *C. gayana* Kunth.

7. ***Chloris montana*** **Roxb**. Perennial tufted erect herb. Distributed in tropics, including hilly districts of Coromandel, Gangetic Plains, and occasionally in plains and forests. Genetic resource for *C. gayana* Kunth. Used medicinally too.
8. ***Chrysopogon aciculatus*** **(Retz.) Trin.; syn. *Andropogon aciculatus* Retz. (Poaceae/Gramineae)**. A vigorous creeping perennial grass. Native to Asia, Pacific, and Australia, including degraded dry and moist deciduous forests and plains of Gangetic Plains and Western Ghats of India. It is a weedy turfgrass suited for production systems like grazing, as it can stand heavy grazing. Also, used in erosion control and folklore medicine. Genetic resource for *C. fulvus* (Spreng.) Chiov. (*Guria grass*).
9. ***Chrysopogon hamiltonii*** **(Hook. f.) Haines**. A Perennial grass with robust culms. Endemic to Bihar and tropical India. Genetic resource for *C. fulvus*.
10. ***Chrysopogon fulvus*, Guria grass**. A perennial highly variable grass with culms densely tufted. Native to tropical Asia, including lower Himlayas, moist deciduous forests of tropical hilly areas up to 1830 m alt., in gravelly lands of Aravalli hills of Rajasthan and central plateau. Used as fodder and erosion control. Cultivated.
11. ***Cynodon dactylon*** **(L.) Pers**. *Dhoob* grass, an enormously variable perennial grass, creeping by means of stolons and rhizomes. Native of India, now widely naturalized between 45° N and 45° S lat., as weed. Grows wild and cultivated as turfgrass for fodder, forage, lawns, and erosion control. It has therapeutic value and, in Ayurveda, considered one of the ten medicinally potential herbs that constitute the group of *Daspushpam*. Presents enormous natural morphological, physiological, and chromosomal variation that can be used in its genetic improvement.
12. ***Desmodium triquetrum*** **(L.) DC. (Leguminosae)**. Erect subshrub. Native to Indo-Malesia to Pacific and China, including semi-evergreen and moist deciduous forests, plains of central and eastern Himalayas and South India. Have medicinal potential, as it contains a wide variety of free radical scavenging compounds.
13. ***Desmodium heterophyllum*** **(Willd.) DC**. Prostrate, perennial, free branching herb. Distributed in Indo-Malesia and China, including degraded moist deciduous forests and plains of Western Ghats and native pastures of India. Used as forage and for erosion control. Significant interspecific variability in cultivated species with *D. intortum*, green leaf type, and *D. uncinatum*, silver leaf type.

In addition, species like *Desmodium kulhaitense* C.B. Clarke, an erect shrub, endemic to Eastern Himalayas (Sikkim) has been described with less variability and needs further investigation.

14. ***Dichanthium annulatum*** **(Forssk) Stapf. (Poaceae/Gramineae).** It is an erect tufted perennial grass with long awn spikelet. Native to tropical Africa and Asia including Indian grasslands. Indian variant withstood heavy grazing and gives a good standing hay. Popular pasture grass, known by several vernacular names.

15. ***Dichanthium tuberculatum* (Hack.)**. Perennial, erect culms grass, distributed from Africa through India to Southeast Asia and Australia. In India, found particularly in upper Gangetic Plains, Malwa Plateau, Central Highlands, and Maharashtra (Fig. 8.1).
16. ***Digitaria compacta* (Roth ex Roem. & Schult.) Veldkamp; syn. *D. cruciata* var. *esculenta* Bor. (Poaceae/Gramineae)**. It is a grass species native to India and Indo-China. Found wild in Eastern Himalayas and northeastern hills. *D. cruciata* var. *esculenta* cultivated in the Khasi Hills of northeast India, used as a glutinous flour for making bread or porridge, and known as *raishan*. Natural variability is the genetic resource.
17. ***Digitaria longiflora* (Ratz.) Pers**. Annual, or short-lived perennial, slender with culms 10–40 cm tall. Paleotropically distributed in grasslands and open areas, including India. Potential gene source for cultivated fonio *Digitaria* spp.
18. ***Digitaria sanguinalis* (L.) Scop**. Annual with decumbent or prostrate culms. Native to temperate and warm regions of the world, including Western Himalayas and other parts of India. Used as forage. Genetic resource for cultivated *Digitaria* spp.
19. ***Digitaria tomentosa* (J. Koenig ex Rottler) Henrard**. Native to Asia-tropical and South India (Bijapur in Deccan). Genetic resource for cultivated *Digitaria* spp.
20. ***Diplachne fusca* (L.) P. Beauv (Poaceae/Gramineae)** *Kamal* is a semiaquatic rhizomatous perennial, tuft, and erect grass. Native to temperate and warm regions of the world, including upper Gangetic Plains, Bengal, Orissa, Andhra Pradesh (AP), Tamil Nadu, and Kerala along coast. Relished, when young. Salt-tolerant, alkaline soils are reclaimed by growing it.
21. ***Eragrostis curvula* (Schvad.) Nees (Poaceae/Gramineae)**. It is densely tufted perennial grass, native to Africa. Naturalized in other regions, including India. At young stage, palatable and is used for grazing and hay purpose. Also, for erosion control and revegetation. Extremely drought tolerant.
22. ***Heteropogon contortus* (L.) Beauv. ex R. & S. (Poaceae/Gramineae)**. A densely tufted, perennial, highly palatable grass. Native to the tropics and subtropics of Africa, south Asia, and parts of Oceania, including India, in all arid and semi-arid regions and North Himalayas up to an elevation of 2000. Cultivated in pastures.
23. ***Indigofera hirsuta* Linn. (Fabaceae/Leguminosae).** It is an erect or spreading annual. Attains a height up to 1.5 m. Distributed in tropical Africa and America, northern Australia, and south Asia, including India all over the plains. Gives good quality green manure.
24. ***Iseilema holei* Haines (Poaceae/Gramineae)**. A perennial with robust culms. Distributed in moist patches of middle and lower Gangetic Plains, tropical areas of MP, Maharashtra, Gujarat, Kerala, Tamil Nadu, AP, and Telangana from sea level to 760 m alt. A wild relative of cultivated *Iseilema laxum* Hack. (*Machuri*).
25. ***Lasiurus sindicus* Henr; syn. *L. hirsutus* (Poaceae/Gramineae)**. *Sewan* grass, it is an erect, tufted and branched perennial grass, attaining a height of about 1.25 m. Native to Africa and India, where it is found particularly in arid zones

of Rajasthan, extending to the parts of Haryana and Punjab. Cultivated in pastures.

26. ***Medicago lupulina* L. (Fabaceae/Leguminosae)**. Black medick, an annual or short-lived perennial self-regenerating legume. Native of Eurasia, northern Africa, and Asia, including Indo-Gangetic Plains. Besides, forage, cultivated as an ornamental, revegetator, bee supporter (honey), and soil improver. Closely related to *M. sativa* L. and to clovers (*Trifolium*) and sweet clover (*Melilotus*).
27. ***Medicago monantha* (C.A. Mey.) Trautv; syn. *Trigonella monantha* C.A. Mey.** Single-flowered medick, an herb with curved fruits and large amount of variability forcing taxonomists to refer it as complex. Native of Asia, from Afghanistan to northern plains of India, up to 1800 m. Cultivated. Genetic resources for *M. sativa*.
28. ***Melilotus indicus* (L.) All. (Fabaceae/Leguminosae)**. *Ban Methi/Senji* or Indian sweet clover is an annual herb with erect or spreading hairless stem and yellow flower. Native to North Africa, Europe, and Asia (India). Used as forge and grown as revegetator, bee supporter (honey), soil improver, and folklore medicine. Related to *M. albus* Medik.
29. ***Mucuna atropurpurea* (Roxb.) DC. ex Wight & Arn. (Fabaceae/ Leguminosae)**. Large twining shrubs, rich in crude protein with most of the essential amino acids, fatty acid such as palmitic and linoleic acid and some minerals. Found in evergreen forests of Peninsular India and Sri Lanka
30. ***Mucuna bracteata* DC. Ex Kurz**. Climbing shrub. Native of Asia, originated in the forest of Tripura, distributed in North Eastern hills of India, and common in Khasi Hills. Has phenolic acids to combat insects and cattle. Grown as cover crop in Kerala, in rubber plantations to improve nitrogen levels. Allied to *M. pruriens*.
31. ***Mucuna gigantea* (Willd.) DC**. A twining climbing lianas. Distributed in Indo-Malesia, Australia, and Pacific Islands, including southern Western Ghats and Tamil Nadu. Poisonous.
32. ***Mucuna monosperma* Wight**. Negro bean, a large climber with branches covered with rusty velvety hairs. Native to Indo-Malesia region, including evergreen forests of Eastern Himalayas and Western Ghats.
33. ***Mucuna pruriens* (L.) DC.; syn. var. *hirsuta* Wight & Arn**. Vigorously growing annual trailing vine, native to Americas (Sorenson and Johannessen 2004.), neutralized throughout tropics, including to south Asia (India) and Malaysia. Variability found among wild types in Peninsular India, humid areas of Western Ghats, and sporadically in the subtropical/temperate Himalayas. Cultivated as fodder, fallow, and green manure crop. Seeds used in Ayurveda medicine as source of L-DOPA. Variety *utilis* (Wall. ex Wight) L.H. Bailey.; syn. *Mucuna capitata* Roxb.; *Stizolobium deeringianum* Bort., is distributed in *Tarai* areas of Western and Eastern Himalaya and Chota Nagpur. It is an underutilized pulse, nutritionally comparable to soybean and other pulses, in terms of protein, lipid, minerals, etc. Grown as ornamental, for honey production, soil improver, and forage crop. Member of primary gene pool of *M. pruriens* with desirable features.

34. ***Pueraria phaseoloides*** **(Roxb.) Benth. (Fabaceae/Leguminosae)**. Tropical kudzu, deep-rooted perennial climbing shrub, distributed along margins of cultivated lands in tropical Asia and Pacific. A promising cover/forage crop or a mixture in pastures. Genetic resource for kudzu, *P. lobata*.
35. ***Pueraria sikkimensis*** **Prain**. Plants are strongly woody climbers. Native to India and Bhutan. Distributed in *terai* and subtropical forest slopes of Eastern Himalayas. Genetic resource for *P. lobata*.
36. ***Sehima nervosum*** **(Rottl.) Stapf. (Poaceae/Gramineae)**. A perennial 1 m tall grass with numerous tillers forming dense tufts. Native to Africa and Arabia, extending to undulating areas of MP, UP, Gujarat, Rajasthan (Arvalli), Maharashtra, Karnataka, Andhra, and Tamil Nadu from sea level to 1830 m alt., in India and Australia. Used as fodder/forage.
37. ***Sesbania concolor*** **Gillett. (Fabaceae/Leguminosae)**. Unarmed shrub, 1–2 m tall, native to Yemen and Indian Subcontinent (Kathiawar peninsular region, Gujarat, Jammu and Kashmir, Pakistan). Related to cultivated *S. sesban*; *S. grandiflora*.
38. ***Vetiveria zizanioides*** **(L.) Nash; syn.** ***Chrysopogon zizanioides*** **(L.) Roberty (Andropogoneae)**. Vetiver or *Khus*, an erect perennial, densely tufted, awnless grass. Native of tropical and subtropical Asia. In India found in wet, damp marshy areas of Gangetic Plains, Brahamputra valley, extending toward the plains of Punjab and Rajasthan. Indian variants are known for sweet, earthy, woody balsamic highly priced essential oil used in perfumary and folklore medicine. Could be utilized/grazed by cattle in young stage, boundary raising, and erosion control.

In addition, species like *Tephrosia jamnagarensis* Santapau., a rare endemic species naturally distributed to semi-arid region of India, particularly the Kathiawar Peninsula, is used as green manure and can be a genetic resource.

8.3 Perspective

Genetic improvement efforts in forage crop have been very limited except for the coarse cereals and minor millets used as both forage and fodder and the temperate crops of European and central Asian origin. The major yield gains in most other crops have been achieved by improving the agronomy and the cultural practices. A great scope exists to increase forage productivity in these crops by strengthening the genetic improvement activities, including utilization of the wild relatives of cultivated species. In this regard efforts are needed for collection, characterization, evaluation (against main yield-reducing factors, such as abiotic and biotic stresses), and conservation of wild species germplasm. Further to facilitate use, a thorough understanding should be developed on species relationships, chromosomal constitution, genome structure of various species that are part of the cultivated species gene pool with identification of putative parentage and phylogenetic relationships, exploring the possible breeding strategies, and extent of possible genetic introgression through recombination breeding. In case of polidy differences, the nature of polyploidy must be understood for manipulation of the breeding program accordingly. In case of postfertilization incompatibility barriers, embryo rescue is most effective and successful technique.

For evaluation of wild species germplasm suitable screening methodologies may need to be developed at field, greenhouse, and laboratory level or combination of these for conclusive information. They may need to be screened in vitro at two levels: (i) seedling level and (ii) tissue or cell level. For example, in vitro studies have shown significant interspecific and intraspecific variation among legume and clover species for salt tolerance.

The strategies for various crops would vary as per the problems encountered in combination of the species. Many of techniques such as embryo rescue, micropropagation, androgenic haploid plant production, and creation of novel variations may be required to help at one or more steps than the conventional breeding methods. For example, the embryo-rescue technique has been well exploited in *Lolium–Festuca* complex for production of hybrids. *Lolium–Dactylis* hybrids and many interspecific hybrids in *Trifolium* have also been developed by this method. Interspecific hybrids have been developed successfully in other crops as well, like berseem.

References

Arora RK, Mehra KL, Hardas MW (1975) The Indian gene center, prospects of exploration and collection of herbage grasses. Forage Res 1:11–12

Kanodia KC, Nanda PC (1978) *Cenchrus rajasthanensis* new species from western Rajasthan India. Geobios (Jodhpur) 5(4):156–159

Sorenson JL, Johannessen CL (2004) "Scientific evidence for pre-Columbian transoceanic voyages" Sino-Platonic papers, Department of East Asian Languages, and Civilizations. University of Pennsylvania, Philadelphia, p 273

Walter KS, Gillett HJ (eds) (1998) 1997 IUCN Red List of Threatened Plants. Compiled by the World Conservation Monitoring Centre. IUCN – The World Conservation Union, Gland, Switzerland and Cambridge, UK. lxiv + p 862

Vegetables 9

9.1 Introduction

Most population of the Indian Subcontinent is predominantly vegetarian. The local tribes and communities have domesticated a large amount of plant species as vegetables and/or use them as alternative food, involving several genera, such as *Cucumis*, *Coccinia*, *Luffa*, *Momordica*, *Trichosanthes*, *Solanum*, *Abelmoschus*, *Amorphophallus*, etc. Additionally, wild species diversity of certain exotic crops has either been extended or introduced to the subcontinent, such as *Amaranthus*, *Dioscorea*, etc. Predominant wild species diversity in the Indian Subcontinent occurs for tropical cucurbitaceous, solanaceous, and leguminous vegetable crops and some others, like okra. In addition, wild species diversity is also found in several leafy and rooty, tuberous, and bulbous vegetable crop species, many of whom are directly harvested from the natural habitat by local tribal communities. The complexity of species diversity using different plant parts as vegetables justifies the classification of vegetable crops into three groups, fruity vegetable, leafy vegetable, and root and tuber crops.

Among the cucurbitaceous vegetables, a number of wild taxa occur in genera *Citrullus*, *Cucumis*, *Luffa*, *Momordica* (Fig. 9.1), and *Trichosanthes*. Recently, Renner and Pandey (2013) have listed 94 accepted species, including ten endemics. Important ones are, namely, *Citrullus colocynthis*, *Cucumis hardwickii*, *C. setosus*, *C. hystrix*, *Luffa graveolens*, *L. acutangula* var. *amara*, *L. cylindrica*, *L. tuberosa*, *L. echinata*, *Momordica cochinchinensis* (syn. *M. macrophylla*), *M. subangulata*, *M. cymbalaria* (syn. *M. tuberosa*), *M. dioica*, *M. denticulata*, *M. balsamina*, *Siraitia sikkimensis*, *Trichosanthes anamalaeiensis*, *T. bracteata*, *T. cordata*, *T. cucumerina*, *T. cucumeroides*, *T. dioica*, *T. lobata*, *T. khasiana*, *T. ovata* (syn. *T. truncata*), *T. tricuspidata* (syn. *T. nervifolia*), *T. perottetiana*, and *T. wallichiana*, based on the checklist of Chakravarty (1982). In *Luffa*, most of the species occur in disturbed lands, forest openings, etc. *Luffa acutangula* var. *amara* occurs in Peninsular India and is the wild type related to the cultivated ridge gourd. *Luffa echinata* occurs in the Western Himalayas, the upper Gangetic plains, and Central India, and

A.K. Singh, *Wild Relatives of Cultivated Plants in India*,
DOI 10.1007/978-981-10-5116-6_9

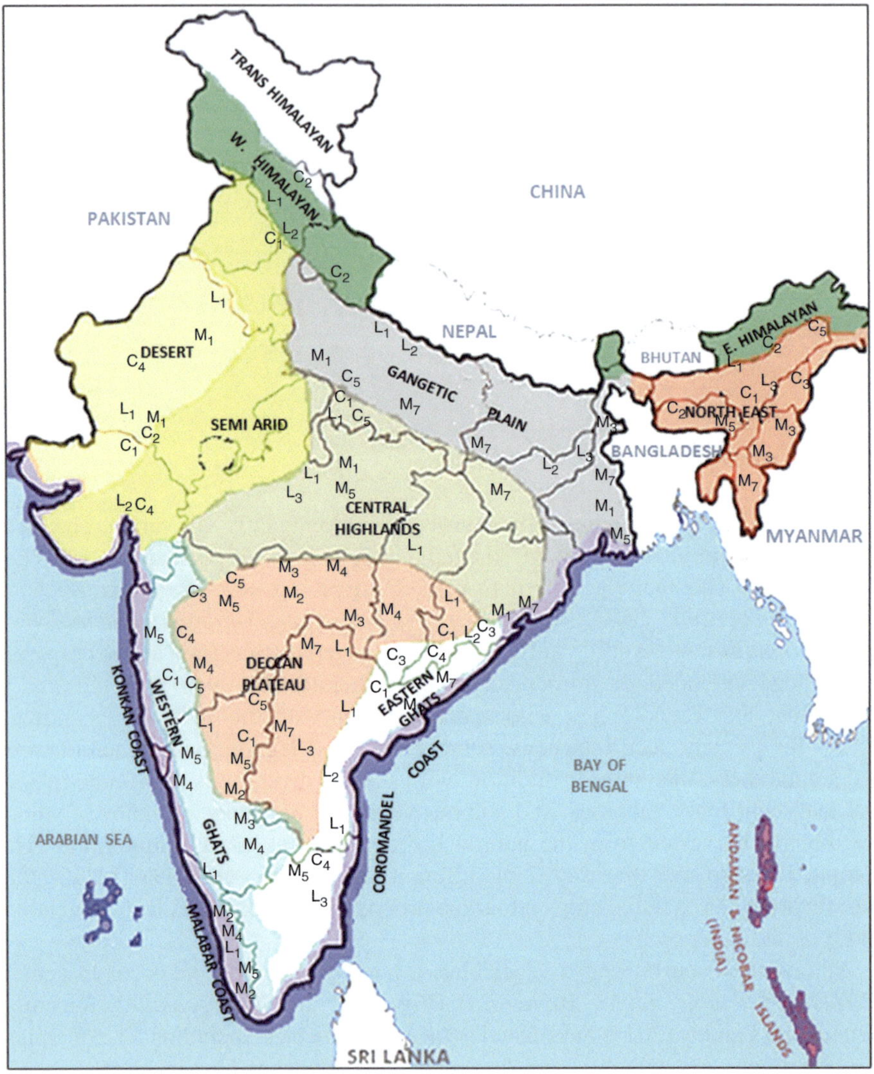

Fig. 9.1 Biogeographical regions associated with distribution of wild species of fruity cucurbitaceous vegetables, *Cucumis*, *Luffa*, *Momordica*
C1 = *Cucumis callosus*; C2 = *C. hardwickii*; C3 = *C. hystrix*; C4 = *C. prophetarum*; C5 = *C. trigonus*; L1 = *Luffa acutangula*; L2 = *L. echinata*; L3 = *L. graveolens*; M1 = *Momordica balsamina*; M2 = *M. charantia muricata*; M3 = *M. cochinchinensis*; M4 = *M. cymbalaria*; M5 = *M. dioica*; M6 = *M. sahyadrica*; M7 = *M. subangulata*

L. graveolens (considered a wild progenitor of *L. hermaphrodita*) in Maharashtra, Bihar, and northward to Sikkim and West Bengal (Fig. 9.1). In *Momordica*, *M. balsamina* occurs in semi-arid northwestern plains and only sporadically elsewhere in upper Gangetic region and in the northern parts of Western and Eastern Ghats.

Momordica dioica and *M. cochinchinensis* occur wild/semi-wild in the Gangetic plains extending eastward. *Momordica cymbalaria* is restricted to the Deccan Plateau, Maharashtra, and southward, with only sporadic occurrence in the eastern peninsular region. *Momordica subangulata* and *M. cochinchinensis* occur largely in the northeastern region (Fig. 9.1). *Trichosanthes* is the most species-rich genera with 22 species occurring in India, and the major zones of species concentration are (a) along the Malabar coast in Western Ghats and (b) low and medium elevation zones (up to 1500 m) in Eastern Ghats and northeastern region. Contiguous regions are major areas of distribution of the wild types of potential importance. *Trichosanthes cucumerina* occurs in the northeastern region, along with semi-wild *T. dioica*. A widely distributed species, *T. bracteata*, occur in eastern India, extending to the south and sporadically in the Himalayas (1500 m). *T. cordata* (related to *T. bracteata*) occurs in the peninsular region, extending to northeastern plains and hills. *Citrullus colocynthis* exhibits much variation in northwestern arid and semi-arid plains, whereas in *Cucumis*, the wild species related to cultivated cucumber (*Cucumis sativus*), *C. hardwickii* occur mostly in the Himalayas or hills. *C. callosus*, distributed in Indo-Malayan region, is found throughout the Peninsular India including the hills. Localized variability of *Cucumis setosus* is restricted to Maharashtra and eastern plains, while *C. hystrix* extends its range from eastern plains to northeastern hills in Assam, Tura Range in Meghalaya, and Mishmi hills, and, in the drier northwestern plains, *C. prophetarum* (Fig. 9.1).

In non-tuberiferous *Solanum* species, the maximum diversity has been observed in South India, foothills of Himalayas, and the northeastern region (Fig. 9.2). *Solanum torvum*, *S. indicum*, *S. insanum*, *S. surattense*, *S. khasianum*, and *S. pubescens* are the widely distributed species in these areas. *Solanum melongena* complex has three species, namely, the *S. melongena, S. incanum* (syn. *S. coagalans*), and *S. melongena* var. *insanum.* Phylogenetically, they are very closely related. More variability in *S. incanum* occurs in northwest plains of India; in *S. insanum* in eastern peninsular tract; and in primitive cultivated form, the potangi type of *S. melongena*.

In leguminous vegetables, the humid tropical belt of Western Ghats is reported to have wild species of *Canavalia*, particularly in the lowland forests along the coast. Between the two distinct species found, the *C. cathartica* (syn. *C. virosa*) and *C. obtusifolia*, the latter is related to cultivated sword bean, *C. gladiata*. Another climbing large flowering shrubby bean, *C. stocksii* Dalz., a native of Mahabaleshwar (Maharashtra), is possibly a rare variant of cultivated *C. ensiformis*, which has now been treated as part of *C. cathartica*. The northeastern region (Manipur and Mizoram) is another area of diversity for both wild and cultivated forms of sword bean.

Diversity occurs in India for the wild relatives of Okra (*Abelmoschus esculentus*). Nine species and two varieties, namely, *A. angulosus*, *A. tuberculatus*, *A. manihot*, *A. moschatus*, *A. ficulneus*, *A. esculentus*, *A. crinitus*, *A. caillei*, and *A. tetraphyllus* var. *tetraphyllus*, *A. tetraphyllus* var. *pungens*, occur in India. The nearest wild related species to cultivated okra is *A. tuberculatus*, which sporadically occurs in fields/field margins. *Abelmoschus tuberculatus* is confined to north/

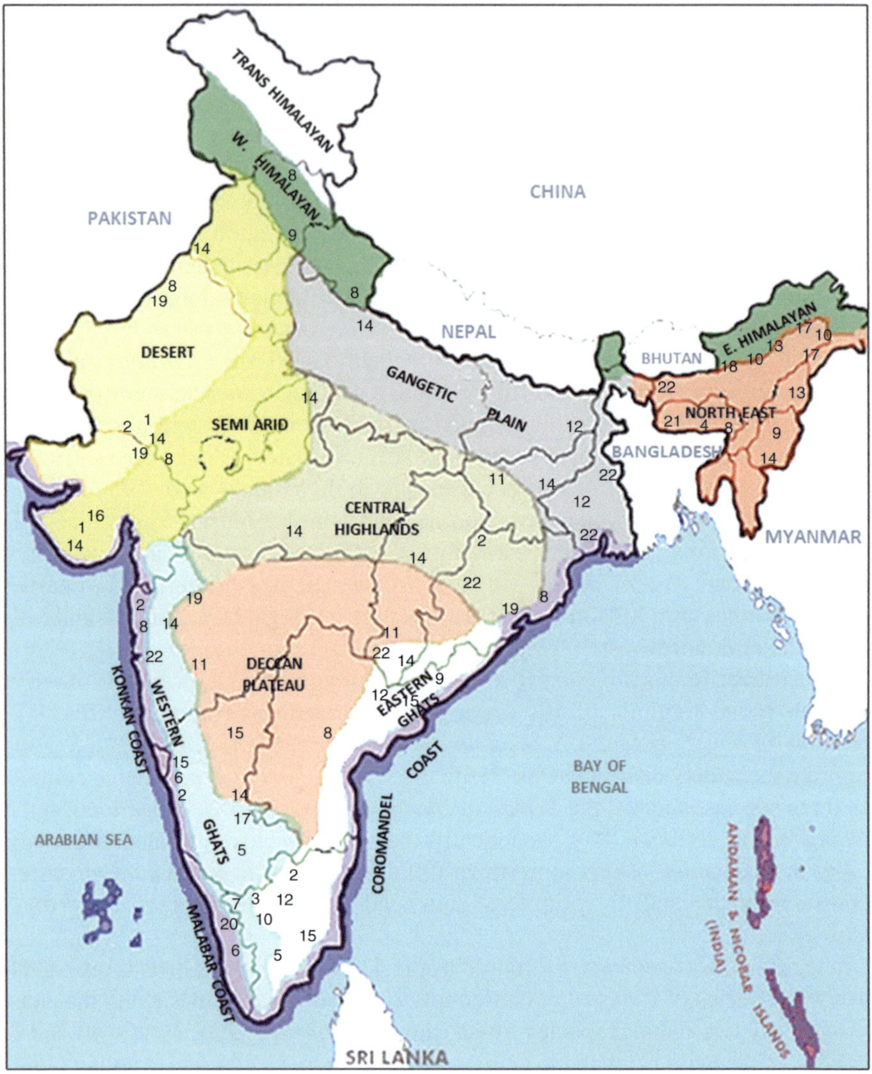

Fig. 9.2 Biogeographical regions associated with distribution of wild *Solanum* species
1 = *Solanum albicaule*; 2 = *S. anguivi*; 3 = *S. barbisetum*; 4 = *S. dubium*; 5 = *S. erianthum*; 6 = *S. giganteum*; 7 = *S. grandiflorum*; 8 = *S. incanum*; 9 = *S. khasianum*; 10 = *S. kurzii*; 11 = *S. mammosum*; 12 = *S. melongena insanum*; 13 = *S. myriacanthum*; 14 = *S. nigrum*; 15 = *S. pubescens*; 16 = *S. purpureilineatum*; 17 = *S. sisymbriifolium*; 18 = *S. spirale*; 19 = *S. torvum*; 20 = *S. vagum*; 21 = *S. viarum*; 22 = *S. surattense*

northwestern plains, often along field borders and except for its tuberculate hard fruits (up to 8 cm long) looks unmistakably like the cultivated okra. Variability in *A. pungens* and *A. tetraphyllus* occurs in the sub-Himalayan range, the former in northeastern belt and the latter in the northwestern *terai* region. Wild *Abelmoschus* species are also available in the Western Ghats (three of the six spp.): *A. angulosus*, a polymorphic taxon with greater variation in Karnataka and Nilgiri hills, and also *A. manihot* (including *A. pungens*) and *A. ficulneus*.

Among leafy vegetables, northern plains are the major area of variability, where wild types of *Amaranthus*, *Chenopodium*, *Rumex*, and *Malva* are found as common weeds. Predominant ones are *Amaranthus* spp., particularly *A. spinosus*, *Malva rotundifolia*, *M. sylrestris*, and *Chenopodium album.*

In case of tuberous vegetables, which may be the plant roots or non-root tubers of modified plant stem, such as rhizomes, corm, or bulb, the subcontinent is blessed with greater variability in tuberous root type such as *Amorphophallus* and *Dioscorea*, modified stem type such as *Colocasia*, and bulb such as *Allium* spp. About 50 species of *Dioscorea* are found wild, including four cultivated, largely in the west, east, and northeast regions. Other tuberous species are represented by *Amorphophallus paeoniifolius*, *Alocasia macrorrhiza*, *Colocasia esculenta*, *Dioscorea alata*, *D. rotundata*, *D. esculenta*, *D. bulbifera* var. *sativa*, *Ipomoea batatas*, *Manihot esculenta*, and *Xanthosoma sagittifolium*. Besides these crops, there are several minor tuber crops, namely, *Canna edulis*, *Maranta arundinacea*, *Moghania vestita*, *Psophocarpus tetragonolobus*, *Pachyrrhizus erosus*, and *Solenostemon rotundifolius*, found in western peninsular region. In bulbs, around 30 species represent *Allium*, which is naturally distributed in India. The cultivated species are *Allium cepa* var. *cepa*, *A. cepa* var. *aggregatum*, *A. cepa* var. *viviparum*, *A. fistulosum*, *A. tuberosum*, *A. sativum*, *A. ampeloprasum* var. *porrum*, and *A. schoenoprasum*, whereas *A. carolinianum*, *A. chinense*, *A. consanguineum*, *A. humile*, *A. przewalskianum*, *A. stoliczkii*, *A. stracheyi*, *A. victorialis*, and *A. wallichii* occur wild in the Himalayan region. The production of hybrid between leek and garlic demonstrates the scope interspecific hybridization (Yanagino et al. 2003).

9.2 Wild Species Found in India

9.2.1 Fruity Vegetable

1. ***Abelmoschus angulosus* Wall. ex Wight & Arn. (Malvaceae)**. Prickly annual shrub. Native to moist deciduous, semi-evergreen, and shola forests of tropical Asia, including the Western Ghats, up to 1600 m, and forests of Gujarat, Maharashtra, Kerala, and Tamil Nadu. Source of resistance to yellow vein mosaic virus (YVMV) and mites and tolerance to low temperatures and light frost
2. ***Abelmoschus caillei* (A. Chev) Stevels**. Annual, erect hardy plant with large leaves, thick stem, and unique-shaped fruits. Native of Africa and naturalized to the Indian Subcontinent in Western Ghats. It appears close to *A. esculentus* (L.) Moench and *A. manihot* ssp. *tetraphyllus*. Segregating populations involving *A. caillei* and ssp. *tetraphyllus* have been found in Karnataka (Velayudhan et al. 2007). Source of resistance to YVMV and shoot and fruit borer.

3. ***Abelmoschus crinitus* Wall.; syn. *A. cancellatus* (L.f.) J.O. Voigt**. An erect, perennial herb, around 1.5 m tall. Habitat to deciduous forests and wastelands of China, Indo-Malaysia, and India, including Northwest Himalayas, upper Gangetic Plains, and Peninsular India, extending down to the south. Source of resistant *Cercospora* blight.
4. ***Abelmoschus ficulneus* (L.) Wight & Arn**. Stout annual weedy herb, up to 1.5 m tall (Fig. 9.3a). Native to Africa and tropical Asia, including India, extending from Jammu and Kashmir to southern Western Ghats. Yields a white fiber for twine and possesses medicinal properties and a genetic resource.
5. ***Abelmoschus manihot* (L.) Medik**. It is a perennial plant, growing up to 2 m tall. Native of Asia, in India found in tropical and subtropical zones, consisting of middle and lower Gangetic Plains and Southern Deccan Plateau. Grown as ornamental, vegetable, and fiber crop and used for gum/resin. Source of resistant to YVMV. Van Borssum Waalkes, in 1966, classified morphological variation into two subspecies, **ssp. *manihot* (L.) Medik**., mainly consisting of cultivated erect herb types with smooth stems (wild and weedy forms found in coastal plains of Western Ghats and West Bengal), and **ssp. *tetraphyllus* (Hornem.) Borss. Waalk.,** with prickly habit. Within ssp. *tetraphyllus*, Van Borssum Waalkes described two different botanical varieties: variety *pungens* (Roxb.) Hochr., which is distributed in Asia, including tropical Himalayas, middle and lower Gangetic Plains, and northeastern India and a source for resistance to enation leaf curl virus, and variety *tetraphyllus* (Roxb. ex Hornem.) Hochr., which is distributed in sub-Himalayan region, Upper Gangetic, and northwestern plains. Additionally, a large seeded variety *megaspermus* Hemadri is found in Central India (Gujarat, Maharashtra, and Madhya Pradesh) on shady hill slopes and foothills.
6. ***Abelmoschus moschatus* Medik**. Musk okra, an aromatic (essential oils) and medicinal, tropical perennial weedy shrub, valued for its scented seed, is used for flavoring. Native to Asia, including central Himalayas, Eastern Ghats, and Andaman table Islands, the morphological variation has given rise to subspecies *moschatus* and ssp. *tuberosus* (Span.) Borss. Waalk (ornamental), found in Karnataka, Kerala, and Tamil Nadu. Cultivated.
7. ***Abelmoschus tuberculatus* Pal & Singh**. Herb or undershrub, found in semi-arid regions of North and Northwestern India. A close relative of *A. esculentus* but differs by its strigose hairs on the stem and shorter capsules. A spontaneous hybrid between *Abelmoschus tuberculatus* and *A. esculentus* has been reported (Nair and Kuriachan 1976). Tolerant to YVMV and fruit borer.
8. ***Canavalia cathartica* Thours; syn. *C. virosa* (Roxb.) Wight & Arn. (Fabaceae/Leguminoseae)**. Perennial climber or trailer (Fig. 9.3b), native to Africa and Asia, including northeastern region, Eastern Ghats of Peninsular India, and Australia. Grows even in mangroves. A neglected wild bean, consumed by Malayali tribes in Tamil Nadu as staple food after soaking seeds in water. Anti-poisonous. Occasionally cultivated. Source of nutritional traits for jack bean *C. ensiformis*.

Fig. 9.3 Representative wild relatives of economically important vegetable crops: (**a**) *Abelmoschus ficulneus*, (**b**) *Canavalia cathartica*, (**c**) *Cucumis hardwickii*, (**d**) *Solanum khasianum*, (**e**) *Trigonella emodi*, and (**f**) *Allium tuberosum*

9. ***Canavalia rosea* (Sw.) DC.; syn. *C. obtusifolia* DC., *C. maritima* (Aubl.) Thouars**. Coastal jack bean, a trailing herbaceous vine. Pantropical distribution including the coastal habitats of western Peninsular India and Andaman Islands. A rare variant of cultivated *C. ensiformis* gene pool. Based on probable affinities with *C. gladiata*, it belong to secondary gene pool of sword bean. Highly salt tolerant.
10. ***Citrullus colocynthis* Schrad. (Cucurbitaceae)**. A drought hardy desert vine. Native to northern Africa to Mediterranean Basin, extending and naturalizing in fallow land of arid and semi-arid zone of northwestern plains of India. Seeds are rich in oil and can be source of biofuel and also used medicinally. It is a member of secondary gene pool of watermelon. Natural hybrid between *C. lanatus* and *C. colocynthis* has been reported (Singh 1978). Potential source for drought tolerance and pest resistance.
11. ***Cucumis callosus* (Rottl.) Cogn.; syn. *C. trigonus* Roxb.; *C. melo* L. subsp. *agrestis* (Naudin) Pangalo**. A feral vine found in semi-arid regions of most of the Indian Subcontinent, particularly in northwest and central plains, and Eastern Ghats. Progenitor and member of primary gene pool of *C. melo* (John et al. 2013). Reported to possess genes for resistance to fruit fly and leaf-eating caterpillars. *Turbinatus* types possess edible fruits.
12. ***Cucumis hardwickii* Royle; syn. *C. sativus* L. var. *hardwickii* (Royle) Alef**. An annual monoecious climber (Fig. 9.3c) found in the foot hills of Himalayas and those of Gangetic Plains and other hills. Member of primary gene pool of *C. sativus*, being a wild sympatric taxa having $2n = 14$, with which it hybridizes readily and considered its progenitor. It is cold tolerant and resistant to powdery mildew.

13. ***Cucumis hystrix* Chak**. A rare scandent herb with white hirsute stem, climbing by tendrils. Native to Asia including northeastern hills of India, with 2n = 24. Genetic resource, as member of tertiary gene pool of *C. sativus*, based on successful interspecific hybrid obtained between *C. sativus* and *C. hystrix* (Chen et al. 1997) and of *C. melo* based on distant affinities (Sebastian et al. 2010). An amphidiploid species (*Cucumis* x *hytivus* Chen and Kirkbride, 2n = 38) was produced by chromosome doubling of hybrid (Chen and Kirkbride 2000). These facts indicate the possibility of gene introgression from *C. hystrix* to both cultivated species of *Cucumis* with 2n = 24 and 2n = 14.
14. ***Cucumis prophetarum* L**. A monoecious gray-green perennial trailer, habitat to dry areas of semi-evergreen forests of arid and semi-arid Indus and northwestern plains and parts of Peninsular India. Genetic resource, as member of secondary gene pool of West Indian gherkin *C. anguria* and tertiary of *C. melo* and *C sativus* (Sebastian et al. 2010). Used in treatment of inflammation in Indian medicine system.
15. ***Cucumis ritchiei* (Clarke) H Schaef**. Prostrate or climbing hirsute herbs. Habitat to hedges and bushes in Peninsular India. Genetic resource, member of tertiary gene pool of *C. anguria* and *C. melo* (Sebastian et al. 2010).
16. ***Cucumis setosus* Cogn**. A rare, endemic, wild, or semidomesticated vegetable restricted to Maharashtra and adjoining states of India (Gangetic plains, Bundelkhand). Member of primary gene pool of *C. sativus*, distinct, without bitter principle (John et al. 2014).
17. ***Dolichos uniflorus* Lam.; syn**. ***Macrotyloma uniflorum* var. *uniflorum* (Fabaceae/Leguminosae)**. Horse gram or *kulthi*, growing wild as annual herb in Western Ghats, Khandala, southward. The variability in these wild types is part of primary gene pool of horse gram, used for food and fodder.
18. ***Lablab purpureus* (L.) Sweet.; syn. *L. niger* Medik (Fabaceae/Leguminosae)**. An herbaceous, twining plant. Native of Africa, escaped to the wild in India. Available variants have been classified into long-podded var. *lignosus* (Linn.) Prain, synonym of cultivated *Lablab purpureus*, hyacinth bean or *seim*, found in eastern coast of Andhra Pradesh. Small-podded, drought hardy types, var. *typicus* Prain is restricted to scrub jungles. Besides being a vegetable, it is used as pulse, fodder, soil improver, etc.
19. ***Luffa acutangula* var. *amara* (Roxb.) Clarke. (Cucurbitaceae)**. Monoecious, climbing annual with prominently ribbed ovary. Native to moist deciduous forests all over India. Medicinal, fruits used in vata, kapha anemia, asthma, leukoderma, tumors, etc. Member of primary gene pool of cultivated ridge gourd, *L. acutangula*.
20. ***Luffa echinata* Roxb**. It is a spreading climbing herb of tremendous medicinal value (antiulcer). Native to Africa and Indian Subcontinent, distributed all over, particularly northwestern Himalayas and upper Gangetic plains. Genetic resource, member of secondary gene pool of ridge gourd, *L. acutangula*, and sponge gourd (*L. cylindrica*), with whom it is crossable, but hybrid is complete sterile (Singh 1990).

21. ***Luffa graveolens* Roxb**. Native to India, widely distributed in north-central and eastern plains and Eastern Ghats, extending up to Tamil Nadu; sporadically occur elsewhere in Eastern Himalayas. Genetic resource, as member of secondary gene pool of ridge gourd and sponge gourd, with whom it is crossable, but produces hybrid with complete pollen sterility (Dutt and Roy 1990).
22. ***Momordica balsamina* L. (Cucurbitaceae)**. Balsam apple, a tendril-bearing monoecious annual vine with red spindle-shaped ripe fruit. Native to Africa and Asia. In India adapted to dry sandy soil as an invasive species in northwestern Indo-Gangetic Plains. A medicinal and nutraceutical plant. Cultivated as an underutilized vegetable.
23. ***Momordica charantia var. muricata* (Willd.) Chakrav**. Wild free-living monoecious annual climber with small round fruit than cultivated *M. charantia*. Native to Africa, tropical Asia, and Australia. In India, it is found in many parts of the subcontinent, including Malabar region. Possesses many medicinal benefits. Part of primary gene pool of *M. charantia* with high crossability and pollen fertility (Bharathi et al. 2012).
24. ***Momordica cochinchinensis* Spreng; syn. *M. macrophylla* Gage**. A tropical dioecious species of South and Southeast Asian origin. In India, distributed in peninsular and northeastern region in humid tropical forests and openings. Fruit traditionally used as vegetable and in medicine. Rich in nutrients including carotenoids, fatty acids, vitamin E, polyphenol compounds, and flavonoids. All dioecious species are closely related (Bharathi et al. 2012) and appear to be a member of secondary gene pool of *M. dioica*.
25. ***Momordica cymbalaria* Fenzl. Ex Hook.f.; syn. *M. tuberosa* (Roxb.) Cogn**. It is a monoecious perennial vine with tuberous roots. Native to Africa and Indian Subcontinent in eastern peninsular tract of Madhya Pradesh, Andhra Pradesh, Karnataka, and Tamil Nadu. Charantin in its leaves and fruits is responsible for medicinal properties. Traditionally used for the treatment of diabetes mellitus and as an antiovulatory agent. Member of secondary gene pool for *M. charantia* and tertiary for *M. dioica* (Bharathi et al. 2012).
26. ***Momordica dioica* Roxb. Ex Willd**. Climbing dioecious herbs with tuberous roots, native to Indo-Malayan region and deciduous and semi-evergreen forests and plains of Western Ghats of India. Cultivated for fruit.
27. ***Momordica subangulata* Blume subsp. *renigera* (Wall. ex G. Don) W.J.de Wilde**. A perennial dioecious species with ovoid fruit having flattened spines. Found in Northeast India. It is a tetraploid and has been successfully crossed with *M. dioica* (Bharathi et al. 2011) producing sterile triploid. Therefore, it is a tetraploid species close to cultivated dioecious species as part of the secondary gene pool (Bharathi et al. 2012).

Another dioecious perennial species *Momordica sahyadrica* earlier considered *M. dioica* is described from a very narrow habitat of Western Ghats (Kattukunnel and Antony 2008).

28. ***Nesphostylis bracteata* (Baker) D. Potter & J.J. Doyle; syn. *Dolichos bracteatus* Baker (Fabaceae/Leguminoseae)**. A perennial climber with purple flowers. Endemic to Indian Peninsula and Western Ghats (largely Konkan Hills). The young pods are eaten; rare and allied to hyacinth bean and *Lablab purpureus*.
29. ***Solanum anguivi* Lam.; syn. *S. indicum* L**. An annual, diffusely branched 1.5 m tall undershrub, prickly with sharp, yellowish spiny berries. Native of Africa, naturalized in tropical and South India (Tamil Nadu). Traditional ethnomedicine; roots are anti-inflammatory, and fruits are used in treatment of diabetes/respiratory disorders. Progenitor of scarlet eggplant (Sakata and Lester 1997) and member of tertiary gene pool of eggplant. Source of resistance to eggplant diseases (*Fusarium* wilt).
30. ***Solanum erianthum* D. Don**. Potato/tobacco tree or *Ban Tamakhu*, a fast-growing evergreen shrub/small tree. Native of Americas and naturalized to tropical Australia and Asia, including the tropical and subtropical India, Andaman's forests, and wastelands. Grown as ornamental and listed as cultivation escape. Eaten in the south, a potential medicinal plant.
31. ***Solanum khasianum* C.B. Clarke**. A bushy annual or short-lived perennial plant (Fig. 9.3d). Found in Western Himalayas, northeastern region, and Eastern Ghats, but common to Meghalaya. A close relative of *S. melongena* (Brinjal) and member of its secondary gene pool with resistance to shoot and fruit borer. Berries are important source of the glycoalkaloid solasodine, a raw material to produce steroidal drugs.
32. ***Solanum kurzii* Brace ex Prain**. A rare perennial shrub species with white-blue flowers and egg-shaped fruits. Native to Himalayas and northeastern region. Used in folk medicine.
33. ***Solanum melongena* Linn var. *incanum* L. and var. *insanum***. Variety *insanum* is an armed subshrub 75 cm tall. Distributed over tropical Asia, including India on roadsides, wastelands, homesteads, in the eastern and southern parts. It is the wild progenitor of the cultivated eggplant, *S. melongena,* and produces fully fertile hybrid. Member of primary gene pool and source of resistance to shoot and fruit borer. Used in Ayurvedic formulations. The distribution range of the other variant *incanum* L. is from Africa to India. It is a member of secondary gene pool and being used as a source of variation for phenolic content and drought resistance as well as to develop ILs (introgression lines).
34. ***Solanum nigrum* L**. Black nightshades or *Mokoi*, an annual weed, erect, much branched, tall up to 0.5 m, with white flowers and fruits, berry, edible. Native to Africa, Europe, and Asia, including India, as weed. Genetic resource with disease resistance (bacterial wilt) for potato (Jansky 2000). Used in folklore medicine.
35. ***Solanum pubescens* Willd**. An unarmed pubescent shrub, habitat to dry deciduous forests, wayside, and shola border of Peninsular India, Karnataka, Kerala, and Tamil Nadu. Needs further investigations.

36. ***Solanum purpureilineatum* Sabnis & Bhatt**. A new species reported from Kathiawar, Gujarat, India (Sabnis and Bhatt 1970). Needs to be bio-prospected.
37. ***Solanum sisymbriifolium* Lam**. Sticky nightshade, a 1 m-tall densely prickly perennial plant with small edible fruits. Native to South America, naturalized in Eastern Himalayas, India. Member of tertiary gene pool of eggplant. Source of resistance to pest and diseases of eggplant, such as shoot and fruit borer (Collonnier et al. 2001, 2003). Also cultivated as ornamental.
38. ***Solanum spirale* Roxb**. A small fruiting shrub found in mid-elevation of 500 to 1900 m in Paleotropical areas of Asia, extending from China to India, Southeast Asia, and Australia. Cultivated in backyard gardens as vegetable and used in local medicine. In addition to Etioline, solaspiralidine, a glycoside, was isolated from its roots (Ripperger 1996).
39. ***Solanum torvum* Sw.; syn. *S. rudepannum* Dunal**. Turky berry, a bushy, erect, and spiny perennial plant 2–3 m tall. Native to Americas and widely naturalized, including in the Indian tropics and the eastern plains. Source of resistance to *Verticillum* wilt, bacterial wilt, and pest. Member of tertiary gene pool of eggplant (Bletsos et al. 1998). Used as root stock for eggplant (Gisbert et al. 2011). Used as vegetable and folklore medicine.
40. ***Solanum vagum* Heyne *ex* Nees**. An unarmed spreading shrub with globose berries, red when ripe, endemic to grasslands of southern Western Ghats.
41. ***Solanum viarum* Dunal.; syn. S. *khasianum* var. *chatterjeeanum***. The tropical soda apple or *Khasi kateri* is a perennial shrub with golf ball-sized fruit. Native to South America, invasive and naturalized all over India as an exotic noxious weed. Genetic resource, as a part of the secondary gene pool of tomato and tertiary gene pool of eggplant (Collonnier et al. 2001). Medicinal source of solasodine.
42. ***Solanum virginianum* L.; syn. *S. surattense* Burm. f., *S. xanthocarpum* Schard. & H. Wendl**. Yellow-fruit nightshade or *Kateli*, a wild prostrate prickly undershrub. Distributed from India to Polynesia, common up to 2000 m in the warmer parts, particularly of South India. Member of tertiary gene pool of eggplant with disease resistance (Collonnier et al. 2001). Potential source of solasodine and used to treat cough, asthma, and chest pain, considered one of the *Dasamula* of Ayurveda.

Several exotic wild species of African and American origin are cultivated (*Solanum giganteum* Jacq., *S. mammosum* L.) in India, whereas some have naturalized either in specific or many regions, depending on adaptation. For example, *Solanum albicaule* Kotschy ex Dunal and *S. dubium* (Kantakaari) native of Africa are recorded from Saurashtra and Rajasthan and Maharashtra, respectively, whereas *Solanum barbisetum* Nees. and *S. grandiflorum* Ruiz & Pav., native of South America, are naturalized in evergreen forests of Nilgiri Hills and Coimbatore in Tamil Nadu, while *S. myriacanthum* Dunal., native of Mexico and South America, is in Nagaland and Arunachal Pradesh. Some are being used in traditional medicine system and need further investigation for assessing their potential as genetic resource.

43. ***Trichosanthes bracteata* Lam.; syn. *T. tricuspidata* Lour. (Cucurbitaceae)**. A large woody climber, fruits with hard rind, red when ripe. Common on trees and hedges, throughout India, particularly in peninsular region and northeast; var. *bracteata* occurs largely in eastern India, in the Himalayas up to 1500 m, near cultivation lands. Distribution extends southward up to Andaman Islands. Medicinally, one teaspoon of mature seed powder, once a day, is taken by tribes to develop sterility. *Trichosanthes anamalaiensis* Bedd. is now considered its variant.
44. ***Trichosanthes cordata* Roxb**. Robust climber, habitat to foot hills of Himalayas from Garhwal to Sikkim and Northeast Hill region, up to 500 m. Doubtful entity, needs further investigations.
45. ***Trichosanthes cucumeriana* L**. Monoecious climbing herb with fruits 3–7 x 2.5–5.5 cm. Native to Asia and Australia, including India, particularly Gangetic Plains and Eastern Himalayas. Ancestor of cultivated snake gourd *T. anguina* L., with which it is conspecific and freely cross-compatible (Singh and Roy 1979), with smaller fruit size. Part of primary gene pool of cultivated snake gourd.
46. ***Trichosanthes dioica* Roxb**. Pointed gourd/*parwal*. Native to tropical Asia, wild or semi-wild populations occur in Brahmaputra Valley, plains of Assam, and North India, with greater resilience to stresses. Wild forms are part of primary gene pool of cultigens.
47. ***Trichosanthes majuscula* (C.B. Clarke) Kundu**.; **syn. *T. dunniana***. Robust climber, confined to Meghalaya, Khasi-Jaintia-Garo hills. Requires further study. Appears close to *T. wallichiana* except for larger leaves.
48. ***Trichosanthes himalensis* Clarke.; syn. *T. pilosa* Lour.; *T. ovigera* Blume**. Japanese snake gourd, a perennial climbing plant with stem that can sprawl or climb by tendrils. Reported in India from Meghalaya and Eastern Himalayas between 600 and 1500 m. It is cultivated in China for its edible fruit and medicinal roots.
49. ***Trichosanthes kerrii* Craib; syn. *T. tomentosa* Chakr**. Robust climber. Reported from Northeastern Hill region. Needs further study.
50. ***Trichosanthes lepiniana* (Naud.) Cogn**. Tendril climber with robust stem. Habitat to roadsides, evergreen broad-leaved forests, and groves in Deccan Peninsula and Kerala.
51. ***Trichosanthes nervifolia* L.** Scandent climbing herb with white flowers, habitat to semi-evergreen forests and plains in Southwest India, Malabar, and Sri Lanka.
52. ***Trichosanthes truncata* Clarke.; syn. *T. ovata* Cogn**. A perennial climbing vine with egg leaves and dioecious flower. Reported from Northeast Hill region. Needs further study.
53. ***Trichosanthes wallichiana* (Ser.) Wight**. A climber, reported from moist deciduous forests of Northeast India. Close ally of *T. bracteata* from which membranous leaves with basal black-dotted glands separated it.

9.2.2 Leafy Vegetables

54. ***Chenopodium album* L. (Amaranthaceae)**. A fast-growing weedy annual herb, distributed from Europe to Eastern Asia as weed. Found in North India abundantly during winters, harvested from nature, and consumed as leafy vegetable/food crop. It has a very complex taxonomy and is divided in numerous micro-species, subspecies, and varieties. It hybridizes with several *Chenopodium* species, such as *C. foliosum* (Moench) Asch., and many others found in Himalayan region (Singh 2015). Cultivated on a very small scale as leafy vegetable or grain.
55. ***Ipomoea aquatica* Forssk. (Convolvulaceae)**. A semiaquatic, tropical herb that grows wild in water or on moist soil. Native to Africa, Asia, and Australia, including India. Invasive. Leaves vary from sagittate to lanceolate. Cultivated in tropics for vegetable and fodder.
56. ***Lactuca cooperi* Anthony (Compositae/Asteraceae)**. A perennial herb, endemic to Eastern Himalayas (Sikkim) in alpine and sub-alpine meadows and on exposed hill slopes. Genetic resource for *L. sativa* L., the lettuce. Under threat, endangered.
57. ***Lactuca remotiflora* DC.; syn. *L. intybacea***. An annual or biennial erect herb, up 25–60 cm tall. Leaves mostly radical, obovate, orbicular, or oblong. Distributed on dry sandy locations of Indus and Gangetic plains and Indian peninsula. A rare medicinal plant.
58. ***Lactuca runcinata* DC.; syn. *L. heyneana* DC**. An erect herb, with leaves radical and cauline. Common weed in many parts of India, Jammu and Kashmir, Rajasthan, Maharashtra, and Tamil Nadu. Antibacterial.
59. ***Lactuca serriola* L.; syn. *L. scariola* L**. An annual erect, prickly herb up to 1.5 m tall. Leaves sessile pinnatifid. Native to Africa, Europe, and Asia, including Indian Himalayan region. Referred as prickly lettuce or compass plant, because under the Sun, the upper leaves twist around to hold their margins upright. Ornamental, culinary, and folklore medicine. Closest wild relative and member of primary gene pool of *L. sativa*, the lettuce (Koopman et al. 2001).
60. ***Malva pusilla* Sm.; syn. *M. rotundifolia* L. (Malvaceae)**. Dwarf Mallow, an annual/biennial herb. Native to Europe and Asia. An invasive species, naturalized in Western Himalayas, northern plains, and the Deccan Peninsula. Blessed with medicinal properties. Vegetable is extracted from the wild. Successful interspecific hybridization within genus *Malva* suggests possibility of genetic introgression.
61. ***Malva sylvestris* L.** Blue mallow, an annual, biennial, or perennial plant, standing 1 m tall with showy bright mauve-purple flowers, with dark veins. Native of Africa, Europe, and Asia, including Western Himalayas up to 2400 m, Bihar, and Deccan Peninsula, in fields, hedgerow and fallow lands. Ornamental, young leaves used as vegetable and medicine. Extracted from the wild as well as cultivated.
62. ***Malva verticillata* L.** Whorled mallow, an annual/biennial plant growing to 1.7 m. Leaves cordate, downy pubescent, with pink flowers, crowded in clusters.

Native to China and naturalized in Himalayas up to 3600 m and in Nilgiri Hills. Stem and leaves are eaten as vegetable, and roots are used for whooping cough and the ash of leaves in scabies.

63. ***Nasturtium officinale* R. Br. (Brassicaceae)**. Watercress is aquatic or terrestrial noxious perennial weed with significant genetic diversity that can be used in improvement. Native to Europe and Asia including Indian Subcontinent. One of the oldest leafy vegetable and folklore medicine with diverse chemical activities. Cultivated as leafy vegetable, rich in vitamins and minerals.
64. ***Rumex acetosella* L. (Polygonaceae).** The red sorrel is a perennial weedy herb that can reach 0.5 m in height. A garnish for green salad, tart-flavoring agent, and a curdling agent for cheese. Native to Africa, Europe, and Asia, including India. Found as weed in Eastern Himalayas (Sikkim), Nilgiris, and Palani hills.
65. ***Rumex dentatus* Linn.** Indian dock, an annual herb 20–60 cm tall. Native in Europe and Asia, including Indian Subcontinent, common as roadside weed in Northwestern Himalayas and northern and central plains. More variability in hills, extending to eastern region; forms from plains are akin to *R. nigricans*. Plant infusion rich in flavonoids, phenolic, reducing sugars, total sugars, and antioxidant activity.
66. ***Rumex patientia* L.** An herbaceous perennial, native to Europe and Asia, including India. Widely distributed in Western Himalayas, from Kashmir to Kumaun, and Northeast with lot of genetic variation. Consumed as spinach in high-altitude cold desert of Ladakh. Several health-promoting compounds, including phenols, are found, besides antioxidant activity.
67. ***Rumex vesicarius* L.** An annual, native of Africa and Asia, including many parts (Punjab) of Indian Subcontinent. Most popular *Rumex* reported from Chhattisgarh. Cultivated. Used as vegetable and in folklore medicines as source of vitamins, minerals, proteins, fiber, carotenes, and flavonoids with health benefits. Vital source of phytochemical antioxidants to prevent damage associated with free radicals.
68. ***Trigonella cachemiriana* Cambess. (Fabaceae/Leguminosae).** Prostrate or spreading much branched herb. Native of Asia, including Himachal Pradesh, Kashmir, and Punjab in India. Needs bioprospecting from genetic resource point of view of fenugreek.
69. ***Trigonella emodi* Benth**. Himalayan fenugreek, a perennial glabrous erect herb 20–60 cm tall (Fig. 9.3e). Native to Western Asia and Himalayas, up to 3000 m, and plains of Punjab and Uttar Pradesh. *T. emodi* var. *podperae* Sirj., syn. *T. podperae* (Širj.) Vass., is endemic to Ladakh. It is an erect herb with many branches from the base. Not much work done, needs bioprospecting.
70. ***Trigonella fimbriata* Benth**. Perennial erect herb. Native to Northwestern Himalayas, including Kashmir, Himachal Pradesh, northern Punjab, and Uttarakhand.
71. ***Trigonella uncata* Boiss. & Noë; syn. *T. glabra* Thunb. subsp. *uncata* (Boiss. & Noë) Lassen**. Not a climbing annual herb. Native to Asia, including West Asia and Indian Subcontinent. Common in Gujarat, Rajasthan. Needs collection, characterization, and evaluation of genetic potential.

72. ***Trigonella gracilis* Benth**. *Tinpaate Jhaar*, perennial trailing glabrous 15–45 cm-long herb. Native to Indian Subcontinent: Kashmir, Himachal Pradesh, Punjab, and Uttarakhand. It is used as a green fodder and a genetic resource for fenugreek.
73. ***Trigonella occulta* Delile**. Densely branched annual herb with glabrous stem, pinnately trifoliate leaf. Found in moist places of Indus plains, Gangetic Plains, and in Koraput region. Leaves used as vegetable and green fodder and seeds in diarrhea.
74. ***Trigonella polycerata* L.; syn. *Medicago polycerata* (L.) Trautv**. It is glabrous or slightly hairy herb. Native to Indian Subcontinent, found in Western Himalayas, Indus, and Gangetic Plains. It has medicinal potential.

9.2.3 Root, Tuber, and Bulb Vegetables

75. ***Allium carolinianum* DC. (Amaryllidaceae/Liliaceae)**. A 60 cm-tall herb, with narrow, flat leaves, shorter than the scape, which produces egg-shaped bulbs up to 25 mm across. Native to Asia, including Indian Subcontinent in upper reaches of Western Himalayas, Ladakh. Used for flavoring and as vegetable.
76. ***Allium cepa* var. *aggregatum*; syn. *A. ascalonicum* L**. Shallot or *Ek-dana-lasun* is a small, clustered onion with intense flavor. Native to Central or Southwest Asia, from where it spreads to the Mediterranean and Indian hills. Expressing great variability for bulb. Mostly cultivated and used as condiment in Indian cuisine and known as an Ayurveda herb. Variety *viviparum* (Metzg) Alef., member of the primary gene pool of *Allium cepa*, is also cultivated; wild form does not occur.
77. ***Allium chitralicum* Wang & Tang**. An herb with narrow leaves, rose-colored flowers, and egg-shaped bulbs. Native to upper reaches of Afghanistan, Pakistan, Northwestern Himalayas, and Ladakh. Needs collection, characterization, and evaluation of genetic potential.
78. ***Allium griffithianum* Boiss**. Up to 40 cm-tall herb with ovoid bulb, coriaceous outer coats, fibrous grayish black; membranous inner coats. Native to high mountains of Asia, including Indian Subcontinent (Himalayas and Khasi hills).
79. ***Allium humile* Kunth**. An 8–30 cm-tall plant with flat, fleshy, linear leaves, about 5 mm wide, scape is up to 15 cm, with slightly compressed narrow, cylindrical bulbs. Grows naturally on slopes at high elevations in the Western Himalayas.
80. ***Allium hypsistum* Stearn**. A bulbous herb 20 cm tall, bulbs densely clustered, tufted, cylindrical, and elongated, 1 cm in diameter, and situated on short rhizome. Native to Nepal, extending to Eastern Himalayas. Used as food and folk medicine.
81. ***Allium macleanii* Baker**. It is a perennial herb 1 m tall, with a spherical umbel crowded with many purple flowers. It has tuberous bulbs. Native to mid- and

West Asia and Indian Subcontinent (Himalayas). Cultivated in Afghanistan and used in Unani and Ayurveda medicine.

82. ***Allium prattii* Wright**. A 60 cm-tall herb with solitary or clustered subcylindrical bulbs. An Asian species of wild onion, native to Central Himalayas (Assam, Nepal, Sikkim, Bhutan) and China. Important genetic resource, needing collection and characterization.
83. ***Allium rhabdotum* Stearn**. Native to Indian Subcontinent (Bhutan and Sikkim). Member of tertiary gene pool of onion (Fritsch and Friesen 2002).
84. ***Allium roylei* Stearn**. A plant with ca. 40 cm tall scape. It has an egg-shaped bulb ca. 30 mm across. Endemic to high altitudes of Afghanistan and Himalayas in India. Source of resistance to powdery mildew, leaf blight, etc., as member of secondary gene pool of onion and Welsh onion (Chuda and Abrams 2009). Leaf blight introgressed into *A. cepa* (de Vries et al. 1992). Used as flavoring agent. Endangered.
85. ***Allium rubellum* M. Bieb**. A lovely, bright rare species, diploid (2n = 16). Native to Asia, including Western Himalayas on exposed hillocks in the Indian Subcontinent with much variation. A triploid form was collected from Pathankot. Used locally as flavoring agent in curries and as vegetable. Semidomesticate, grown in kitchen gardens. Overexploitation is shrinking its populations.
86. ***Allium schoenoprasum* L**. *Chives*, a cultivated pot herb, allied to garlic, with purple flowers and narrowly ovoid, 2 cm long, 0.5 cm broad, scaly bulb. Native to Asia, Europe, and North America, including Western Himalayas from Kashmir to Kumaon up to 3300 m, in alpine meadows. Member of tertiary gene pool of leek and Welsh onion (Kik 2002; Umehara et al. 2007)
87. ***Allium sikkimense* Baker**. Bulbs are clustered, cylindrical, 0.3–0.5 cm in diameter. Native to Sikkim, Tibet, Bhutan, Nepal, India, and parts of China in forest margins, scrub, slopes, and meadows; 2400–5000 m. An ornamental onion with large true light-blue flowers. Possesses ethnobotanical use and a genetic resource.
88. ***Allium stracheyi* Baker**. *Jambu* is a perennial herb about 35 cm tall with rosy flowers. Native to Tibet to Western Himalayas and upper Gangetic Plains. Used as vegetable, spice, condiment and seasoning, and occasionally as medicine (Pandey et al. 2008). Vulnerable.
89. ***Allium tuberosum* Roxb**. Perennial herb with tender leaves having light taste of garlic (Fig. 9.3f), used to flavor vegetables, salads, omelets, etc. Bulbs cylindrical, 4–6 cm long, light brown on drying. Distributed in Himalayas, northeastern hills, and Shanxi, China. Sporadic, up to 1800 m, wild on exposed hills, etc. Cultivated as ornamental and vegetable. Member of primary gene pool of Chinese chives (Blattner and Friesen 2006).
90. ***Allium victorialis* L**. Alpine leek, a perennial, broad-leaved species of wild onion with cylindrical bulb 4–6 cm long, coat net-veined, fibrous. Native to Asia, Europe, and North America, including Northwestern Himalayan region of India. Cultivated.

91. ***Allium wallichii* Kunth**. Himalayan onion herb 0.6 m tall with linear-to-oblong lanceolate leaves. Distributed over Kumaon Himalaya to Northeast Himalayas (Sikkim, Darjiling) to Tibet and parts of China. Used for medicinal purposes and as a spice, threatened due to overharvesting.

In addition, *Allium gilgiticum* Wang & Tang., an herb with cylindrical bulbs, endemic to upper reaches of Western Himalayas in Gilgit and Ladakh, known from type collections gathered in 1930 (Karthikeyan et al. 1989). Rare and quite possibly extinct and needs recollection and further investigation.

92. ***Alocasia acuminata* Schott**. A terrestrial herb, small to medium sized, slightly robust, up to 75 cm tall, evergreen. Distributed in forests of Assam, Northeast India, and Indochina region. Used as food and extracted from the wild.
93. ***Alocasia cucullata* (Lour.) G. Don**. Chinese taro, perennial herb producing thick, erect stems and heart-shaped leaf. Native to Asia, including Northeast India and Western Ghats. Used as food, source of starch and vegetable and also folk medicine and ornamental. Extracted from the wild and cultivated.
94. ***Alocasia macrorrhizos* (L.) G. Don; syn. *A. indica* Schott**. Elephant ear taro, a massive aroid, stout perennial; stem 4–5 cm thick, arrow-shaped leaves. Native of rainforests of Malaysia. Naturalized to tropical Asia, including sub-Himalayan tract of East Himalayas, northeastern region, Bengal plains, and peninsular humid tract. Member of primary gene pool of giant taro (Nauheimer et al. 2012). Cultivated as ornamental/vegetable. The stem tuber is peeled, cut into pieces, and eaten. An Ayurveda plant.
95. ***Amorphophallus bulbifer* (Roxb.) Blume (Araceae)**. Prettiest of the voodoo lilies with striking flower, corm globose, up to 15 cm across, and tubercle. Petiole 1 m long, produces bulbils on the top of the petioles, wherefrom the leaflets depart. Native to Asia, including Eastern Himalayas (Sikkim), northeastern hills (Khasi Hills), Eastern peninsular tract, West Bengal, and Kerala, Western Ghats in India.
96. ***Amorphophallus commutatus* (Schott) Engl**. Herbs with corms. Leaves tripartite compound, up to 50 m-long, elliptic leaflets, caudate acuminate; petiole to 50 cm long. Endemic to evergreen forests of Western Ghats, with significant variability of varietal complex. With antioxidant properties and genetic resource.
97. ***Amorphophallus hohenakeri* (Schott.) Engl. & Gehrm**. An herb with corm of 3 cm across, smooth, depressed globose, and dull white inside. Endemic to evergreen forests of Western Ghats (Jaleel et al. 2011). Needs collection and characterization.
98. ***Amorphophallus konkanensis* Hett., S.R. Yadav & K.S. Patil**. An herb with globose or depressed globose, 3–4 cm in diameter tubers. Found on forest clearing, road sides, and hedges of cultivated field in North Western Ghats (Goa, Maharashtra, and Karnataka). Recently reported from Madhya Pradesh.

99. ***Amorphophallus longistylus* Kurz ex Hook. f.** A little known rare species, endemic to India in Andaman and Nicobar Islands. Needs collection and characterization.
100. ***Amorphophallus muelleri* Blume**. A perennial herb producing a single leaf each year from a tuberous rootstock. Triploid, native in Tropical Asia (Myanmar, Thailand, Malaysia, Indonesia, and Andaman Islands). Corm eaten raw or boiled or baked. Valued for glucomannan content and used in the food industry, paper industry, pharmacy, and cosmetics.
101. ***Amorphophallus mysorensis* E. Barnes & C.E.C. Fisch**. An herb, endemic to South Karnataka—Billigirirangan Hills. Needs collection and characterization.
102. ***Amorphophallus paeoniifolius* (Denst.) Nicolson A. *campanulatus* (Roxb.) Bl. ex Decne**. Elephant yam or *Jimikand*, annual herb with large, depressed globose, much-warted tubers, 20–25 cm in diameter. Native of Africa and Asia, including Indian Deccan Plateau. Cultivated as a tropical tuber crop for starch and as vegetable. Corm is considered stomachic and tonic; used in piles and as restorative in dyspepsia and debility.
103. ***Amorphophallus smithsonianus* Sivadasan**. A new species with smooth, glossy, compressed globose or sub-globose, 3.0–4.5 x 3.5–4.0 cm corms. Native to South Western Ghats. Fruits and seeds have medicinal properties. Needs further studies.
104. ***Amorphophallus sylvaticus* (Roxb.) Kunth**. An herb with 5 cm-thick corms. Distributed in semi-evergreen forests of India [Eastern (Bastar) and Western Ghats] and Sri Lanka. Fruits and seeds have medicinal properties.
105. ***Colocasia affinis* Schott (Araceae)**. Dwarf elephant ear, a perennial terrestrial herb with stolon's trailing horizontally, thin, 3.7–40 cm × ca. 4 mm, pale green, branched, and triangular or ovate tubercles, ca. 2 cm in diameter. Extended from China to East Himalayas, Assam, and Indochina. Ornamental. Related to taro.
106. ***Colocasia antiquorum* Schott.; syn. *C. esculenta* (L.) Schott var. *antiquorum*.** An erect herb up to 2 m tall with small main corm and many cormels. Native to Southeast Asia including India, in sub-Himalayan tract, peninsular region, and northeastern region, in water-logged humid tropical belt with greater variability in eastern region. Member of primary gene pool of taro, *C. esculenta* (Ivancic and Lebot 1999).
107. ***Colocasia fallax* Schott**. Silver leaf dwarf elephant ear, a small herb, growing to 0.6 m tall with inconspicuous fragrant white flowers. Extending from China to Eastern Himalayas (Sikkim, Assam, and Khasi Hills) with three forms different in petiole color.
108. ***Dioscorea alata* L. (Dioscoreaceae)**. Known as Greater, water or winged yam is a creep and shrub. Native of Asia, cultivated for desserts and starch, escaped into the wild in Western Himalayas and northeastern region generating variability. Part of primary gene pool of greater yam (Egesi et al. 2002). Medicinal and ornamental.

109. ***Dioscorea bulbifera* Linn.** Air potato, a perennial vine with broad leaves forming bulbils in leaf axil and tubers beneath ground. Native of Africa and Asia, cultivated as food crop, escaped, and naturalized in Western and Eastern Himalayas and most of India. Part of primary gene pool of air potato (Terauchi et al. 1991). Used as folk medicine.
110. ***Dioscorea decipiens* Hook.f**. Climber of up to 12 m, tubers are vertical and cylindrical, one per growing season, with withering tuber from previous season. Tubers are edible. Habitat to mixed deciduous and hilly evergreen forest of Northeast India.
111. ***Dioscorea deltoidea* Wall. ex Griseb.** Perennial, dioecious, glabrous twining climber. Roots are tuberous, and rhizome is horizontal, irregular, and ginger shaped. Native of Asia in forest and open fields of Himalayas and Punjab. Possesses high medicinal value (roundworm, constipation, antimicrobial). Cultivated, but still harvested from the wild, endangered.
112. ***Dioscorea glabra* Roxb.** Glabrous herb with cylindrical tubers. Habitat to evergreen broad-leaved and scrub forests, mountain slopes, valley sides in Southeast Asia, including Northeast Himalayas/India, Bengal, Chota Nagpur, and Andaman. Ayurveda medicinal plant for general tonic. Extracted from the wild.
113. ***Dioscorea hamiltonii* Hook.f.** Plant with angled stem, glabrous. Habitat to Asia, including Eastern Himalayas, Bihar, Chota Nagpur Plateau, and the Western Ghats. Ayurveda plant, *Varahi*, leaves used to treat jaundice and mumps. Part of secondary gene pool of greater yam (Hsu et al. 2013).
114. ***Dioscorea hispida* Dennst; syn. *D. daemona* Roxb.** Twiner with glabrous stems twining to left; leaflets are subequal, 17x12 cm, and obovate, with large tubers. Distributed from India and southern China, through Southeast Asia to New Guinea. Produces intoxicating tubers of medicinal value. Famine food.
115. ***Dioscorea intermedia* Thw.** *Shoddi kalasu*, a climbing herb; stems neither winged nor conspicuously angled. Leaves alternate. Tuber >1, grows horizontally. Distributed in evergreen forests of South India (Kerala) and Sri Lanka. Medicinal.
116. ***Dioscorea kalkapershadii* Prain & Burkill.** Climber with slender stem, sparsely prickly, hirsute, at length glabrescent; leaves alternate, compound, single tuber. Found in Gangetic Plains, Chota Nagpur Plateau, and Southern Western Ghats.
117. ***Dioscorea kumaonensis* Kunth.** Stem somewhat smooth. Tubers elongate to globose, seldom eaten. Distributed in Temperate Himalayas and Northeast India with var. *straminea* and var. *vera*. A new source of antirheumatic drug.
118. ***Dioscorea lepcharum* Prain & Burkill.** It is an edible yam with several tubers produced upon at the end of a long stalk, spreads laterally, descends in soil, recently rediscovered (Saikia, et al. 2014) in the Eastern Himalayas (Sikkim) and northeastern region.
119. ***Dioscorea oppositifolia* L.** *Amilkaakanda*, a climber, terete stem, twining right. Native to moist deciduous, semi-evergreen, and evergreen forests of Myanmar and to the Indian Subcontinent. Ayurveda medicine to reduce swelling.

120. ***Dioscorea pentaphylla* L.; syn *D. jacquemontii* Hook.f.; *D. tripylla* L.** *Vaaraahikanda*, a tuberous climber, native to Asia and Australia including India, from 1800 m in the Himalayas to Andaman Islands. Famine food, Ayurveda medicine. Dry tubers sold.
121. ***Dioscorea prazeri* Prain & Burkill; syn. *D. clarkei* Prain & Burkill.** Climber up to 5 m. Habitat to evergreen forests of Indo-Malayan region, including Central to Eastern Himalayas, northeast region. Ayurveda medicine, tuber antiphthiriac and fish poison. Endangered medicinal plant.
122. ***Dioscorea pubera* Blume; syn. *D. anguina* Roxb.** Plant pubescent, with stem twining to the right having 1–2 tubers narrowly cylindrical, rootless, flesh lemon yellow. Habitat to India, Nepal, and Bhutan in Eastern Himalayas, Northeast and Bastar. Exploited from nature.
123. ***Dioscorea scortechinii* Prain & Burkill.** Large forest climber with compound leaf, tuber one like *D. pentaphylla*. Distributed from Eastern Himalayas, Assam, Bangladesh, and Northeast India to Sumatra.
124. ***Dioscorea spicata* Roth.** A perennial twiner, leaves opposite or subopposite. Found in Indian Subcontinent, including Bangladesh and Sri Lanka. Ayurveda plant with antioxidant with free radical scavenging capacity. Marginally cultivated.
125. ***Dioscorea tomentosa* Koenig ex Spreng.** Climber with stem twining to the left, terete, with flattened and matted hairs. Leaves trifoliate. Habitat to semi-evergreen, moist, and dry deciduous forests of Indian Subcontinent.
126. ***Dioscorea trinervia* Roxb. ex Prain & Burkill.** Distributed in Northeast India, Assam, Bangladesh, Indochina, and Myanmar. Medicinal.
127. ***Dioscorea vexans* Prain & Burkill.** Tuberous climbers with opposite leaves. Endemic and rare species from Andaman and Nicobar Islands. Medicinal, having antifertility properties.
128. ***Dioscorea wallichii* Hook. f.** A yam with stem twining to the right, smooth or scarcely prickly toward the base. Leaves up to 12 × 12 cm, orbicular. Tubers can grow over 1 m. Found in evergreen and moist deciduous forests of Eastern Himalayas and Western Ghats. Tubers are edible and possess medicinal properties along with roots.
129. ***Dioscorea wattii* Prain & Burkill.** Tubers unknown, woody crown is present with numerous rhizome, large capsule, seeds surrounded by wings. Found in Eastern Himalayas, Northeast India, and Naga Hills, rare.
130. ***Dioscorea wightii* Hook.f.** Glabrous climbing shrub, stem slender, terete, twining to the right. Leaves simple, 3–6 × 2–4.5 cm, ovate, acuminate. Endemic to Indian Peninsula and Chota Nagpur Plateau. Listed in the Red List of Threatened Species (Chadburn and Contu 2014). Medicinal.
131. ***Flemingia procumbens* Roxb.; syn. *F. vestita* Benth. ex Baker (Fabaceae/Leguminosae).** Perennial non-climbing shrub with tuberous roots, used as food and drink. Cultivated for juicy tuber, a highly priced vegetable in Meghalaya. Naturally distributed in Himalayas and Northeast Hills with significant genetic variability as source for improvement. Also found in Uttar Pradesh and Tamil Nadu. Used in mixed cropping because of its nitrogen fixation property. Medicinal.

9.3 Perspective

The utilization of wild relatives in crop improvement has been most successful and popular in solanaceous vegetables, particularly *Lycopersicon* and *Solanum*. In these taxa, many wild relatives have been identified with gene conferring resistance to biotic and abiotic stresses. The experience in utilization of these taxa in introgression of desired genes into cultivated species from their wild relatives has been possible because of the positive response of solanaceous species to both conventional and biotechnological manipulations thereby developing genetically engineered interspecific hybrids/cultivars. Therefore, successful utilization of wild species in crop improvement requires amenability to these resources with expertise in biotechnological techniques to overcome hybridization barriers, establishing successful hybrids, besides conventional cytogenetic manipulations to improve genetic recombination. In the Indian context, conventional approaches would be sufficient in the case of eggplant for transfer of resistances from species, such as *S. insanum* and *S. incanum* into *S. melongena*, since there are no crossability barriers. Transfer of desired characters from species, such as those resistant to shoot and fruit borer from *S. khasianum* and root knot nematode from *S. sisymbrifolium* and those resistant to several root diseases from *S. torum*, may require biotechnological approaches because of the cross-ability barriers between these species and the cultivated *S. melongena*.

A large number of wild species related to important vegetable crops have been described based on morphological distinctiveness; however, experimental taxonomic studies are required in certain cases, such as *Trichosanthes* in fruity vegetables, *Lactuca* and *Trigonella* in leafy vegetables, and *Allium* and *Dioscorea* in bulbous and tuberous vegetables, to produce an inventory of species that are genetically distinct in terms of their distinctive features, genetic distance, and phylogenetic relationships, particularly with cultivated species to assess their value as a genetic resource and facilitate conservation, avoiding redundancy and feasibility for introgression of genes with application of appropriate breeding strategies. Such studies shall throw new light on genetic relationship of *Lactuca* and *Trigonella* wild species found in India with their cultivated counterpart and as potential genetic resource.

In other cases, such as *Abelmoschus*, narrow genetic base has been the major concern. Concerted efforts are required for exploitation of available wild genetic resources and to augment genetic diversity from unexplored pockets of the subcontinent as well as from Africa for broadening the genetic base on priority. Useful information has been generated on the economic potential of wild species, particularly on their inherent resistance potential against one or more biotic and abiotic stresses. However, in these cases, greater emphasis is required on their collection for capturing existing variability before they are being lost under the pressure of infrastructure development and climate change, particularly from the eastern and northeastern regions of the Indian Subcontinents. Further, in several cross-combinations, variability has been created through interspecific hybridization, while in others, intermediate forms resulting from natural hybridization do occur. Both types are easy to be exploited for genetic introgression. Therefore, they require immediate collection attention. For example, the variability created by natural hybridization between *A. manihot*, *A. tetraphyllus*, and *A. caillei* (African) needs to be fully understood, collected, and exploited.

In cases like cucurbitaceous vegetables, efforts are needed to collect diverse germplasm in case of the species reported with resistance to various biotic and abiotic stresses, such as *Cucumis*, while in other genera, *Luffa*, *Momordica*, etc., the wild relatives need to be characterized and evaluated, possibly under different agroclimatic conditions, because of their wider distribution throughout the subcontinent.

Also for making full use of genetic diversity available in the wild relatives of vegetable crop species, the application of molecular techniques needs to be encouraged for identification and differentiation of germplasm and study of interspecific differentiation between cultivated species and the wild relatives. These tools should also be encouragingly utilized to overcome crossability barriers and promote utilization of wild species.

References

Bharathi LK, Vinod DAB, Ghosh N, Behera TK, Nak G, Nath V (2011) Cytomorphological and molecular characterization of interspecific F1 hybrid of *Momordica dioica* x *Momordica subangulata* subsp. *renigera* (G. Don) de Wilde. Af J Agric Res 6(13):2982–2990

Bharathi LK, Munshi AD, Behera TK, Vinod, Joseph John K, Das AB, Bhat KV, Sidhu AS (2012) Production and preliminary characterization of inter-specific hybrids derived from *Momordica* species. Curr Sci 103:178–186

Blattner FR, Friesen N (2006) Chapter 10. Relationships between Chinese chive (*Allium tuberosum*) and its putative progenitor *A. ramosum* as assessed by random amplified polymorphic DNA (RAPD) (Docum Domestication), pp 134–142

Bletsos FA, Roupakias DG, Tsaktsira ML, Scaltsoyjannes AB, Thanassoulopoulos CC (1998) Interspecific hybrids between three eggplant (*Solanum melongena* L.) cultivars and two wild species (*Solanum torvum* Sw. and *Solanum sisymbriifolium* Lam.) Plant Breed (New York) 117:159–164

Chadburn H, Contu S (2014) *Dioscorea wightii*. The IUCN red list of threatened species 2014: e.T44392879A44545069

Chakravarty HL (1982) Cucurbitaceae. In: Jain SK (ed) Fascicles of Flora of India, 11. Botanical Survey of India, Calcutta

Chen JF, Kirkbride JH (2000) A new synthetic species of *Cucumis* (Cucurbitaceae) from interspecific hybridization and chromosome doubling. Brittonia 52:315–319

Chen JF, Staub JE, Tashiro Y, Isshiki S, Miyazaki S (1997) Successful interspecific hybridization between *Cucumis sativus* L. and *C. hystrix* Chakr. Euphytica 96:413–419

Chuda A, Abrams A (2009) Aspects of interspecific hybridization within edible Alliaceae. Acta Physiol Plant 31:223–227

Collonnier C, Fock I, Kashyap V, Rotino GL, Daunay MC, Lian Y, Mariska IK, Rajam MV, Servaes A, Ducreux G, Sihachakr D (2001) Applications of biotechnology in eggplant. Plant Cell Tissue Organ Cult 65:91–107

Collonnier C, Fock I, Daunay MC, Servaes A, Vedel F, SiljakYakovlev S, Souvannavong V, Sihachakr D (2003) Somatic hybrids between *Solanum melongena* and *S. sisymbrifolium*, as a useful source of resistance against bacterial and fungal wilts. Plant Sci (Elsevier) 164:849–861

De Vries JN, Wietsma WA, de Vries T (1992) Introgression of leaf blight resistance from *Allium roylei* Stearn into onion (*A. cepa* L.) Euphytica 62:127–133

Dutt B, Roy RP (1990) Cytogenetics of the old-world species of *Luffa*. In: Bates DM, Robinson RW, Jeffrey C (eds) Biology and utilization of Cucurbitaceae. Comstock Public. Assoc, a division of Cornell University Press, Ithaca, pp 134–140

Egesi CN, Pillay M, Asiedu R, Egunjobi JK (2002) Ploidy analysis in water yam *Dioscorea alata* L. germplasm. Euphytica 128:225–230

Fritsch RM, Friesen N (2002) 1 Evolution, domestication and taxonomy. In: Rabinowitch HD, Currah L (eds) *Allium* crop science: recent advances. CABI Publishing, Wallingford, pp 5–30

Gisbert C, Prohens J, Raigón MD, Stommel JR, Nuez F (2011) Eggplant relatives as sources of variation for developing new rootstocks: effects of grafting on eggplant yield and fruit apparent quality and composition. Sci Hortic. doi:10.1016/j.scienta.2010.12.007

Hsu K-m, Tsai J-L, Chen M-Y, Ku H-M, Liu S-C (2013) Molecular phylogeny of *Dioscorea* (Dioscoreaceae) in East and Southeast Asia. Blumea 58:21–27

Ivancic A, Lebot V (1999) Botany and genetics of New Caledonian wild taro, *Colocasia esculenta*. Pac Sci 53:273–285

Jaleel VB, Sivadasan M, Alfarhan AH, Thomas J, Altar AA (2011) Revision of *Amorphophallus* Blume ex Decne. Sect. Rhaphiophallus (Schott) Engl. (Araceae) in India. Bangladesh J Plant Taxon 18(1):1–26

Jansky S (2000) Breeding for disease resistance in potato. In: Janick J (ed) Plant breeding reviews, vol 19. Wiley, Oxford, pp 86–155. ISBN 0-471-38787-8 © 2000

John KJ, Scariah S, Muhammed Nissar VA, Latha M, Gopalakrishnan S, Yadav SR, Bhat KV (2013) On the occurrence, distribution, taxonomy and genepool relationship of *Cucumis callosus* (Rottler) Cogn., the wild progenitor of *Cucumis melo* L. from India. Genet Resour Crop Evol 60:1037. doi:10.1007/s10722-012-9899-2

John KJ, Khedasana R, Muhammed Nissar VA, Scariah S, Sutar S, Rao SR, Nizar MA, Latha M, Yadav SR, Bhat KV (2014) On the occurrence, distribution and taxonomy of *Cucumis setosus* Cogn., an endemic wild edible vegetable from India. Genet Resour Crop Evol 61:345. doi:10.1007/s10722-013-0038-5

Karthikeyan S, Jain SK, Nayar MP, Sanjappa M (1989) Florae Indicae Enumeratio: Monocotyledonae: 1–435. Botanical Survey of India, Calcutta

Kattukunnel JJ, Antony VT (2008) *Momordica sahyadrica* sp. nov. (Cucurbitaceae), an endemic species of Western Ghats of India. Nord J Bot 24:539–542

Kik C (2002) Chapter 4. Exploitation of wild relatives for the breeding of cultivated *Allium* species. *Allium* Crop Sci:89–91

Koopman WJM, Zevenbergen MJ, van den Berg RG (2001) Species relationships in *Lactuca* s.l. (Lactuceae, Asteraceae) inferred from AFLP fingerprints. Am J Bot 88:1881–1887

Nair PG, Kuriachan P (1976) A spontaneous hybrid between *Abelmoschus tuberculatus* Pal & Singh and *A. esculentus*. New Bot 3:48–53

Nauheimer L, Boyce PC, Renner SS (2012) Giant taro and its relatives: a phylogeny of the large genus *Alocasia* (Araceae) sheds light on Miocene floristic exchange in the Malesian region. Mol Phylogenet Evol 63:43–51

Pandey A, Pandey R, Negi KS, Radhamani J (2008) Realizing value of genetic resources of Allium in India. Genet Resour Crop Evol 55(7):985–994

Renner SS, Pandey AK (2013) The Cucurbitaceae of India: accepted names, synonyms, geographic distribution, and information on images and DNA sequences. Phytokeys 20:53–118

Ripperger H (1996) Steroidal alkaloids from roots of *Solanum spirale*. Phytochemistry 43:705–707

Sabnis SD, Bhatt RP (1970) *Solanum purpureilineatum* new species from Gujarat India. Bulletin Bot Survey of India 12(1-4):258–260

Saikia B, Saikia M, Das AK (2014) Rediscovery of *Dioscorea lepcharum* Prain & Burkill [Dioscoreaceae] – from Arunachal Pradesh, India Pradesh, India after a century. Pleione 8(1):178–180

Sakata Y, Lester RN (1997) Chloroplast DNA diversity in brinjal eggplant (*Solanum melongena* L.) and related species. Euphytica 97:295–301

Sebastian P, Schaefer H, Telford IRH, Renner SS (2010) Cucumber (*Cucumis sativus*) and melon (*C. melo*) have numerous wild relatives in Asia and Australia, and the sister species of melon is from Australia. Proc Natl Acad Sci U S A 107:14269–14273

Singh AK (1978) Cytogenetics of Semi-Arid Plants III. A natural interspecific hybrid of *Cucurbitaceae* (*Citrullus colocynthis* × *vulgaris* Schrad). Cytologia 43:569–574

Singh AK (1990) Cytogenetics and evolution in *Cucurbitaceae*. In: Bates DM, Robinson RW, Jeffrey C (eds) Biology and utilization of Cucurbitaceae. Comstock Publc Assoc, a division of Cornell University Press, Ithaca, pp 10–28

Singh AK (2015) Agricultural biodiversity heritage sites and systems in India. Asian Agri-History Research Foundation, Secunderabad, p 467

Singh AK, Roy RP (1979) Analysis of interspecific hybrids in *Trichosanthes* L. Caryologia 32:329–334

Terauchi R, Terachi T, Tsunewaki K (1991) Intraspecific variation of chloroplast DNA in *Dioscorea bulbifera* L. Theor Appl Genet 81:461–470

Umehara M, Sueyoshi T, Shimomura K, Hirashima K, Shimoda M, Nakahara T (2007) Production and characterization of interspecific hybrids between *Allium fistulosum* L. and *Allium schoenoprasum* L. Bull Fukuoka Agric Res Cent 26:25–30

Van Borssum Waalkes J (1966) Malaysian Malvaceae revised. Blumea 14(1):1–231

Velayudhan KC, Nizar MA, John J, Latha M, Abraham Z (2007) A note on taxonomy of the genus *Abelmoschus* in India with special reference to the vulnerable status of *Abelmoschus moschatus* subsp. *tuberosus* (Span) Borss. Plant Genet Resour Newsl 152:77–81

Yanagino T, Sugawara E, Watanabe M (2003) Production and characterization of an interspecific hybrid between leek and garlic. Theor Appl Genet 107:1–5

Fruits and Nuts

10

10.1 Introduction

The climatic diversity of the Indian Subcontinent has generated significant genetic diversity in cultivated tropical, subtropical, temperate, and arid fruits. Rich diversity is also exhibited in the wild relatives of these fruit types. Tropical fruits constitute a major proportion of the spectrum of fruit diversity available with large cultivation of indigenous fruits like mango (*Mangifera indica* L.), banana (*Musa* spp.), citrus fruits (*Citrus* spp.), jackfruit (*Artocarpus heterophyllus* L.), litchi (*Litchi chinensis*), etc., which produces about 66.4% of the annual fruit production. Maximum concentration of wild species has been recorded in *Citrus* and *Musa* (Fig. 10.1) occurring in the northeast region. In *Citrus*, Singh and Chadha (1993) recorded 30 species/botanical varieties, indigenous to India. Shillong Plateau has *Citrus assamensis*, *C. latipes*, *C. macroptera*, and *C. medica*. *Citrus indica* has sporadic distribution around Tura range and Khasi Hills in Meghalaya and in the foothills of Nagaland, where *C. ichangensis* (*cavaleriei*), a cold-tolerant native of China, is also found (possibly a hybrid of *C. latipes*). *Citrus jambhiri* (considered a variety of *C. limon*) is found sporadically in the peninsular hills and is used as rootstock. *Citrus aurantifolia* has a comparatively wider distribution occurring in the sub-Himalayan tract extending eastward to Khasi Hills and southward to Nilgiris. Similarly, in *Musa*, maximum variability occurs in the northeast (Fig. 10.1), in the wild types of *M. balbisiana* and *M. acuminata*. *Musa flaviflora* (*M. thomsonii*) is confined to Manipur and Meghalaya, and *M. nagensium* is endemic to Naga Hills. *Musa sikkimensis* occurs in North Sikkim, West Bengal, Khasi Hills, and Manipur. Among others, *M. superba* occurs in the Western Ghats and Assam. In the foothills of Assam also occur *M. cheesmanii*, *M. mannii*, and *M. velutina*. *Mangifera* wild species are represented by *M. sylvatica*, *M. khasiana*, and *M. camptosperma* from northeast region and *M. andamanica* from tropical areas of wet evergreen forests of Andaman Islands. The northeast region also has diversity in other wild fruits such as *Elaeocarpus floribundus*, *Myrica esculenta*, *Docynia indica*, and *D. hookeriana*. The humid tropical zone of the Western Ghats is another region of concentration of

A.K. Singh, *Wild Relatives of Cultivated Plants in India*,
DOI 10.1007/978-981-10-5116-6_10

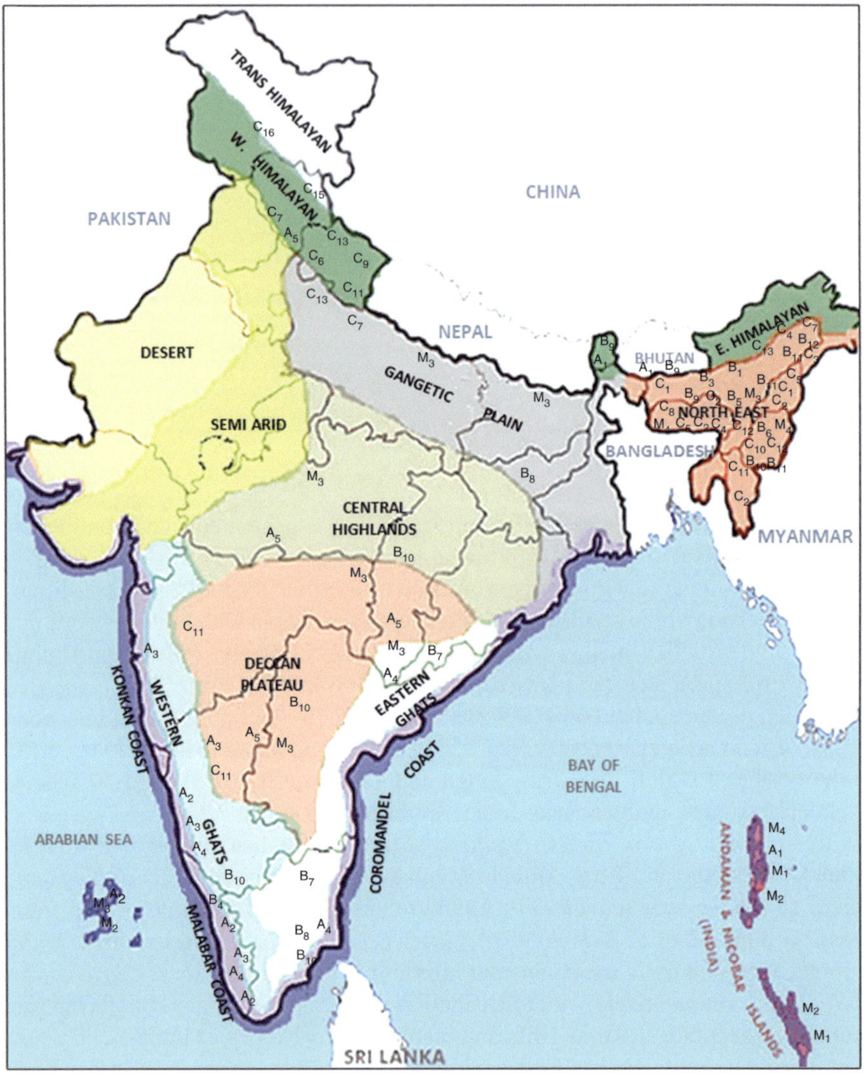

Fig. 10.1 Biogeographical regions associated with distribution of wild species of tropical fruits, *Artocarpus*, *Citrus*, *Mangifera*, and *Musa*

A1 = *Artocarpus chaplasha*; A2 = *A. gomezianus*; A3 = *A. heterophyllus*; A4 = *A. hirsutus*; A5 = *A. lakoocha*; B1 = *Musa acuminata*; B2 = *M. balbisiana*; B3 = *M. cheesmani*; B4 = *M. itinerans*; B5 = *M. mannii*; B6 = *M. nagensium*; B7 = *M. ornata*: B8 = *M. sapientum*; B9 = *M. sikkimensis*: B10 = *M. superba*; B11 = *M. thomsonii*; B12 = *M. velutina*; C1 = *Citrus assamensis*; C2 = *C. aurantifolia*; C3 = *C. cavaleriei*; C4 = *C. hystrix*; C5 = *C. indica*; C6 = *C. jambhiri*; C7 = *C. karna* (*aurantium*); C8 = *C. latipes*; C9 = *C. limonia*; C10 = *C. maxima*; C11 = *C. medica*; C12 = *C. megaloxycarpa*; C13 = *C. reticulata*; C14 = *C. rugulosa*; C15 = *C. sinensis*; C16 = *C. trifoliata*; M1 = *Mangifera andamanica*; M2 = *M. camptosperma*; M3 = *M. indica*; M4 = *M. sylvatica*

wild relatives of cultivated fruits such as *Artocarpus heterophyllus* and *A. lakoocha* and wild forms of *Garcinia indica*, *Mimusops elengi*, and *Euphoria longan* (*Dimocarpus longan*). In case of *Garcinia*, there are two major centers of diversity in the Indian Subcontinent; the northeast region, extending from Assam up to Myanmar; and the Western Ghats. There are 36 species reported from India, out of which 15 are endemic. Of these, seven are endemic to the Western Ghats, six to the Andaman and Nicobar Islands, and four to the northeast region. Comparatively wider distribution in peninsular tracts are of species of *Diospyros* (temperate, diffused to all over continent), *Syzygium*, and *Vitis*. *Spondias pinnata* also exhibits much variability in this tract.

The cultivation of temperate fruits in the subcontinent extends from Northwest Himalayas, starting from Jammu and Kashmir to the subtropical plains in the north, to Arunachal Pradesh in the Eastern Himalayas. These fruits are also grown in the hilly regions of South India around 10^0N and 75°E. Apple (*Malus domestica* Borkh.), pear (*Pyrus communis* L.), peach [*Prunus persica* (L.) Batsch], plum (*Prunus domestica* L.), almond (*Prunus amygdalus* Batsch.), apricot (*Prunus armeniaca* L.), cherry (*Prunus avium* L.), and walnut (*Juglans regia* L.) are commercially cultivated in the Northwest Himalayan region (Jammu and Kashmir, Himachal Pradesh, and Uttarakhand). As per Vavilov (1951), temperate fruits have originated in three broad centers, Chinese center; Central Asiatic center, including the region extending from Hindu Kush to Kashmir; and Near-Eastern center, extending from Northeastern Iran to Caucasus and central Turkey. It is believed that some species of the temperate fruit genera such as *Malus*, *Pyrus*, *Prunus*, *Rubus*, etc., escaped to the Himalayas and over a period of few thousand years got neutralized and evolved into new species and varieties. Randhawa (1987) presented a comprehensive description of the wild temperate fruit species native to the Himalayas. Subsequently, several reviews (Rana et al. 2003) described the genetic diversity in temperate fruits, primarily in Northwest Himalayan region and relatively of lesser degree in the Northeast Hill region. Among the wild relatives of temperate fruits, species of *Pyrus*, *Prunus*, *Rubus*, *Sorbus*, and *Ribes* occur in the Himalayas. Some of these such as *Prunus prostrata* and *P. tomentosa* are confined to the Western Himalayas; *P. jenkinsii* occurs in the eastern belt, while few other species are widely distributed in the Himalayan zone, such as *P. cornuta* and *P. napaulensis*, allied to *P. padus* and *P. acllminata*. Likewise, in *Pyrus*, *P. communis* occurs largely in Kashmir; *P. kumaoni* is localized in the Western Himalayas, but *P. pashia* and *P. baecala*, besides the Western Himalayas, extend eastward into subtropical zone of the Northeast Hills. *Pyrus pyrifolia* is naturalized in the Khasi Hills, also found in Nilgiris, and used as a rootstock. *Sorbus aucuparia* occurs in the Western and the Eastern Himalayas, and *S. vestita* extends its distribution range from Eastern Himalayas to northeastern subtropical belt. The species of *Rubus* also exhibit similar distribution patterns; *R. fruticosus* is confined to Western Himalaya and *R. lineatus* to the Eastern Himalayas; *R. lanatlts*, *R. lasiocarpus*, *R. moluccanus*, *R. niveus*, and *R. reticulatus* are widespread in the Himalayas; and *Rubus ellipticus* and *R. lasiocarpus* extend to south in peninsular hills. In *Ribes*, *R. nigrum* occurs in Western Himalayas, and *R. gracilis* and *R. acliminatum* in Eastern Himalayas.

The hot arid and semi-arid region spread over the Northwest India occupies around 12% of countries land, covering the states of Punjab, Haryana, Rajasthan, Gujarat, Maharashtra, Andhra Pradesh, and Karnataka. These regions have comparatively few fruit species, but there is wide diversity within species, though confined to small pockets. Many fruits, such as *aonla* (*Emblica officinalis* Gaertn.), *bael* (*Aegle marmelos* L.), *ber* (*Ziziphus mauritiana* Lam.), *chironji* (*Buchanania lanzan* Spreng.), *jamun* [*Syzygium cumini* (L.) Skeels], *karonda* (*Carissa carandas* L.), *ker* [*Capparis decidua* (Forsk.) Edgew], *khejri* (*Prosopis cineraria* (L.) Druce], *lasoda* (*Cordia myxa* Roxb.), *phalsa* (*Grewia asiatica* L.), *pilu* (*Salvadora persica* L.), date palm (*Phoenix* L.), and wood apple (*Limonia acidissima* L.), are indigenous and present significant genetic diversity. Comparatively fewer species occur in the northern and northwestern plains and in the Aravalli hills, such as *Carissa congesta*, *Capparis decidua*, *Grewia asiatica*, and *Ziziphus* spp. (*Z. mauritiana*, *Z. nummularia*). Cultivated *Ziziphus mauritiana* exhibits maximum variability in the semi-arid plains; wild types such as *Z. oenoplia* and *Z. rugosa* occur in humid peninsular region. Recently *Ziziphus jujuba* Mill. subsp. *spinosa* (Bunge) Peng, Li & Li has been recorded from the Kashmir Himalaya (Akhter et al. 2013). *Cordia* (*lasoda*) native to India, in addition to cultivated *C. myxa*, is represented by *C. gharaf* growing wild in sub-Himalayan tract and Central India and *C. vestita* in tropical Himalayas. Variation in *Ficus palmata*, the Indian fig (related to *F. carica*), is also represented in the northwestern zone. Other wild *Ficus* species, *F. benghalensis*, *F. glomerata*, *F. hispida*, and *F. lucescens*, occur throughout India, *F. auriculata* in sub-Himalayan region and Northeast Hills and Eastern India, *F. heterophylla* in hot parts, *F. hirta* in the tropical Himalayas, and several more in Himalayan and sub-Himalayan region. Many species of *Syzygium* grow wild, such as *S. zeylanicum* in the Western Ghats, *S. malaccense*, *S. uniflora* in South India and West Bengal, *S. fruticosum*, *S. densiflora*, and many more in different regions of the subcontinent. *Koranda*, *Carissa carandas*, grown widely, has other species, *C. inermis* Vahl., growing wild in Deccan Peninsula; *C. spinarum* L. in Punjab, Jammu and Kashmir, and Western Uttar Pradesh; and *C. paucinervia* in dried tracts of India. *Grewia* is represented by 40 wild species, besides the cultivated *G. subinaequalis*. They are *G. damine* Graerth in Rajasthan; *G. elastica* Royle in Himalayas; *G. flavescens* Juss in Gangetic Plains and central and South India; *G. hirsuta* in warm regions of the subcontinent ascending to sub-Himalayas; *G. rothi* in peninsular region; *G. sapida* in northwestern India, Assam Hills, and Eastern Ghats; *G. sclerophylla* in tropical Himalayas; *G. tenax* in Rajasthan and adjacent areas; *G. tiliaefolia* Vahl. throughout India; and *G. villosa* Willd. in hotter parts of northwestern and Central India. The diversity in wild relatives and types in fruit tree species appears to be high in all the phytogeographical zones, except the Gangetic Plains with comparatively less variability.

The wild relatives of dual-purpose fruit species, such as mulberry (*Morus*), date (*Phoenix*), and khejri (*Prosopis*), have been listed under industrial, commercial, and agroforestry crop groups, respectively, as per their primary use.

10.2 Wild Species Found in India

10.2.1 Temperate Fruits

1. ***Actinidia strigosa* Hook.f. & Thomson (Actinidiaceae).** Scandent shrubs, up to 8 m tall with edible berries. Distributed in broad-leaved forests of Eastern Himalayas (Darjeeling and Sikkim). Wild relative of kiwi fruit.
2. ***Fragaria daltoniana* Gay. (Rosaceae).** A wild strawberry native to the Himalayas. Its fruit has a poor flavor and is of no commercial value. Frost tolerant. Member of tertiary gene pool of strawberry (Staudt 1989).
3. ***Fragaria nilgerrensis* Schlecht.** Perennial wild strawberry. Fruit has poor flavor with no commercial value. Native to evergreen forests of Peninsular India and Eastern Himalayas. Resource for quality. Part of tertiary gene pool of strawberry (Staudt 1989).
4. ***Hippophae salicifolia* D. Don (Elaeagnaceae).** A fast-growing deciduous 15 m-tall tree. Rich natural source of multivitamins, A, B1, B12, E, and K, polyphenols, etc., accounting for its pharmaceutical, nutraceutical, and cosmetic potential. Cultivated as ornamental. Native to Asia, including cold arid region of Ladakh. Wild relative of sea buckthorn (*Elaeagnus rhamnoides*).
5. ***Hippophae tibetana* Schltr.** Perennial woody deciduous 15 m-tall tree, fruits rich in vitamin C. Habitat to riverbeds in cold arid region of Ladakh. Close relative of sea buckthorn, considered as subspecies. Medicinal.
6. ***Malus baccata* (L.) Borkh. (Rosaceae).** Siberian crab, a deciduous tree narrowly upright native to Asia, extending from Siberia to Himalayas and Northeast Hill region. Presents significant genetic variability with var. *himalaica* (Maxim.) Schneid., in the region between 1800 and 3000 m. Part of secondary gene pool of apple (Zhukovsky 1965) and source of resistance to diseases (Hokanson et al. 2001), cold tolerance, and used as graft stock.
7. ***Malus sikkimensis* (Wenz.) Koehne.** Sikkim crabapple, a rare deciduous 7 m-tall tree, endemic to Sikkim, Bhutan, and Western China. Ornamental, produces white and pink flowers and dark red fruit. A triploid. Part of secondary gene pool of apple (Forte et al. 2002) and source for disease resistance and dwarfing gene. Threatened due to loss of habitat.
8. ***Prunus arborea* var. *montana* (Hook.f.) Kalkman (Rosaceae).** Deciduous tree distributed in mountainous evergreen or dense humid forest from 200 to 500 m altitude of northeast region. Tree trunk is used to make house pillars and in general construction. The tree has ornamental potential.
9. ***Prunus bracteopadus* Koehne.** Distributed in Eastern Himalayas, Assam, and Uttarakhand, closely related to *P. napaulensis*.
10. ***Prunus cornuta* (Wall. ex Royle) Steud.; syn. *Padus cornuta* (Wall. ex Royle) Carrière.** Himalayan bird cherry, a deciduous medium-sized tree, distinct from *Prunus* species in flowers, being in racemes. Fruit is an ovoid-globose drupe, 7–9 mm long, blackish brown when ripe. Native of Asia, including Himalayas, Uttar Pradesh, West Bengal, and Tamil Nadu hills. Source of disease resistance for sweet cherry and as graft stock (Pandey et al. 2008).

11. ***Prunus cerasoides* D. Don.; syn. *P. carmesina* H. Hara; *Cerasus cerasoides* (Buch. -Ham. ex D. Don) S.Y. Sokolov.** A deciduous cherry tree up to 15 m tall. Native of Asia, including Indian Subcontinent (Himalayas, Khasi Hills, Eastern and Western Ghats). Used as fruit, for flavoring and wood. Member of tertiary gene pool of ornamental and sweet cherries, source of disease resistance and graft stock for sweet cherry (Ohta et al. 2005, 2007; Pandey et al. 2008).
12. ***Prunus imanishii* Kitam.; syn. *P. rufa* var. *imanishii* (Kitam.) Ghora & Panigrahi; *P. trichantha* Koehne; *Cerasus trichantha* (Koehne) C.L. Li & S.Y. Jiang.** A 2–10 m-tall tree, distributed in sparse forests margins, thickets, and mountain slopes from China to Eastern Himalayas (Sikkim). Wild relative of *Prunus* with cold hardiness.
13. ***Prunus jacquemontii* Hook. f.** Dwarf flowering cherry/almond, small deciduous shrub' with more ornate rosy pink flowers. Native to Asia, including Himachal Pradesh, Jammu and Kashmir, Uttarakhand, and Punjab. Tertiary genetic resource of plum, Japanese plum, and myrobalan plum (Chin et al. 2014).
14. ***Prunus jenkinsii* Hook. f. & Thomson; syn. *Laurocerasus jenkinsii* (Hook. f. & Thomson) Browicz.** A deciduous tree ca. 15 m tall. Native of Asia, including Indian Subcontinent (Eastern Himalayas, Northeast Hills, and Assam). A potential fruit, semidomesticated in India.
15. ***Prunus* prostrata Labill.; syn. *Cerasus prostrata* (Labill.) Ser.** Mountain cherry, a hardy alpine shrub found naturally above 2000 to 4000 m in Western Himalayas. The fruit is edible but not preferred by humans, ornamental. Part of secondary (tertiary?) gene pool of Japanese and myrobalan plum (Bouhadida et al. 2007), used as graft stock, and potential source of disease resistance. Member of tertiary gene pool of apricot and plum (Chin et al. 2014).
16. ***Prunus rufa* Hook. f.** Himalayan cherry, a deciduous tree ca. 6 m tall. Ornamental. Native of India (Sikkim, West Bengal) Bhutan, and Myanmar. Variety *imanishii* listed threatened (Archive.is/7ytuq 2001).
17. ***Prunus tomentosa* Thunb.** A deciduous shrub with pink flowers. Native to Asia from China to Northwest Himalayas, including Kashmir and Ladakh. Cultivated for fruit and flower, also harvested from wild. Graft stock for apricot, peach, and plum (Hartmann and Neumüller 2009); part of secondary gene pool of peach and tertiary gene pool Japanese and myrobalan plum and sweet cherry (Aradhya et al. 2004; Chin et al. 2014).
18. ***Prunus undulata* Buch. -Ham. ex D. Don; syn. *P. wallichii* Steud, *Laurocerasus undulata* (Buch. -Ham. ex D. Don) M. Roem.** A deciduous shrub. Distributed in Central and Eastern Himalayas; Northeast Hill region, particularly Khasi and Mishmi hills and Sikkim; Uttar Pradesh; and West Bengal. Also, cultivated.
19. ***Punica granatum* Linn. (Punicaceae).** A shrub with multiple stems, commonly grows 2–5 m tall. Semi-wild in Western Himalayas, naturalized in hills of Himachal Pradesh and Jammu and Kashmir, around 1200–1500 m. Cultivated as ornamental and for fruit.

20. ***Pyrus jacquemontiana*** **Decne. (Rosaceae).** Resembles *P. pashia* but for its smooth young fruit. Found in hills of Punjab. Based on SSR markers, groups with *P. pyrifolia*, *P. pashia*, *P. serotine* (Rana et al. 2015). Appears part of primary gene pool of *P. pashia*.
21. ***Pyrus pashia*** **Buch-Ham. ex D. Don.** Wild Himalayan pear, a medium-sized deciduous tree with small and oval-shaped crown with ovate, finely toothed leaves and white flowers. Native to the mid-hills (700 and 2000 m), from Western Himalayas (India) to Vietnam. Common rootstock for Asian pear. Frost tolerance and part secondary gene pool Asian pear (Zheng et al. 2008). *P. pashia* var. *kumaoni* (Decne.) Stapf. is reported from Himalayas and Northeast Hills.
22. ***Pyrus polycarpa*** **Hook. F.** *Deingloporam*, wild in Shillong at 1300 m and in Somarin at 1350 m altitude, Meghalaya. It bears small globose fruits in cluster. Good rootstock for pear. Field resistant to powdery mildew and fire blight.
23. ***Pyrus pyrifolia*** **(Burm. f.) Nakai.; syn.** ***P. serotina*** **Rehder.** *Nashpati* or Asian pear, a deciduous tree ca. 10 m tall with large, juicy bronze skinned, sweet flavor with very low acid fruits. Native of East Asia, naturalized in Khasi and Nilgiri hills. Cultivated in mid hills. Preferred rootstock for pear. Belong to primary gene pool of Ussurian pear and secondary of pear (Westwood and Bjornstad 1971).

In addition, *Pyrus khasiana* Hook.f. *Soh Phoh Khasi* is reported from the northeast region, used as rootstock for pear. This is provisionally accepted as species; however, availability is unknown.

24. ***Ribes glaciale*** **Wall. (Grossulariaceae).** A deciduous shrub ca. 3 m tall. Habitat to forests between 2600 and 4400 m from Afghanistan to China, including India (Himachal Pradesh, Jammu and Kashmir, Uttarakhand, Sikkim, and Assam). The sour-tasting red fruit is about 7 mm in diameter, eaten raw or cooked, and medicinal. Genetic resource for black currant. Used for breeding immunity to *Cronartium ribicola* in *R. orientale*, the Oriental Gooseberry.
25. ***Ribes nigrum*** **L.** Blackcurrant, erect aromatic shrub, up to 2 m tall. Naturally habitat to cold temperate regions, extending from Himalayas (Western Himalayas, Kunawar to Kashmir, up to 3600 m) to West China to Northern Asia and Europe. Part of primary gene pool of black currant and tertiary of gooseberry (Hjamarsson and Wallace 2007). Cultivated.
26. ***Ribes takare*** **D. Don; syn.** ***R. acuminatum*** **Wall. ex G. Don.** A deciduous shrub, distributed in Bhutan, China, India (Kashmir), Myanmar, and Nepal with varietal variability. Primary genetic resource for black currant.
27. ***Rubus ellipticus*** **Smith (Rosaceae).** Yellow Himalayan raspberry, an evergreen shrub ca. 4.5 m tall. Native to China and the Indian Subcontinent. Invasive, common in Himalayas from Shivalik range to Sikkim, Khasi Hills, and Peninsular India. Delicious wild fruits from mid-hills of Himachal Pradesh with significant genetic diversity. Cultivated. Part of primary gene pool of red raspberry (Hall et al. 2009). Shade tolerant.

28. ***Rubus fockeanus*** **Kurz.** Creeping herb with pubescent stems, simple pinnately trifoliate leaves. Extending from China to Central to Eastern Himalayas (Sikkim).
29. ***Rubus ghanakantae*** **Rolla Rao et Joseph.** A new species described from Arunachal Pradesh (Rao and Joseph 1970). Endemic and threatened taxa.
30. ***Rubus kumaonensis*** **Balakr.; syn. *R. reticulatus* Wall. ex Hook.f.** A deciduous shrub growing to 0.1 m by 1 m. Distributed from SE China to temperate Himalayas, from Kumaon to Sikkim between 2000 and 2400 m. Resembles *Rubus rugosus*.
31. ***Rubus lanatus*** **Wall.; syn. *R. glandulifera* N.P. Balakrishnan.** It is a deciduous shrub with hermaphrodite flowers. It is found from Western and Central Himalayas, from Kumaon to Sikkim between 900 and 2500 m.
32. ***Rubus lineatus*** **Reinw.** It is a deciduous shrub growing to 3 m with clusters of red fruit, extending from China to Northeast India (Sikkim 1800–2700 m); much variable. Also, ornamental. Member of secondary gene pool of blackberry (Li et al. 2012). Natural hybrids available with *R. splendidissimus* Hara.
33. ***Rubus nepalensis*** **(Hook. f.) Kuntze; syn. *R. nutantiflorus* H. Hara.** An evergreen raspberry, growing to 0.2 × 1 m with small versatile edible fruit. Aggressive ground coverer. Endemic to Indian Subcontinent (Nepal and Indian Himalaya). A plant of future.
34. ***Rubus niveus*** **Thunb.; syn. *R. lasiocarpus* Smith.** Mysore raspberry, a straggling thorny shrub with purplish stem and sweet with fine blend of acid fruit, excellent for dessert. But have very short shelf life, not more than 24 h in ambient conditions. Native to Asia from Afghanistan through India, China to Taiwan and the Philippines, and to Sri Lanka. Invasive. Source of high yield. Part of secondary gene pool of black and red raspberry and tertiary of blackberry (Finn et al. 2002).
35. ***Rubus paniculatus*** **Sm.** It is a deciduous shrub growing to 3 m., Native to China and the temperate Himalayas from Kashmir to Khasi Hills with varietal diversity. Natives eat the fruit and use it in puddings and pies.
36. ***Rubus rosifolius*** **Sm. ex Baker.** It is a prickly subshrub with leaves resembling rose, called rose-leaf bramble. Widely distributed from Africa to Asia to Pacific, including rainforest of the Himalayas and Khasi Hills of India. Part of secondary gene pool of red raspberry (Hall et al. 2009); source of yield. Leaves used as herbal tea. Cultivated; fruit is sweet and flavored when grown on good soil moisture.
37. ***Rubus rugosus*** **Sm.; syn. *R. moluccanus* L.** It is a scrambling shrub or climber providing fresh edible fruits. Widely distributed throughout Southeast Asia and the Pacific, including tropical Himalayas from Nepal to Eastern Himalayas and Northeast Hills and Myanmar to Andaman and Nicobar.
38. ***Rubus wardii*** **Merr.** Extensive trailing or scrambling shrubs with red fruit. Found in mixed forests, valleys, and rocky slopes of Southeast China to East Himalayas (Sikkim).
39. ***Sorbus arachnoidea*** **Koehne (Rosaceae).** Tree 5–10 m tall with distinctive stout twigs and fruits that are initially crimson, becoming pink to white flushed pink globose and fleshy.

40. ***Sorbus himalaica* Gabrielian.; syn. *S. foliolosa* (Wall.) Spach.** Shrub or tree with red or pink globose fleshy fruit. Native to temperate (China) and tropical Asia, including montane conifer forests in Eastern Himalayas. Like *S. arachnoidea*, but branchlets slenderer; leaflets proportionately shorter and broader. Often grow with *S. microphylla*. Extracted from wild for economic use.
41. ***Sorbus lanata* (D. Don) Schauer.** Known as hairy rowan, is a small deciduous tree 10 m tall with lobulated leaf margin white-woolly flower clusters, distributed temperate Himalayas from Afghanistan to Kashmir to Kumaon. Extracted from wild for economic use.
42. ***Sorbus microphylla* (Wall. ex Hook. f.) Wenz.** Small-leaf rowan, a shrub or small tree with compound leaves and globose, 8–12 mm, white or pink fruit. Common in Fir, Hemlock, and Spruce forests of Himalayas from Himachal to Arunachal Pradesh.
43. ***Sorbus vestita* (Wall. ex G. Don) Lodd.; syn. *S. cuspidata* (Spach) Hedl.; *Pyrus vestita* Wall.** Himalayan whitebeam. A deciduous tree 6–10 m tall with large bold foliage, distributed from temperate China to Northwest Himalayas to Nepal and Myanmar. Grown as ornamental for beautiful foliage.

10.2.2 Tropical and Subtropical Fruits

44. ***Ampelocissus araneosa* (Dalz.) Gamble; syn. *Vitis araneosus* Dalz. & Gibs. (Vitaceae).** A slender climber up to 5 m long with 1–4 seeded berry fruit. Endemic to semi-evergreen forests of Peninsular India at high altitude, though rare. Roots are cooling and astringent. Listed Ayurveda plant (Nadkarni 1996).
45. ***Ampelocissus tomentosa* (Heyne ex Roth.) Planchon; syn. *Vitis tomentosa* Heyne ex Roth.** Hairy wild grape, a climbing shrub with branchlets and leaves densely woolly. Endemic to semi-evergreen and moist deciduous forests of South Western Ghats, extended throughout India and Sri Lanka.
46. ***Artocarpus chaplasha* Roxb.; syn. *A. chama* Buch. -Ham. (Moraceae).** A deciduous tree up to 30–40 m tall. Fruiting syncarp yellow when young. Drupes ellipsoid. Native of Northeast India (Bhutan and Sikkim), lower Myanmar, and the Andaman and Nicobar Islands. Genetic resource for jackfruit.
47. ***Artocarpus gomezianus zeylanicus* Jarrett.** Medium- to large-sized tree with syncarp sub-globose, uneven, fleshy, greenish yellow. Distributed in semi-evergreen and moist deciduous forests of Southwest India and Sri Lanka, including Andaman and Lakshadweep Islands.
48. ***Artocarpus heterophyllus* Lam.** Jackfruit, a tree with the largest tree-borne fruit of the world. Originated in the southwestern rain forests of Indian Subcontinent (Western Ghats). Natural variability found in India is part of primary gene pool of jackfruit (Wangchu et al. 2013). Widely cultivated fruit; wood and seeds are also used.
49. ***Artocarpus hirsutus* Lam.** An evergreen tree, ca. 50 m tall with globose or ovoid, echinate, yellow 6–7.5 cm across fruit. Endemic to evergreen forests up

to 900 m of South Western Ghats, Tamil Nadu, and Bastar. Part of tertiary gene pool of breadfruit (Zerega et al. 2010).

50. ***Artocarpus lakoocha* Roxb.** Monkey jack, a medium to large deciduous tree with a spreading crown, dropping leaves, and nearly round or irregular 5 × 12.5 cm fruit. Distributed in deciduous forests of East Asia, including Western sub-Himalayan tract and humid Peninsular India. Widely cultivated and used as medicinal plant by tribes. Found to be antioxidant, antiglycation, and oxyresveratrol.
51. ***Cayratia pedata* Gagnep. var. pedata; syn. *Vitis pedata* (Lam.) Wall. ex Wight (Vitaceae).** A weak climber with rusty tomentose 1 × 1 cm, 2–4 lobed berries. Native to moist deciduous and semi-evergreen forests of Indo-Malayan region, including Eastern Himalayas and Eastern and Western Ghats.
52. ***Cissus javana* DC.; syn. *C. discolor* Bl., *Vitis discolor* Dalzell (Vitaceae).** Creeping or climbing shrub with one seeded ca. 4 × 3 mm berries. Habitat to moist deciduous and semi-evergreen forests of Indo-Malayan region from southern China to Eastern Himalayas and Eastern and Western Ghats of India. Related to *V. vinifera.*
53. ***Cissus repanda* Vahl; syn. *Vitis repanda* (Vahl) Wight & Arn**. *Pani Bel*, a woody deciduous climber with one seeded ca. 7 × 10 mm berries. Distributed from southern China to Eastern Himalayas and Eastern and Western Ghats of India.
54. ***Cissus repens* Lam.; syn. *Vitis repens* (Lam.) Wight & Arn.** A tropical rainforest creeping vine, native Indo-Malayan region, including Indian Subcontinent (Eastern Himalayas, Eastern and Western Ghats). Distantly related to *V. vinifera*.
55. ***Citrus assamensis* Dutta & Bhattach. (Rutaceae).** A diploid species with 2n = 18. The meiotic behavior indicates it being a possible structural hybrid. Endemic to Assam, southern slopes of Khasi Hills and Shillong Plateau. Part of tertiary gene pool of *Citrus* crop (Hazarika 2012); source for waterlogging tolerance (Sharma et al. 2004).
56. ***Citrus aurantifolia* (Christm.) Swingle**. Indian lime, a shrubby thorny tree with leaves like bitter orange, *C. aurantium*. Dwarf types are also found. Probably originated in East India. Progenitor of limequat, Tahiti lime, and tetraploid giant key lime (Curk et al. 2016). Fruit used for flavoring, beverage and essential oil, and medicine.
57. ***Citrus cavaleriei* H. Lév. ex Cavalerie; syn. *C. ichangensis* Sw.** A 2–3.5 m-tall evergreen shrub. Native to China, also seen in Northeast Hills (Khasi Hills, Nagaland). Part of tertiary gene pool of *Citrus* crop (Grosser et al. 1996); source of cold tolerance and graft stock.
58. ***Citrus hystrix* DC.; syn. *C. macroptera* var. *annamensis* Yu. Tanaka.** Kaffir lime, a small tree known for its aromatic leaves. Native to China and tropical Asia, including Indian Subcontinent (Meghalaya, Northeast Hills). Part of tertiary gene pool of *Citrus* crops (Bayer et al. 2009). Source of pest resistance. Used in flavoring, beverage, and essential oils.

59. ***Citrus indica* Tan.** Wild orange, native to Tura range, Meghalaya. Also, sporadically distributed in Khasi range in Assam and near Dimapur, Nagaland. A progenitor of cultivated orange. Part of tertiary gene pool of *Citrus* crops (Bayer et al. 2009). Source of disease resistance. Endangered because of poor regeneration capacity.
60. ***Citrus jambhiri* Lush.** Rough Lemon, a 3–6 m-tall tree, is a hybrid species evolved from *C. reticulata* x *C. medica*, like Rangpur or mandarin orange. Originated in northwestern India (Pathankot). Cold hardy, source of drought resistance (Sharma et al.2004) used as rootstock with many cultivars. Considered a variety of *C. limon* (L.) Burm. F., the cultivated lemon, which itself has hybrid origin.
61. ***Citrus karna* Raf. (=*Citrus* x *aurantium* L., pro sp.).** A vigorous tree with medium to medium-large fruits. An indigenous species of India, distributed in Eastern Himalayas and Gangetic Plains. Reported to be hybrid (*C. reticulata* x *C. medica*) and appears close to *C. limon* x *C. aurantium* too. Source of graft stock. Cultivated, medicinal, and yields essential oils.
62. ***Citrus latipes* (Sw.) Tanaka.** Khasi papeda, confused with *C. hystrix*. Native of Shillong Plateau and Khasi Hills, up to 1800 m. Part of tertiary gene pool of *Citrus* crops (Penjor et al. 2013). Cold tolerant, potential disease resistance and graft stock.
63. ***Citrus limonia* Osbeck.** *Surkh nimboo* or Rangpur lime, a hybrid of *C. reticulata* x *C. medica* (mandarin orange and lemon) having very acidic fruit with orange peel. Originated in India, only cultivated. Most used source of rootstock for commercial *Citrus* and disease resistance. Like *C. limon*, which involves different species in parentage.
64. ***Citrus maderaspatana* Hort. ex Tan.** A hybrid species involving *C. latipes*, originated in South India. Source of graft stock.
65. ***Citrus maxima* (Burm.) Merr; syn. *Citrus grandis* Osbeck.** Pomelo or *chakotara,* a medium-sized tree largest among *Citrus*, native of South and Southeast Asia, including eastern region of India. Progenitor of the grapefruit (*C.* x *paradisi*), lemon, and the tangelo (*C. reticulata*) (Garcia-Lor et al. 2013). Cultivated.
66. ***Citrus medica*** L. Citron or *Bara Nimbus* is one of the four true species, a tree with large fragrant fruit with thick rind. Native to India, in areas bordering with Myanmar, foothills of Himalayas, and the Western Ghats. Progenitor of combava (Luro et al. 2011), lemon, and lime (Garcia-Lor et al. 2013) and rough lemon (Li et al. 2010b). Cultivated for fruit and ornamental purpose. Also, used for flavoring and therapeutic value.
67. ***Citrus megaloxycarpa* Lush.** Possibly a hybrid of *C. maxima* x *C. limon*, cultivated in northeastern India and Uttar Pradesh. A highly acidic citrus cultivar of Manipur. The earliest reference to this fruit was made in Babar's Memoirs (1519) (Woodford 2005).
68. ***Citrus nakoor* Singh.** A new species reported from South India in 1967 (Singh 1967). Needing collection, characterization, and evaluation.

69. ***Citrus pennivesiculata* Tan.** Native of India, mostly wild in South India and the southern parts of the Central India. Needing further exploration and investigation.
70. ***Citrus reshni* Hort. ex Tan.** Cleopatra mandarin is a small, erect-branching miniature tree with handsome, bright green foliage and small but very flavorful fruit. Originated in India. Gene sources for salt tolerance in graft stock for *Citrus* (Bhusal et al. 2002).
71. ***Citrus reticulata* Blanco var. *crenatifolia*.** Mandarin is a small evergreen tree with fruit resembling other oranges. A true species from Indochina region, including northwestern India. Progenitor of lemon, mandarin, sweet orange, tangelo, and tangor *Citrus* (Garcia-Lor et al. 2013). Source of graft stock and winter hardiness. Cultivated, eaten raw or in fruit salad.
72. ***Citrus rugulosa* Hort. Ex. Tan.** An indigenous species reported from the foothills of northwestern India, Uttarakhand. Flavonoid analysis showed high percentage of naringin. Endangered.
73. ***Citrus* x *sinensis* (L.) Osbeck.; syn. *C. aurantium*.** An evergreen ca. 9 m-tall tree, belongs to sweet orange group, includes the commonly cultivated sweet orange, blood oranges, and navel oranges. Progenitor of grapefruit (Scora et al. 1982), mandarin (Froelicher et al. 2011), and tangor (Garcia-Lor et al. 2013) graft stock. Cultivated, used in folklore medicine for essential oils and lipids.
74. ***Citrus trifoliata* L.; syn. *Poncirus trifoliata* (L.) Raf.** A deciduous shrub or small tree, native of China, recorded in Kashmir Valley (Aslam et al. 2010), probably an escape from cultivation into Himalayas. Readily hybridizes with all *Citrus* species, reflecting closeness for use as a source of dwarfing, graft stock, hardiness, and disease resistance. Part of tertiary gene pool (Rao et al. 2011).

Except for the four-basic species, most other *Citrus* species originated consequence to natural hybridization between the basic species and between basic and subsequently established such species in wild or cultivation. This reflects the potential of interspecific breeding in genetic improvement of cultivated *Citrus* crops and should be a major emphasis in *Citrus* crops improvement.

75. ***Cordia gharaf* (Forsk.) Ehrenb. ex Asch.; syn. *C. rothii* Roem. & Schult.** Small trees with crooked trunk 1–1.2 cm-long, ovoid, acute fruit, distributed in northwest and central plains. Used as minor desert fruit. Related to cultivated *lasura* (*C. myxa* L.).
76. ***Dimocarpus longan* Lour.; syn. *Euphoria longan* (Lour.) Stued. (Sapindaceae).** A medium-sized evergreen tree with small, round, sweet, and edible fruits. Native to Asia, including Eastern Himalayas, Northeast Hills, and Andaman and Nicobar Islands. Cultivated for fruit/ornamental. Belongs to same family as *Litchi chinensis*, fruits slightly later.
77. ***Diospyros barberi* Ramas. (Ebenaceae).** A tree 25 m tall with 4 cm across berry fruit. Endemic to the Western Ghats (Agasthyamalai and South Sahyadri Hills). Listed vulnerable by IUCN (WCMC 1998a).

78. ***Diospyros bourdillonii* Brand**. An evergreen dioecious tree endemic to evergreen forests of South Western Ghats and Central Sahyadri. Known as tribal medicines for diabetes. Wild relative of persimmon fruit/timber trees, *D. kaki* and *D. peregrina.*
79. ***Diospyros chloroxylon* Roxb.** Green ebony persimmon, a deciduous tree with cherry-like palatable fruit. Distributed in dry evergreen to dry deciduous forests of Maharashtra, Andhra Pradesh, Karnataka, and Tamil Nadu.
80. ***Diospyros lotus* L.** Date-plum, a deciduous 9 m-tall tree. Fruits used as food and beverage base. Distributed in Himalayas and Northeast Hill region. Gene sources for graft stock. Possess medicinal properties. Also, cultivated.
81. ***Diospyros malabarica* (Desr.) Kostel.; syn. *D. malabarica* var. *siamensis* (Hochr.) Phengklai; *D. peregrina* (Gaertn.) Gürke.; *D. embryopteris* Pers.** Indian persimmon, black-and-white ebony, is a long-lived slow-growing small- to medium-sized evergreen tree. Native to tropical Asia, including Indian Subcontinent. In Bastar region, it is habitat to shady wet places and near streams. Yields gum/resin/tenin/dyestuff. Bark and unripe fruit are used in Ayurveda, called *Tinduka* in Sanskrit. Also, reported to have antidiabetic and antimicrobial activities.
82. ***Diospyros montana* Roxb.** A deciduous ebony 20 m-tall tree, distributed in dry evergreen to dry deciduous forests of Indo-Malesia to Australia, including the Western Ghats. Tender shoots are cooked and eaten. It has medicinal properties useful in treatment of tumors. The wood is used for making agricultural implements.
83. ***Diospyros nilagirica* Bedd**. Evergreen dioecious tree, endemic to wet evergreen forests, generally between 800 and 1600 m of South Western Ghats (between Periyar regions and Anaimalai in South Sahyadri Hills). Listed in threatened (Archive.is/7ytuq 2001).
84. ***Diospyros paniculata* Dalz.** Panicled ebony, dioecious up to 15 m-tall tree, with berry fruit. Endemic to wet evergreen/semi-evergreen forests of Peninsular India.
85. ***Diospyros pruriens* Dalz**. A small- or mid-sized 10 m-tall dioecious understory tree with ovoid-conical berry fruit endemic to wet evergreen forests of Western Ghats up to 1000 m.
86. ***Diospyros racemosa* Roxb**. A dioecious ca. 20 m-tall tree with racemose inflorescences and ovoid-ellipsoid, top flattened, berry fruit. Habitat to evergreen or semi-evergreen forests of the Western Ghats and Koraput. Known for medicinal properties.
87. ***Diospyros sulcata* Bourd.** Dioecious tree, grows up to 27 m. Inflorescences bear up to three flowers. The fruits are ovoid or roundish. Endemic to evergreen forests of South Western Ghats. Critically endangered.
88. ***Diospyros sylvatica* Roxb.** A mid-sized dioecious ca. 35 m-tall tree, distributed in evergreen or semi-evergreen forests of Indo-Malesia region, including Peninsular India and Sri Lanka up to 1400 m. It produces antitermitic root extract. Limited cultivation in peninsular region.

89. ***Diospyros tomentosa*** **Roxb.; syn.** ***D. exsculpta*** **Buch. -Ham.** A large tree originated in India, habitat to sub-Himalayan tract from Ravi to Nepal and dry deciduous forests of Peninsular India. Called black ebony of North India. Fruit edible.
90. ***Eriobotrya angustissima*** **Hook.f. (Rosaceae).** An evergreen shrub. Native of Southeast Asia, including the northeast region (Meghalaya) of India. Wild relative of Loquat, *E. japonica.*
91. ***Eriobotrya bengalensis*** **Kurz.** Medium-sized tree. Native to Asia, including low land and middle forests (1000 to 3000 m) of northeast region of India. Wild relative of loquat, *E. japonica.*
92. ***Eriobotrya dubia*** **(Lindl.) Decne.; syn.** ***Photinia dubia*** **Lindl.** A medium-sized tree, native of humid forests of Himalayas (Kumaon to Bhutan) of India. Locally called *Jure kaphal.* Wild relative of Loquat, *E. japonica.*
93. ***Eriobotrya hookeriana*** **Decne.** Spreading 10 m-tall tree with 1–2 seeded pome. Native to dry hill slopes of Eastern Himalayas (Sikkim), Bhutan, and Nepal.
94. ***Eriobotrya petiolata*** **Hook. f.** A 15 m-tall tree with yellow bark. Common in upper hills on steep hillsides in warm broad-leaved forests of Eastern Himalayas.
95. ***Garcinia andamanica*** **King. (Clusiaceae).** A rare 6–12 m-tall tree species with globose or oval berries. Have two varieties. Endemic to evergreen forests of Andaman Islands. Needs evaluation as genetic resource.
96. ***Garcinia atroviridis*** **Griff.** *Asam Gelugur* is a large rainforest tree, native to Indochina, Malay, and northeast region of India (Arunachal Pradesh, Assam). Cultivated, used in vegetable salad for sourness. The fruit is sliced, dried and then stewed, and used in curries and soups. Also, possess medicinal properties.
97. ***Garcinia cambogia*** **(Gaertn.) Desr.; syn.** ***G. gummi-gutta*** **(L.) Roxb.** Malabar tamarind or *Matka imli* is an evergreen tree endemic to the humid tracts of Western Ghats and Sri Lanka. Cultivated for its dried rind (pericarp of the under ripe fruits) used as condiment. Known for weight loss and medicinal properties. Wild forms are genetic resource for cultivated type.
98. ***Garcinia cowa*** **Roxb. ex Choisy.** Cowa is an evergreen 8–12 m-tall tree, habitat to mixed forests on hills or in valleys of Northeast Hills and Eastern and Western Ghats. Harvested from the wild for its edible fruits and leaves. Also, cultivated. Used as genetic resource for rootstock. Possess medicinal properties.
99. ***Garcinia hombroniana*** **Pierre.** Seashore mangosteen, a rare, little-known fruit tree, native to Malaysia, extending into coastal area of Nicobar Islands. An important component of *Garcinia* gene pool, as rootstock and for genetic improvement of mangosteen. Allied to mangosteen and the eastern Himalayan *G. paniculata.* Known for antioxidant and antimicrobial activities.
100. ***Garcinia indica*** **Choisy.** *Kokum*, a fruit-bearing tree with culinary, pharmaceutical, and industrial uses. Wild types occur in humid evergreen forests of the Western Ghats, presenting immense genetic diversity. Cultivated for

acidic, juicy fruits, used as a staple souring/flavoring agent. Kokum squash/drink is popular for improving digestion and cooling the body during summers.

101. ***Garcinia kurzii* Pierre.** A shrub, endemic to Andaman (South Andaman) and Nicobar (Central Nicobar) Islands. Fruits small berries, up to 3 cm in diam., edible.
102. ***Garcinia lanceifolia* Roxb.** It is an important medicinal plant, endemic to Assam, Northeast Hills and extending to Eastern and Western Ghats. Various ethnic communities of Northeast India have used this to treat various disorders like dysentery, dyspepsia, and biliousness. Also, eaten raw or made into pickles.
103. ***Garcinia morella* (Gaertn.) Desr.** Evergreen moderate sized up to 18 m-tall tree, habitat to evergreen and semi-evergreen forests of Myanmar, North East India, and Western Ghats. Also, cultivated and used as food and Ayurveda medicinal plant. Help moderate loss of weight.
104. ***Garcinia pedunculata* Roxb. ex Buch. -Ham.** Popularly known as *bar thekera* in Assamese, an evergreen tree endemic to the Southeast Asia, including Eastern Himalayas and northeast region of India. Related to *G. mangostana*. Large edible fruit that is a substitute to lemon in Khasi Hills. Sliced fruits are much valued and used for preparing delicacies.
105. ***Garcinia sopsopia* (Buch-Ham) Mabberley.** 12–20 m-tall tree, distributed in evergreen forests of Eastern Himalayas and lower hills of Nepal, Bhutan, Sikkim, and Northeast Hills. Aril (an extra seed covering) pulpy with odor and sour taste like mangosteen is eaten. Often cultivated. It has been recommended as a suitable rootstock for mangosteen.
106. ***Garcinia spicata* Hook.** Medium-sized evergreen tree, native to evergreen and semi-evergreen forests of India (Eastern and Western Ghats and in Northeast Hills) and Sri Lanka. Fruit broadly oblong, yellowish, smooth, with bad odor, pulp sweet, acidic, and bitter. Sometimes cultivated, as an ornamental for its interesting, glossy foliage and colorful fruits.
107. ***Garcinia wightii* T. Anders.** Rare understory shrub to small tree with globose or sub-globose berries, endemic to low elevation of evergreen and semi-evergreen forests of South Western Ghats. Listed in threatened species (Archive.is/7ytuq (2001).
108. ***Garcinia xanthochymus* Hook.f. ex T. Anderson; syn. *G. tinctoria* Dunn.** A shrub or a small erect tree, native to temperate and tropical Asia, including humid tropical forests of Eastern Himalayas and Eastern and Western Ghats. Known for medicinal properties. Cultivated. Source of graft stock for cultivated mangosteen.

In Addition, there are *Garcinia* species, such as *G. cadelliana* King., a small tree, reported only from tropical evergreen forests of South Andaman and known only by type collection made long back by George King in 1884 and listed critically endangered (Nayar and Sastry 1987–1988); *G. jelinckii* Kurz., distributed in Great Nicobar and appears threatened; and *G. kingii* Pierre ex Vesque., a small tree reported endemic to island eco-region of India, with no specimen. Listed endangered in the IUCN Red List of Threatened Species 1998 (WCMC 1998c). They need further exploration and characterization before consideration.

109. ***Mangifera andamanica* King. (Anacardiaceae)**. It is a huge tree with oval-shaped fruits borne in clusters, endemic to Andaman and Lakshadweep Islands. The peel of the fruits was observed to be thin, orange colored. Distantly related to *M. indica* and phylogenetically needs reconsideration for inclusion in subgenus *Mangifera* (Dinesh et al. 2015).
110. ***Mangifera camptosperma* Pierre**. It is a tall tree with sparse foliage endemic to Andaman and Lakshadweep Islands. Fruits flat and round. The fruit pulp was found to be very fibrous and nonedible. The pulp recovery is 23%, with hardly any edible pulp. Phylogenetically close related to cultivated *M. indica* along with rare species *M. griffithii* and part of the subgenus *Mangifera* (Dinesh et al. 2015).
111. ***Mangifera indica* L.** A tree occurring wild in peninsular tract of India, in evergreen forests; also in northeast and *terai* region with greater variation in the former. Domesticated around 2000 BC, with cultivated varieties getting introduced to other warm regions of the world. Hence, wild types are part of primary gene pool of cultivated types. Source of graft stock (Campbell and Ledesma 2015).
112. ***Mangifera sylvatica* Roxb.** *Himalayan mango*, a large evergreen tree that can grow up to 45 m tall, largely found in hilly forests of northeast region, in Shillong Plateau; elsewhere sporadic; also in Andaman Islands. Part of primary gene pool and potential source of graft stock for mango (Bompard 1993). Medicinal too. Another species, *M. khasiana* Pierre, of doubtful occurrence have been reported from India.
113. ***Musa acuminata* Colla**. Desert banana is an evergreen stoloniferous perennial, native to Southeast Asia, including cleared evergreen forests of Northeast Hill region and South Western Ghats of India. Domesticated around 8000 BC. A wild progenitor contributing 'A genome' to the present-day dessert bananas (Perrier et al. 2011). It is part of tertiary gene pool of cultivated hybrid bananas (Boer and Ella 2000). It is highly variable having 6–9 subspecies. The yellow variety known as the Cavendish banana used for diverse purposes.
114. ***Musa balbisiana* Colla.** It is a perennial herbaceous plant, native to Eastern India (Northeast Hill region in cleared forests), northern Southeast Asia, and southern China. It is one of two species with *M. acuminata* that are wild progenitors of modern bananas and plantains, contributing B-genome (Perrier et al. 2011). It is a member of tertiary gene pool of cultivated hybrid bananas (Heslop-Harrison and Schwarzacher 2007). It has been the source of improved vigor and tolerance to biotic and abiotic stresses and used in breeding program for incorporation of these traits. Not widely cultivated because fruit contains seed.

115. ***Musa cheesmani* Simmonds.** It is a large, robust, and fast-growing wild diploid seeded banana forming tight groups; seeds are flattened and bigger than those of *M. balbisiana.* Endemic to India, widely distributed in Arunachal Pradesh, Assam, Manipur, and Nagaland mountains up to 2500 m. Based on molecular affinity, it belongs to tertiary gene pool of cultivated *Musa* (Singh et al. 2015).
116. ***Musa itinerans* Cheesman**. Diploid (2n = 22) species with long rhizome and pink fruit, commonly called Yunnan banana. Dispersed across Southeast Asian continental part to Northeast India. Resembles *M. acuminata* and may be a variant; occurring in the northeastern India. It belongs to tertiary gene pool of banana, based on affinities to A- and B-genome of *Musa* (Li et al. 2010a).
117. ***Musa mannii* H. Wendl. ex Baker.** Indian pink dwarf banana (*M. assamica*), native to China to Eastern Himalayas and hilly tracts of Assam. Almost extinct in its habitat, but grown as ornamental. Adapts well to both tropical and warm temperate climates and grows best in protected, filtered sun or shade. It has intraspecific varietal diversity and based on affinities to A- and B-genome of *Musa* belongs to tertiary gene pool of banana (Li et al. 2010a).
118. ***Musa nagensium* Prain**. A very beautiful plant with dark purple-red pseudostem, waxy around the shoulders. Native to Eastern Himalayas, Naga Mountains, and Assam (seeds like *Ensete gillettii*). Taxonomy of the species has been uncertain. Recently, variety *hongii* Häkkinen is reported, as a new addition to the flora of Arunachal Pradesh (Gogoi 2013). Based on affinities to A- and B-genome of *Musa*, it belongs to tertiary gene pool of banana (Li et al. 2010a).
119. ***Musa ornata* Roxb.** It is a small species of ornamental value with green foliage having pink tones throughout. Native to South India and Koraput region, now widely cultivated in tropics as ornamental. Has food and medicinal value. Belongs to section *Rhodochlamys* with 2n = 22. It hybridizes freely with several other *Musa* species. Consequently, many hybrids are derived from it in the tropical regions of America, and interspecific breeding is valuable option in genetic improvement. It belongs to tertiary gene pool of banana, based on probable crosses with edible *Musa* (Silva et al. 2001).
120. ***Musa sapientum* L.; syn. *Musa x paradisiaca*.** It is a 5–9 m-tall perennial herb with tuberous rhizome, hard, long pseudo-stem. It is a cultivated triploid hybrid resulted from cross between *M. acuminata* x *M. balbisiana* (AAB) or *M. acuminata* alone developing a wide range of cultivars. Natural hybrid of this is found in Chota Nagpur and Kaveri Basin. Cultivated in tropical and subtropical regions, for nutritious fruit and medicine for many treatment, including diabetes mellitus.
121. ***Musa sikkimensis* Kurz**. Darjeeling banana is a beautiful herb, about 4 m tall with yellowish-green foliage, reddish tinged pseudo-stem, and sweet edible fruits. Native to Sikkim, North Bengal, Khasi Hills, and Bhutan at high altitude. It is a diploid, with useful traits such as cold tolerance, drought, etc. It belongs to tertiary gene pool of banana, based on affinities to A- and B-genome of *Musa* (Li et al. 2010a).

122. ***Musa superba*** **Roxb.; syn.** ***Ensete superbum*** **(Roxb.) Cheesman.** *Jungli kela* is an erect plant with unbranched globose leaf sheaths formed into a pseudostem of 50–100 cm diameter. Seeds like *M. nagensium*. Habitat to moist deciduous and semi-evergreen forests of Eastern Himalayas, Khasi Hills, Manipur, Peninsular India, and Western Ghats. Soft stem cooked as vegetable and has medicinal properties.
123. ***Musa thomsonii*** **(King ex Baker) A.M. Cowan & Cowan; syn.** ***M. flaviflora*** **N.W. Simmonds**. A plant with slender green stems and edible fruit. Native to Himalayas and Northeast India (Assam, Khasi Hills, Manipur). Corm juice used for home remedy. The taxonomic identity is uncertain. May be a variant of *M. sapientum* and *M. paradisiaca* interspecific complex. Cold hardy.
124. ***Musa velutina*** **Wendl. ex Drude**. The hairy or pink banana is an ornamental perennial dwarf banana with red mid-ribbed shiny green leaves and strikingly bright pink, red, and orange flowers. Fruits are 3 in. long, pink, and fuzzy with seed. It is endemic to Eastern Himalayas and Assam, confused with *M. mannii*/*M. assamica*. Part of tertiary gene pool (Silva et al. 2001).
125. ***Parthenocissus semicordata*** **(Wall.) Planch.; syn.** ***P. himalayana***, ***Vitis himalayana*** **(Royle) Brandis (Vitaceae).** A deciduous climbing shrub, native of Himalayas. It has small fruits, which look like grapes and are dark blue/black when ripe. Related to *V. vinifera* family.
126. ***Vitis flexuosa*** **Thunb.; syn. var.** ***parvifolia*** **(Roxb.) Gagnep,** ***V. parvifolia*** **Roxb. (Vitaceae).** A deciduous climber, native range extends from Asia temperate to tropical region, including Himalayas, up to 600–1800 m. Part of tertiary gene pool of grapes (Tröndle et al. 2010).
127. ***Vitis lanata*** **Roxb.; syn.** ***V. heyneana*** **DC.,** ***V. rugosa*** **Wall.** A woody deciduous climber, with edible sweet fruit, found throughout the sub-Himalayan tract, ascending to 1500 m. Member of tertiary gene pool of wine grape. Source of late ripening, induced due to excessive rains at the ripening time, and resistant to disease.
128. ***Vitis lanceolaria*** **(Roxb.) Wall.; syn.** ***Tetrastigma lanceolarium*** **(Roxb.) Planch.** A large woody climber with white flower and small berries, distributed in Indo-Malayan region, including Kumaon to Sikkim, North Bengal, Assam, and Western Peninsula. Fruits edible.
129. ***Vitis pallida*** **Wight & Arn.; syn.** ***V. adnata*** **(Roxb.) Wall.** ***Cissus adnata*** **Roxb.** Large woody climbers with sulcate branches. Berries obovoid, bluish black. Distributed in northeast region, Eastern Ghats, and Deccan Peninsular region. Listed under medicinal plants.

10.2.3 Arid and Semi-arid Fruits

130. ***Aegle marmelos*** **L. Correa (Rutaceae).** Bael, a deciduous ca. 12 m-tall tree. Native to tropical Asia, including Indian Subcontinent with much variability (var. *mahurensis* Zate.). Marginally cultivated around temple and homesteads

as a sacred tree. Fruits, source of food and Ayurveda medicines. Also, yields wood, gum, resin, and essential oil.

131. ***Baccaurea courtallensis*** **(Wight) Muell-Arg. (Euphorbiaceae).** An evergreen ca. 15 m-tall tree. The capsule fruits are edible. Endemic to evergreen and semi-evergreen forests of Peninsular India. Distantly related to castor. Medicinal plant.
132. ***Capparis pachyphylla*** **Jacob. (Capparaceae)**. Shrubs to small tree, fruits not known. Endemic to subtropical evergreen forests of Arunachal Pradesh, India. Listed, vulnerable in IUCN Red List of Threatened Plants (WCMC 1998b). Related to *kair*, *C. decidua*.
133. ***Carissa spinarum*** **L.; syn.** ***C. congesta*** **Wight. (Apocynaceae).** Wild *karanda* or bush plum, a spiny large shrub. Native to Africa and Asia, including degraded slopes in the plains of Gangetic Plains, northwestern India, and the Western Ghats. Related to *Karanda*, *C. carandas*. Fruit edible, used in Ayurveda.
134. ***Cordia myxa*** **L. (Boraginaceae).** It is a medium-sized broad-leaved deciduous tree, native to West Asia and all over India, except hills, with variability in northwestern and central plains. Two forms are available in northwestern region, which are part of primary gene pool. An underutilized fruit, listed under Ayurveda plants.
135. ***Cordia gharaf*** **(Forsk.) Ehrenb. ex Asch.; syn.** ***C. rothii***, ***C. sinensis*** **Lam.** Small tree found in dry deciduous forests of Africa, Arabia, and India. Fruits contain gum "*gond*" a sticky substance. Wild relative of *C. myxa*.
136. ***Ficus auriculata*** **Lour. (Moraceae).** Giant Indian fig, a perennial evergreen small spreading tree with big round leaves. Native to Asia, including India in outer Himalayas, from Punjab to Assam, Khasi Hills, Chota Nagpur, and Orissa. Fruits are edible and used to make jams, juices, and curries. Medicinal. Related to common fig, *F. carica*.
137. ***Ficus nemoralis*** **(Wall. ex Miq.); syn.** ***F. neriifolia*** **Sm.** *Dudhilo* is an up to 15 m-tall tree with hairless, leathery leaf blades of varying shapes. Native to Asia, including India in higher altitudes of Himalayas. Popular as bonsai plant.
138. ***Ficus palmata*** **Forsk.** Wild Himalayan fig, a small tree native to Africa and Asia, including mid-Himalayas, though common to northwestern India and Rajasthan. Fruits are edible, medicinal as laxative and to treat bladder and lung diseases. Genetic resource of graft stock for fig, *F. carica*.
139. ***Flacourtia montana*** **Graham (Flacourtiaceae).** A thorny 8 m-high tree. Native to India as understory tree endemic to evergreen to semi-evergreen forests of the Western Ghats. Wild relative of Indian Plum, *F. indica* (Burm.f.). Yields timber too.
140. ***Grewia damine*** **Gaertn.; syn.** ***G. salviolia*** **Mast. (Malvaceae).** A shrub to small tree, habitat to dry low land areas of sub-Himalayan tract, Central India, and the Western Ghats. Ayurveda plant, wild relative of *Phalsa*, *G. subinaequalis* DC.; syn. *G. asiatica* Mast.

141. ***Grewia oppositifolia* Buch. -Ham. ex Roxb.; syn. *G. optiva* J.R. Drumm. ex Burret.** *Bhimal* or *Bihul* a deciduous small tree, native to Indian Subcontinent (Himalayas, Upper Gangetic Plains, and Nepal). A multipurpose tree, yielding edible fruit, fiber, and timber, gathered from wild. Often planted by farmers as part of agroforestry as valuable fodder tree. Tolerant to frost.
142. ***Grewia tenax* Forsk.** *Gangu kanger* or *gango*, a shrub up to 2 m tall with orange fruits usually eaten raw, leaves boiled and eaten. Native to Africa, West Asia, and Indian Subcontinent (Central and Western India and Deccan Plateau). It is being domesticated as source of food, fodder, fiber, fuelwood, timber, and a range of traditional medicines.
143. ***Grewia villosa* Willd**. Shrub to small tree, native to Africa, West Asia, and Indian Subcontinent, including dry deciduous forests of Central/Western India, extending into Deccan Plateau and westward to Maharashtra and Kerala. Sweet acidic fruit eaten by poor. Root is employed in diarrhea.
144. ***Holboellia latifolia* Wall. (Lardizabalaceae).** *Gomphal*, a monoecious, evergreen twining or erect to 4 m-tall shrub, native to East Asia, including India, throughout Himalayas. Also, cultivated as ornamental and for edible fruiting carpels.
145. ***Manilkara zapota* (L.) P. Royen; syn. *Achras zapota* L.; *Mimusops hexandra* Roxb. (Sapotaceae).** Sapote or *Chikoo*, native of Central America, cultivated in tropics, including India. Escaped and naturalized in Gujarat and Central and Western India with variability for fruit size.
146. ***Mimusops elengi* L. (Sapotaceae).** *Maulsari* (Spanish cherry) is a medium-sized evergreen tree, native to Indo-Malayan region, including semi- and evergreen forests of Northeast and Eastern Ghats and Andaman and Nicobar Islands. Cultivated in North India, Western Peninsula, and South India. Known in Ayurveda, a potential medicinal plant.
147. ***Myrica esculenta* Buch. -Ham. ex D. Don; syn. *M. nagi* Hook.f. non-Thumb. (Myricaceae).** *Kaiphal* or box berry, a small tree native to the hills of northern, Northeast India and Nepal with much variability. The fruits are perishable and eaten raw and used in making pickle; bark in powdered form is used to treat dysentery and several other ailments. Potential ethnomedicinal plant. Overexploitation has threatened it to extinction.
148. ***Salvadora oleiodes* Decne.** *Pilu* is a small bushy evergreen tree producing new leaves and small yellow fruit. Native to West Asia and Indian Subcontinent's northwestern plains. Used as fodder, and reserved/protected for grazing in local peasant villages. Related to cultivated *S. persica* L.
149. ***Syzygium alternifolium* (Wight) Walp. (Myrtaceae).** A small to large, evergreen rare tree with fruit varies from 1.5 to 2.5 cm in diameter. Endemic to southern Eastern Ghats of India. Known for medicinal properties. Possible genetic resource for cultivated *Syzygium* spp.
150. ***Syzygium beddomei* (Duthie) Chithra**. A large tree, endemic to southern Deccan Plateau (Tamil Nadu) and South Western Ghats evergreen forests. Listed in the Gazette of India (1932) and listed critically endangered in IUCN Red List (WCMC 1998d).

151. ***Syzygium caryophyllatum* (L.) Alston**. Evergreen shrub to small tree with black globose berry, found along margin of evergreen or semi-evergreen forests or in open formations in the Western Ghats. Medicinal with antimicrobial, antioxidant, and anticancer activities. Sometimes cultivated for its spicy flower buds. Listed endangered in IUCN Red List 1998 (WCMC 1998f).
152. ***Syzygium claviflorum* (Roxb.) Wall ex Cowan & Cowan**. Small sub-canopy tree, native of Bengal. Recorded recently from Agasthyamalai Biosphere Reserve of the Western Ghats, Andaman and Lakshadweep Islands, and Southeast Asia. Used as a food and source of wood. Sometimes cultivated for its edible fruits.
153. ***Syzygium cumini;* (Linn.) Skeels.; syn. *Eugenia jambolana* Lam. (Myrtaceae).** Black plum or *jamun*, a large evergreen tree, native to Africa and Asia, including all over India, mainly in peninsular region; sporadically elsewhere with lot of variability. Part of primary gene pool for use in genetic improvement cultivated types. Cultivated for fruit (eaten raw), medicinal value (against diabetes), wood, etc.
154. ***Syzygium densiflorum* Wall. ex Wight & Arn.; syn. *S. arnottianum* (Wight) Walp**. A large evergreen canopy tree, endemic to high elevations of evergreen forests of South Western Ghats, between 1500 and 2300 m. Fruit, a berry, oblong ovoid, dark purple, fleshy. Needs further evaluation.
155. ***Syzygium gardneri* Thw.; syn. *Eugenia gardneri* (Thw.) Bedd.** Large canopy to emergent evergreen tree found in evergreen forests of the Western Ghats. Fruit, a berry 5–8 mm across, ovoid, rarely globose, purple. Noted for antimicrobial activity of the leaf oil. Needs further evaluation.
156. ***Syzygium manii* (King) Balakr**. It is endemic to India, found in middle Andaman Island eco-region in semi-evergreen forest. Listed critically endangered in IUCN Red List 1998 (WCMC 1998g) due to habitat loss. Needs collection and conservation.
157. ***Syzygium nervosum* A. Cunn. ex DC.; syn. *S. operculatum* (Roxb.) Nied.** It is a many-branched, evergreen tree growing up to 15 m tall. Native to Asia and Australia, including South India and Chota Nagpur Plateau. The bark is harvested from wild as source of tannins, while fruits are eaten raw, made into drinks, and preserved. Sometimes cultivated.
158. ***Syzygium salicifolium* (Wight) J. Graham; syn. *S. heyneanum* (Duthie) Gamble**. *Kath-Jamun*, small 6 m-tall tree found in evergreen forests of India, including Gangetic Plains, peninsular region, and the Western Ghats. Needs further evaluation.
159. ***Syzygium samarangense* Merr. & Perry.; Syn. *Eugenia javanica* Lam.** Java apple is a tropical ca. 12 m-tall tree with highly ornamental, juicier, and flavorful fruit, suitable for eating out of hand from wild. Native of Greater Sunda Islands, Malay Peninsula, extending to Andaman and Nicobar Islands and South India. Cultivated under agroforestry in East Godavari, Andhra Pradesh. Pharmacological aspects known. Closely related to *S. aqueum.*
160. ***Syzygium zeylanicum* (L.) DC.** An evergreen 12–20 m-tall tree. Found on banks of streams in evergreen forests, also in scrub jungles of Indo-Malayan

region, including Kerala. Grows wild in coastal regions. Fruit is sweet, aromatic, and edible; wood is used for fuel and boat making, leaves and roots in traditional medicine.

Additionally, *Syzygium andamanicum* (King) Balakr., known by type collection from Andaman Islands and listed critically endangered in IUCN Red List 1998 (WCMC 1998e) because of logging, and *S. balsameum* (Wt.) Wall. ex Walp.; syn. *Eugenia balsamea* Wight., reported from northeast, need collection and evaluation of genetic potential.

161. ***Ziziphus glabrata* Heyne ex Roth. (Rhamnaceae).** A small tree with globose obovoid drupe fruit. Native to dry deciduous forests and along streams in foothills, Peninsular and Northeast India, and Western Ghats. Dried fruits are edible. Cultivated.
162. ***Ziziphus mauritiana* Lam**. It is a spiny, quick-growing evergreen shrub to small 15 m-tall tree, starts producing fruits within 3 years. Native to Asia, including Himalayas and semi-arid northwestern plains of India with much variability. The fruit are quite nutritious, rich in vitamin C, eaten raw, pickled, or used in beverages. Also, yields fuelwood and medicine and can be potential host for lac insect, *Kerria lacca*. Part of primary gene pool of cultivated type.
163. ***Ziziphus nummularia* (Burm.f.) Wight and Arn**. *Jharber*, a 2 m-tall shrub with distinct leaves having pubescence on the adaxial surface. Native of West Asia and Indian Subcontinent, particularly the arid and semi-arid regions of northwest plains and Central India and Deccan Peninsula with much variability in Thar Desert. Used as food, fodder, fuel, medicine, and genetic source of rootstock.
164. ***Ziziphus oenopolia* (L.) Mill**. *Makkaya,* a climbing shrub with small edible fruits. Native to Asia and Australia, including Gangetic Plains, peninsular tract, and Eastern and Western Ghats of India.
165. ***Ziziphus oxyphylla* Edgew**. A large deciduous shrub to small tree with red, oval-shaped, single-seeded drupe fruit. Habitat to waste places of dry clay in Western Himalayas. Used as folk medicine for antimicrobial and cytotoxic properties.
166. ***Ziziphus rugosa* Lam.; syn. *Z. glabra* Roxb**. *Suran*, a small tree native to Asia, including exposed, dry slopes of hills and mountains and in deciduous and semi-evergreen forest edges of Western Himalayas, Gangetic Plains, humid tracts, and Western/Eastern Ghats. The ripe fruits are eaten directly with salt, medicinal too.
167. ***Ziziphus truncata* Baltt. & Hallb**. A small shrub with divaricate branches. Endemic to rocks in Kaylana, Jodhpur, Rajasthan (Bhandari 1995). Close ally of *Z. nummularia*. Ripe fruit eaten.
168. ***Ziziphus xylopyrus* Hochst. ex A. Rich.; syn. *Z. abyssinica* Hochst. ex A. Rich.** *Katber*, an evergreen shrub to small tree, native to dry and moist deciduous forests of Indian Subcontinent (Gangetic Plains to Central India to

Western Ghats). It has tremendous medicinal value for treatment of bronchial asthma, thirst and diarrhea and treating ulcer, as an antinociceptive, antidepressant, anticonvulsant, etc. Genetic resource for *ziziphus* spp.

In addition, there are some rare *Ziziphus* species, such as *Z. horrida* Roth. and *Z. williamsii* Bhandari & Bhansali, known by type collections, which need further exploration and research attention.

10.3 Perspective

There are many wild species under various taxa that have been documented in the literature based on herbarium specimen, single-type collection or mis-collection of an allied population, and reported rare and are not available for use, for example, *Citrus nakoor*, *Garcinia cadelliana*, *G. jelinckii*, *G. kingie*, *Syzygium andamsnicum*, *Ziziphus horrida*, *Z. truncate*, *Z. williamsii*, etc. These species required recollection on priority, failing which should be deleted under the taxonomic revision.

In case of temperate fruits, the wild relatives belonging to taxa such as *Malus*, *Pyrus*, *Prunus*, and *Rubus* have been visualized to be escapes from their actual center of origin, like West Asia, Central Asia, China, etc., to Himalayas. Therefore, the population species found in Himalayan region of India are expected to be genetically distinct than the original ones and require further investigation to ascertain their taxonomic status, phylogenetic relationships with original populations from center of origin, with greater emphasis on their collection, characterization, and evaluation for desired traits to facilitate their conservation and use.

In case of tropical fruits, such as *Citrus*, *Musa*, etc., natural interspecific hybrid origin has been visualized and confirmed for a number of species, reflecting the amount of opportunities are available in such genera for genetic improvement of cultivated species with introgression of desired genes through interspecific breeding. This demands greater number of biosystematic studies in other cases, such as *Diospyros*, *Garcinia*, *Vitis*, and more tropical fruits to understand the phylogenetic and cross-compatibility relationships among the wild relatives and between wild and cultivated species; to identify truly genetically distinct species, providing solution for the taxonomic debates and classifying them into possible gene pools; and to facilitate conservation and use of appropriate breeding strategies. The same may apply to certain arid fruits as well, such as *Syzygium* spp., *Ficus* spp., and *Ziziphus*.

Recognizing the opportunities offered by the wild relatives of temperate, subtropical, tropical, and arid fruits, in form of reservoir genetic diversity for most valuable traits and that their natural habitat is being eroded, as indicated by information available on threatened species, the collection and conservation program on wild relatives of fruits should be a priority area, before they are lost. For conservation, because of the perennial shrub or tree habit of most fruits, in situ conservation is one of the most appropriate options. It can be taken up at specific ecological niches as per adaptability, for example, in case of temperate fruits in dry regions of Leh for fruits like dry-type apricot (*Prunus*), *Hippophae*, etc., and in Himachal

Pradesh and Uttarakhand for *Malus*, *Prunus*, *Pyrus*, etc. Following this approach, in case of tropical fruit, the first gene sanctuary for *Citrus* was established in their natural habitat at Garo Hills in Meghalaya, India (Singh 1981).

Many of wild relative of fruit species have vast potential for value addition, particularly in case of those species, which are already being used as minor fruits. For example, many have been reported with multiple uses, such as medicinal plant, ornamental plant, and production of diverse product and processes. In this regard, a large number of wild species are known in the Indian traditional medicine system *Ayurveda* for their medicinal properties or being used as folk medicine by various tribes and communities. Such species require further investigation in areas of phytochemical analysis for biochemical profiling, screening against various pathogens and for pharmaceutical activity, and identification of associated active principals to ascertain their potential and possible regular use in industrial entrepreneurships. Similarly, many species, such as *Musa mannii*, *M. ornata*, *M. velutina*, *Ficus nemoralis* (Bonsai), *Holboellia latifolia*, etc., have either attractive canopy/foliage or flowers or fruits that can be exploited for ornamental purposes in gardens or as an avenue tree in landscaping, agroforestry, etc. In addition, several minor fruits, particularly the arid and tropical fruits, such as *Cordia gharaf*, *Carissa spinarum*, *Garcinia lanceifolia*, *G. pedunculata*, etc., and wild and weedy from *ker*, *lasoda*, and *ber* (*Ziziphus glabrata*), have potential for preparation of processed products, such as powder or dried fruit, and/or sour blended culinary, or mixtures with appropriate balance of sweetness, sourness, bitterness, etc., for commercial exploitation. Further, gum, resin, tannin, and dyes can be extracted from some, such as *Cordia gharaf*, for industrial use. Many species may have out-of-box use, such as many *Morus* species may be used for rearing of silkworm, while other can be host for several other valuable insects, such as *Ficus* and *Ziziphus* spp. for lac insect, *Kerria lacca*. These areas need further research to improve their overall potential value of wild relatives.

References

Akhter C, Dar GH, Khuroo AA (2013) *Ziziphus jujuba* Mill. subsp. *spinosa* (Bunge) Peng, Li & Li: a new plant record for the Indian Subcontinent. Taiwania 58(2):132–135

Aradhya MK, Weeks C, Simon CJ (2004) Molecular characterization of variability and relationships among seven cultivated and selected wild species of *Prunus* L. using amplified fragment length polymorphism. Sci Hortic 103:131–144

Archive.is/7ytuq (2001) Studies on rare and endangered species – BSI, India

Aslam S, Ganaie KA, John AQ, Dar GH (2010) Family rutaceae in Kashmir Himalyas: *Poncirus trifoliata* (L) Raf. – a new record for Flora of Kashmir, India. Academia Arena 2(5):40

Bayer RJ, Mabberley DJ, Morton CM, Cathy H, Sharma IK, Pfeil BE, Rich S, Hitchcock R, Sykes S (2009) A molecular phylogeny of the orange subfamily (Rutaceae: Aurantioideae) using nine cpDNA sequences. Am J Bot 96:668–685

Bhandari MM (1995) Flora of Indian desert. MPS Repros, Jodhpur, p 435

Bhusal RC, Fusao M, Kipkoriony LR (2002) Selection of rootstocks for flooding and drought tolerance in *Citrus* species. Pak J Biol Sci 5(5):509–512

Boer E, Ella AB (eds) (2000) Plants producing exudates. Pl Res SEAs 18:130

Bompard JM (1993) The genus *Mangifera* re-discovered: the potential contribution of wild species to mango cultivation. Acta Hortic 341:69–77

Bouhadida M, Martin JP, Eremin G, Pinochet J, Moreno MA, Gogorcena Y (2007) Chloroplast DNA diversity in Prunus and its implication on genetic relationships. J Am Soc Hortic Sci 132:670–679

Campbell RJ, Ledesma N (2015) Mango cultivars with potential for commercial development. Acta Hortic 1075:33–40

Chin SW, Shaw J, Haberle R, Wen J, Potter D (2014) Diversification of almonds, peaches, plums, and cherries – molecular systematics and biogeographic history of *Prunus* (Rosaceae). Mol Phylogenet Evol 76:34–48

Curk F, Ollitrault F, Garcia-Lor A, Luro F, Navarro L, Ollitrault P (2016) Phylogenetic origin of limes and lemons revealed by cytoplasmic and nuclear markers. Ann Bot 117(4):565–583

Dinesh MR, Ravishankar KV, Nischita P, Sandya BS, Padmakar B, Ganeshan S, Chithiraichelvan R, TVRS S (2015) Exploration, characterization and phylogenetic studies in wild *Mangifera indica* relatives. Am J Plant Sci 6:2151–2160

Finn CE, Swartz HJ, Moore PP, Ballington JR, Kempler C (2002) Use of 58 *Rubus* species in five North American breeding programs-breeder's notes. Acta Hortic 585:113–119

Forte AV, Ignatov AN, Ponomarenko VV, Dorokhov DB, Savelyev NI (2002) Phylogeny of the *Malus* (apple tree) species inferred from the morphological traits and molecular DNA analysis. Russ J Genet 38:1150–1160

Froelicher Y, Mouhaya W, Bassene J-B, Costantino G, Kamiri M, Luro F, Morillon R, Ollitrault P (2011) New universal mitochondrial PCR markers reveal new information on maternal citrus phylogeny. Tree Genet Genomes 7:49–61

Garcia-Lor A, Curk F, Snoussi-Trifa H, Morillon R, Ancillo G, Luro F, Navarro L, Ollitrault P (2013) A nuclear phylogenetic analysis: SNPs, indels and SSRs deliver new insights into the relationships in the "true citrus fruit trees" group (Citrinae, Rutaceae) and the origin of cultivated species. Ann Bot 11:1–19

Gogoi R (2013) *Musa nagensium* var. *hongii* Häkkinen – a new addition to the flora of India. Taiwania 58(1):49–52

Grosser JW, Mourao-Fo FAA, Gmitter FG, Louzada ES, Jiang J, Baergen K, Quiros A, Cabasson C, Schell JL, Chandler JL (1996) Allotetraploid hybrids between *Citrus* and seven related genera produced by somatic hybridization. Theor Appl Genet 92:577–582

Hall HK, Hummer KF, Jamleson AR, Jennings SN, Weber CA (2009) Chapter 2. Raspberry breeding and genetics. In: Janick J (ed) Plant breed reviews, vol 32. Timber Press Inc, Portland, pp 39–353

Hartmann W, Neumüller M (2009) Plum breeding. In: Jain SM, Priyadarshan PM (eds) Breeding plantation tree crops: temperate species. Springer, New York, pp 161–231

Hazarika TK (2012) *Citrus* genetic diversity of north-east India, their distribution, eco-geography and eco-biology. Genet Resour Crop Evol 59:1272–1275

Heslop-Harrison JS, Schwarzacher T (2007) Domestication, genomics and the future for banana. Ann Bot (Oxford) 100:1073–1084

Hjamarsson I, Wallace B (2007) Gooseberry and currant in Sweden: history and cultivar development. Pl Breed Rev 29:145–175

Hokanson SC, Lamboy WF, Szewc-McFadden AK, McFerson JR (2001) Microsatellite (SSR) variation in a collection of *Malus* (apple) species and hybrids. Euphytica 118:281–294

Li LF, Häkkinen M, Tuan YM, Hao G, Ge XJ (2010a) Molecular phylogeny and systematics of the banana family (Musaceae) inferred from multiple nuclear and chloroplast DNA fragments, with a special reference to the genus *Musa*. Mol Phylogenet Evol 57:1–10

Li X, Xie R, Lu Z, Zhou Z (2010b) The origin of cultivated *citrus* as inferred from internal transcribed spacer and chloroplast DNA sequence and amplified fragment length polymorphism fingerprints. J Am Soc Hortic Sci 135(4):341–350

Li WL et al (2012) An outline on research and utilization of blackberry in China. Acta Hortic 926:394

Luro F, Gatto J, Costantino G, Pailly O (2011) Analysis of genetic diversity in *Citrus*. Plant Genet Resour 9:218–221

Nadkarni's KM (1996) Indian Materia Medica, Volume 2. Popular Prakashan, Mumbai, p 614
Ohta S, Katsuki T, Tanaka T, Hayashi T, Sato Y, Yamamoto T (2005) Genetic variation in flowering cherries (*Piunus* subgenus *Cerasus*) characterized by SSR markers. Breed Set 55:415424
Ohta S, Yamamoto T, Nishitani C, Katsuki T, Iketani H, Omura M (2007) Phylogenetic relationships among Japanese flowering cherries (*Pinus* subgenus *Cerasus*) based on nucleotide sequences of chloroplast DNA. Plant Syst Evol 263:209–225
Pandey A, Nayar ER, Venkateswaran K, Bhandari DC (2008) Genetic resources of *Prunus* (Rosaceae) in India. Genet Resour Crop Evol 55:91–104
Penjor T, Yamamoto M, Uehara M, Ide M, Matsumoto N, Matsumoto R, Nagano Y (2013) Phylogenetic relationships of *Citrus* and its relatives based on matK gene sequences. PLoS One 8(4):e62574. doi:10.1371/journal.pone.0062574
Perrier X, De Langhe E, Donohue M, Lentfer C, Vrydaghs L, Bakry F, Carreel F, Hippolyte I, Horry JP et al (2011) Multidisciplinary perspectives on banana (*Musa* spp.) domestication. Proc Natl Acad Sci U S A 108:11311–11318
Rana JC, Verma VD, Yadav SK and Pradheep K (2003) Genetic diversity of wild relatives and minor fruits in the Indian Himalayas. In: Proceedings 7th international symposium Temperate Zone Fruits in the Tropics and subtropics. Dr. YS Parmar University of Horticulture and Forestry, Solan, Himachal Pradesh, India, pp 62–63
Rana JC, Chahota SRK, Rana V, Verma M, Verma N, Sharma B, Tilak R (2015) Genetic diversity and structure of *Pyrus* accessions of Indian Himalayan region based on morphological and SSR markers. Tree Genet Genomes 11:821
Randhawa SS (1987) Wild germplasm of pome and stone fruits. Indian Agricultural Research Institute, Regional Station, Shimla, Himachal Pradesh, India, p 51
Rao RS, Joseph J (1970) *Rubus ghanakantae* Rolla Rao et Joseph - a new species from Arunachal Pradesh. Bull Bot Surv India 12(1–4)
Rao MN, Jaya R, Soneji R, Sahijram L (2011) Citrus. In: Kole C (ed) Wild crop relatives: genomic and breeding resources tropical and subtropical fruits. Springer, Berlin, pp 43–59
Scora RW, Kumamoto J, Soost RK, Nauer EM (1982) Contribution to the origin of grapefruit, *Citrus paradisi* (Rutaceae). Syst Bot 7:170–177
SdeOe S, SouzaJunior MT, Alves EJ, Silveira JRS, Lima MB (2001) Banana breeding program at Embrapa. Cropp Breeding and Applied Biotechnology 1(4):399–436
Sharma BD, Hore DK, Gupta SG (2004) Genetic resources of *Citrus* of north-eastern India and their potential use. Genet Resour Crop Evol 51:411–418
Singh D (1967) Nakoor lime-a new *Citrus*. Indian J Hort 24:84–86
Singh B (1981) Establishment of first gene sanctuary for *Citrus* in Garo Hills. Concept Publishing Company, New Delhi
Singh HP, Chadha KL (1993) Genetic resources of *Citrus*. In: Chadha KL, Pareek OP (eds) Advances in horticulture Vol I. Malhotra Publishing House, New Delhi
Singh WA, Singh NS, Handique AJ, Devi HS (2015) Morpho-taxonomical and molecular assessment of Musa genotypes from north-east India by morphological and inter-retrotransposon amplified polymorphism markers. Plant Syst Evol 301:563–575
Staudt G (1989) The species of *Fragaria*, their taxonomy and geographical distribution. Acta Hortic 265:26
Tröndle D, Schröder S, Kassemeyer H-H, Kiefer C, Koch MA, Nick P (2010) Molecular phylogeny of the genus *Vitis* (Vitaceae) based on plastid markers. Am J Bot 97:1168–1178. doi:10.3732/ajb.0900218
Vavilov NI (1951) The origin, variation, immunity and breeding of cultivated plants. Chronica Bot 13:364
Wangchu L, Singh D, Mitra SK (2013) Studies on the diversity and selection of superior types in jackfruit (*Artocarpus heterophyllus* Lam.) Genet Resour Crop Evol 60:1749–1762
WCMC (World Conservation Monitoring Centre) (1998a) *Diospyros barberi*. The IUCN red list of threatened species 1998: e.T37623A10067158. http://dx.doi.org/10.2305/IUCN.UK.1998.RLTS.T37623A10067158.en

WCMC (World Conservation Monitoring Centre) (1998b) *Capparis pachyphylla*. The IUCN red list of threatened species 1998: e.T33607A9795866. http://dx.doi.org/10.2305/IUCN.UK.1998.RLTS.T33607A9795866.en

WCMC (World Conservation Monitoring Centre) (1998c) *Garcinia kingii*. The IUCN red list of threatened species 1998: e. T33491A9782387

WCMC (World Conservation Monitoring Centre) (1998d) *Syzygium beddomei*. The IUCN red list of threatened species 1998: e.T38819A10150912. http://dx.doi.org/10.2305/IUCN.UK.1998.RLTS.T38819A10150912.en

WCMC (World Conservation Monitoring Centre) (1998e) *Syzygium andamanicum*. The IUCN red list of threatened species 1998: e.T32621A9718483. http://dx.doi.org/10.2305/IUCN.UK.1998.RLTS.T32621A9718483.en

WCMC (World Conservation Monitoring Centre) (1998f) *Syzygium caryophyllatum*. The IUCN red list of threatened species 1998: e.T38036A10094391. http://dx.doi.org/10.2305/IUCN.UK.1998.RLTS.T38036A10094391.en

WCMC (World Conservation Monitoring Centre) (1998g) *Syzygium manii*. The IUCN red list of threatened species 1998: e.T32622A9718633. http://dx.doi.org/10.2305/IUCN.UK.1998.RLTS.T32622A9718633.en

Westwood MN, Bjornstad HO (1971) Some fruit characteristics of interspecific hybrids and extent of self-sterility in *Pyrus*. Bull Torrey Bot Club 98:22–24

Woodford RC (2005) Citrus Classification Biotech Books, New Delhi, p 185

Zerega NJC, Nur Supardi MN, Motley TJ (2010) Phylogeny and re-circumscription of Artocarpeae (Moraceae) with a focus on *Artocarpus*. Syst Bot 35:766–782

Zheng XY, Cai DY, Yao LH, Teng YW (2008) Non-concerted ITS evolution, early origin and phylogenetic utility of ITS pseudogenes in *Pyrus*. Mol Phylogenet Evol 48:892–903

Zhukovsky PM (1965) Main gene centers of cultivated plants and their wild relatives within the territory of the U.S.S.R. Euphytica 14:177–188

Spices and Condiments

11

11.1 Introduction

India has been referred as the land of spices, because of its contribution to the world of the two main spices, the black pepper (*Piper longum* L.) and cardamom [*Elettaria cardamomum* (L.) Maton], originated in the tropical evergreen forests of Western Ghats of South India, where their wild form still exists. The other two equally important spices contributed are ginger (*Zingiber* spp.) and turmeric (*Curcuma* spp.) for which South Asian region, possibly India, is believed to be the center of origin, with maximum number of species diversity (Fig. 11.1). Other spices relevant from wild relatives' point of view in Indian context are *Amomum* Roxb., among herbs, *Trigonella* and *Foeniculum* in seed spices, and *Cinnamomum* and *Garcinia* in tree spices. *Amomum* is represented by 22 species, mostly restricted to Northeast India and South India (Thomas et al. 2010). Their number from South India has risen to eight with description *Amomum fulviceps* Thwaites (Thomas et al. 2009) and *A. nilgiricum* (Thomas et al. 2012a).

About 110 species represent the genus *Piper* (Ravindran et al. 2000). The northeastern region and the Western Ghats are recognized as the two independent centers of diversity (Fig. 11.2a, b). About 16 species are reported from the forests of Western Ghats (Fig. 11.2b), while most remaining species are from northeastern region (Fig. 11.2a) (Ravindran and Babu 1994). Many of the *Piper* species, *P. barberi* Gamble; *P. hapnium* Buch. -Ham.; *P. silentvalleyensis* Ravin, Nair, & Asokan; *P. wightii* Miq.; and *P. schmidtii* Hook.f., have been adversely affected by deforestation and the developments; consequently they are now confined to few locations and may soon be extinct, if not collected and conserved immediately. Only one species represents cardamom, *Elettaria cardamomum*, along with one of its closest relatives, the wild cardamom, *E. ensal* (Gaertn.) Abeyw.; syn. *E. major* Thw.

The original version of this chapter was revised.
An erratum to this chapter can be found at https://doi.org/10.1007/978-981-10-5116-6_22

A.K. Singh, *Wild Relatives of Cultivated Plants in India*,
DOI 10.1007/978-981-10-5116-6_11

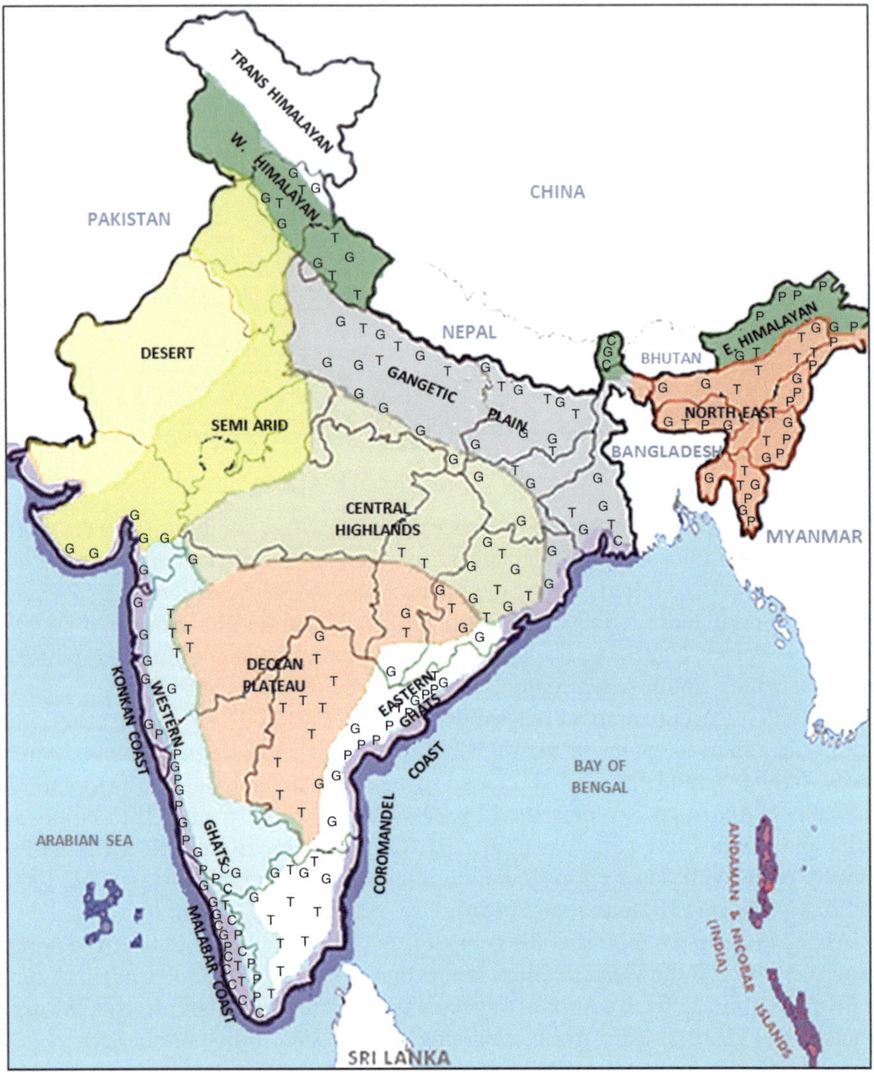

Fig. 11.1 Biogeographical regions associated with the wild species of major spices crops, large cardamom, ginger, pepper, and turmeric
C = cardamom (*cardamomum* spp.); G = ginger (*Zingiber* spp.); P = pepper (*Piper* spp.); T = turmeric (*Curcuma* spp.)

Zingiber is a big genus divided into four sections. Sabu (1991) described eight species from Western Ghats and the adjacent areas. Around 18 *Zingiber* species have been reported from India with variability mainly existing in the northeastern region and South Western Ghats (Kerala). Species like *Zingiber cassumunar* and Z. *zerumbet* are sporadically distributed in the sub-Himalayan tracts. *Z. officinale* var. *rubrum* having pink outer skin of rhizome have been reported recently by IISR. Turmeric, *Curcuma longa* L., is believed to be Indian/Indo-Malayan origin

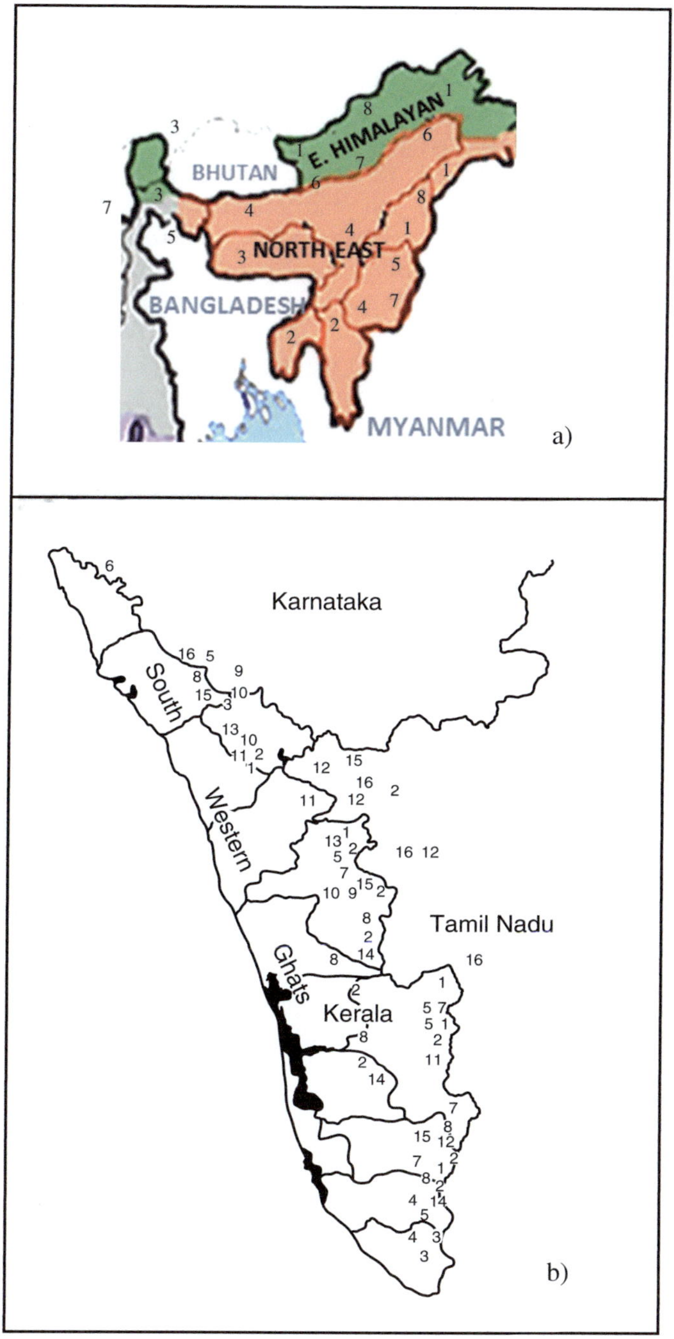

Fig. 11.2 Distribution of wild *Piper* spp. in (**a**) northeast region:
1 = *Piper attenuatum*; 2 = *P. falconeri*; 3 = *P. hamiltonii*; 4 = *P. longum*; 5 = *P. mullesua*; 6 = *P. peepuloides*; 7 = *P. thomsonii*; 8 = *P. wallichii*
(**b**) South Western Ghats of India: 1 = *Piper argyrophyllum*; 2 = *P. attenuatum*; 3 = *P. barberi*; 4 = *P. galeatum*; 5 = *P. hapnium*; 6 = *P. hookeri*; 7 = *P. hymenophyllum*; 8 = *P. longum*; 9 = *P. mullesua*; 10 = *P. nigrum*; 11 = *P. pseudonigrum*; 12 = *P. schmidtii*; 13 = *P. silentvalleyensis*; 14 = *P. sugandhi*; 15 = *P. trichostachyon*; 16 = *P. wightii*

with little information on its ancestors and phylogenetic relationship. The genus *Curcuma* is mainly distributed in the Indo-Malayan region with about 100 species. Baker (1886) described 27 species in the "Flora of British India." He subdivided the genus into three sections, *Exantha*, *Mesantha*, and *Hitcheniopsis*. *Exantha* consists of 14 species including *C. longa* and other economically important species, *C. angustifolia* Roxb., *C. aromatica* Salisb., and *C. zedoaria* Rosc. Sabu (1991) reported 16 *Curcuma* species from South India, out of which nine are endemic (Fig. 10.1). *Curcuma aeruginosa* Roxb., *C. amarissima* Rosc., *C. brog* Val., *C. caesia* Roxb., *C. comosa* Roxb., *C. latifolia* Rosc., *C. soloensis* Val., and *C. sylvatica* Val. are naturally distributed to northeastern region. *Curcuma albiflora* Thw.; *C. aromatica* Salisb.; *C. karnatakensis* Amalraj, Velay., & Mural; *C. kudagensis* Velayudhan et al.; *C. thalakaveriensis* Velayudhan et al.; and *C. malabarica* Velayudhan et al. are found in Karnataka. *Curcuma aurantiaca* Van Zijp, C. *cannanorensis* Ansari et al., *C. coriacea* Mangaly & Sabu, *C. decipiens* Dalzell, *C. ecalcarata* Sivar. & Balach., *C. haritha* Mangaly & M. Sabu, *C. nilamburensis* Velayudhan et al., *C. oligantha* Trimen, *C. raktakanta* Mangaly & Sabu, *C. vamana* Sabu & Mangaly, and *C. vellanikkariensis* Velayudhan et al. are found in Kerala. *C. aromatica* Salisb. and *C. lutea* Ansari & Nair occur both in Karnataka and Kerala. *C. montana* Roxb. is found in Andhra Pradesh, *C. pseudomontana* Graham in Maharashtra and other parts of South India, while *C. amada* Roxb. and *C. longa* are cultivated throughout India (Velayudhan et al. 1999).

Most seed spices cultivated in India are exotic, with limited number of wild relatives confined to remote areas, like Himalayas. Among the tree spices, nutmeg, *Myristica fragrans* Houtt., produces two separate spices, namely, nutmeg (kernel of seed) and mace (aril covering the seed). The species occurring in India are *M. amygdalina* Wall., *M. andamanica* Hook., *M. attenuate* Wall., *M. dactyloides* Gaertn. (syn. *M. laurifolia* Hook.f.), *M. beddomei* King., *M. gibbosa* Hook.f. & Thomson, *M. glabra* Blume, *M. glaucescens* Hook.f & Thom., *M. irya* Gaertn., *M. kingii* Hook., *M. longifolia* Wall. ex Blume, and *M. magnifica* Bedd. Most of these are endemic to the Indo-Malayan region and the forests of Western Ghats. Cinnamon and cassia is obtained from *Cinnamomum verum*, family Lauraceae, which is indigenous to South Western Ghats of India. Cassia cinnamon is obtained from three main sources, the most important being *C. cassia* Bercht. & Presl. (Chinese cassia). Gamble (1925) described 11 species of *Cinnamomum* mainly from the Western Ghats. Kostermans (1983) described 13 species from South India, most of them from the Western Ghats. The species occurring in South India are *Cinnamomum citriodorum* Thw., *C. filipedicellatum* Kosterm. (*C. gracile* Hook.f.), *C. goaense* Kosterm., *C. macrocarpum* Hook., *C. malabathrum* (Burm.f.) Bl., *C. nicolsonianum* Manilal & Shylaja, *C. perrottetii* Meiss., *C. riparium* Gamble, *C. keralaense* Kosterm., *C. travancoricum* Gable, *C. verum* Brecht & Presl, *C. wightii* Meisson, *C. sulphuratum* Nees, *C. heyneanum* Nees, *C. chemungianum* Mohan & Henry, and *C. walaiwarense* Kosterm. All these, except for *Cinnamomum citriodorum*, *C. filipedicellatum*, *C. goaense*, *C. perrottetii*, *C. verum*, *C. sulphuratum*, and *C. walaiwarense*, are endemic to these regions. The species diversity of *Garcinia*, whose rind of the fruits is a common spice, used as substitute of tamarind have been discussed under the tropical fruits. None of *Syzygium* species found in India is related to *S. aromaticum* (clove).

11.2 Wild Species Found in India

1. ***Alpinia galanga*** **Sw. (Zingiberaceae).** Greater Galanga, a perennial tuberous herb with elongate leafy stems and aromatic rootstock. Native to East Asia, including the sub-Himalayan region of India, considered natural home, with greater variability in northeastern region, as undergrowth in evergreen forests. Cultivated, used in local cuisines for flavoring in Southeast Asian countries and medicinally in Ayurveda.
2. ***Alpinia speciosa*** **K. Sch.; syn.** ***A. nutans***. Southeast Asian plant, distributed in sub-Himalayan region, more so in northeastern region of India, as undergrowth in evergreen forests. Cultivated for its shining oblong leaves and arching clusters of white flowers with pink shading and crinkled yellow lips. Rhizomes are a substitute for ginger and for essential oil, used medicinally.
3. ***Amomum aromaticum*** **Roxb. (Zingiberaceae).** Bengal cardamom, a rhizomatous perennial herb with leafy stems, up to 1.2 m high, native to Eastern Himalayas, Sikkim, North Bengal, Khasi Hills, as undergrowth in evergreen forests with significant genetic variability. Used for flavoring and in medicine. The seeds yield essential oil containing cineole as the principal constituent. Cultivated.
4. ***Amomum cannicarpum*** **(Wight) Benth. ex Baker.** An herb, narrowly endemic to evergreen forests of Western Ghats (Thomas et al. 2012b). Essential oil with antimicrobial activity is obtained from the fruits. Genetic resource for cardamom. Reported to be endangered.
5. ***Amomum corynostachyum*** **Wall.** Reported from Assam, Bangladesh to Myanmar in tropical moist forests. Needs more investigations.
6. ***Amomum fulviceps*** **Thwaites.** An herb with red, white flower, recently reidentified from India. Distributed in tropical moist forests of South India and Sri Lanka (Thomas et al. 2009).
7. ***Amomum maximum*** **Roxb.; syn.** ***A. dealbatum*** **Roxb.** A medium-sized, vigorous ginger, with white fragrant flowers found in understory of tropical and warm temperate Asia, including East Himalayas. Newly recorded from the Western Ghats. The seed pods are used like black cardamom, and the unopened flower heads are used in cooking.
8. ***Amomum muricatum*** **Bedd.** Herb with leafy stem and brown capsules, endemic to evergreen forests of South Western Ghats (Anamalai Hills). It is a clustered cardamom, with medicinal properties.
9. ***Amomum pterocarpum*** **Thwaites; syn.** ***A. microstephanum*** **Baker.** Winged-fruit ginger, a rhizomatous herb, growing in hilly regions of moist evergreen forests of South India. Common in Sri Lanka.
10. ***Amomum subulatum*** **Roxb.** Black cardamom, an herbaceous perennial 1.5 m tall plant. Native to East Asia, including the damp under growth of evergreen forests, forested valleys of Eastern Himalayas. The pods are used as spice in cuisines for camphor like odor, which gives a strong smoky flavor, as substitute for true green cardamom and masticatory. The essential oil is extracted from the seed and rind, used in Ayurveda medicine. Cultivated.

11. ***Bunium bulbocastanum* L.; syn. *B. persicum*; syn. *Carum bulbocastanum* W. Koch. (Apiaceae/Umbelliferae)**. Black cumin/*Kala zira*, a perennial herb, native to Northern Africa and Europe. Also, recorded in Northwestern Himalayas with much variability, between 2000 and 3500 m, particularly in alpine stony meadows with grasses. Dried fruits are used as a culinary spice for flavoring in North India and Central Asia. Ayurveda plant with antioxidation and antiglycation properties. Related to cumin (*Cuminum cyminum*). Cultivated in Jammu and Kashmir.
12. ***Carum carvi* L. (Apiaceae/Umbelliferae).** *Shia-jira*, a biennial plant, native of North Africa, Europe, temperate Asia, and Indian Subcontinent, in northern Himalayan plains and hills of Kashmir to Kumaon with variability in alpine meadows up to 3600 m. Cultivated, an established source of essential oils, known for antispasmodic, carminative activities. Member of primary gene pool of caraway.
13. ***Caulokaempferia sikkimensis* (King ex Baker) K. Larsen; syn. *Kaempferia sikkimensis* King ex Baker. (Zingiberaceae)**. Orchid Ginger, a rare beautiful dwarf ginger with dark green paddle-shaped foliage and very attractive red and white flowers. Native to Eastern Himalayas. Cultivated. Needs further studies.
14. ***Cichorium intybus* Linn. (Compositae).** A perennial herb with stunning blue flower head. Native of Africa, Europe, and Asia, including Northwestern Himalayas, Gangetic Plains, and Deccan Plateau of Indian Subcontinent. Cultivated for edible leaves and roots and as ornamental. Used as food additive/flavoring and in medicine. Member of secondary gene pool of chicory, *C. endivia* (Cappelle et al. 2007).
15. ***Cinnamomum cassia* (Nees & T. Nees) J. Presl.; syn. C. *aromaticum* Nees (Lauraceae).** It is an evergreen tree originated in southern China and widely cultivated. Also, found in Western Ghats. Cassia bark is used as a flavoring agent for confectionery, desserts, pastries, and meat; specified in many curry recipes. Cassia oil is a vital ingredient for making traditional Easter biscuits and offers a range of health benefits. Close relative of Ceylon cinnamon (*C. verum*).
16. ***Cinnamomum heyneanum* Nees.** A small evergreen ca.12 m tall tree with ellipsoid berry fruit. Endemic to the Western Ghats, along the banks of streams in semi-evergreen forests. A poorly known species, threat data deficient (WCMC 1998).
17. ***Cinnamomum macrocarpum* Hook.f.** An understory ca.15 m tall tree with smooth, pale brown bark, habitat to evergreen forests. Endemics to South Western Ghats. Rare in South and Central Sahyadris. Antioxidant and antimicrobial.
18. ***Cinnamomum malabatrum* (Burm.f.) J. Presl.** A 15 m tall tree with smooth, brown bark, pustular having aromatic smell. Endemics to South Western Ghats (South and Central Sahyadris), as common understory tree in evergreen and semi-evergreen forests and plains. Possess pharmaceutical property.
19. ***Cinnamomum obtusifolium* Nees; syn. *C. bejolghota* (Buch. -Ham).** Shrub with often aromatic bark and leaves, distributed from China to Eastern Himalayas, Assam, Andaman Nicobar, and Lakshadweep Islands. Extracted

from wild and cultivated. Medicinally important. Related to Indian bay leaf, *tejpat*, *C. tamala*.

20. ***Cinnamomum perrottetii*** **Meissn.** A ca. 8 m tall tree with odorless smooth, brown bark. Endemic as understory tree in evergreen forests of South Western Ghats (Anamalai and Nilgiris). Leaf essential oil characterized. Rare.
21. ***Cinnamomum sulphuratum*** **Nees.** *Matta dalchini* or *Pingadalchini* is a tree ca. 8 m tall with smooth bark. Endemic to evergreen and shola forests of Western Ghats (South and Central Sahyadris) and Northeast India. Leaf and stem bark oil has benzyl benzoate as the major constituent. Highly traded medicinal plants.
22. ***Cinnamomum travancoricum*** **Gamble.** A rare small understory ca. 8 m tall tree. Endemic and widely distributed in the higher elevation of South Western Ghats. Essential oil has been found to have antifungal activities.
23. ***Cinnamomum zeylanicum*** **Blume; syn *C. verum* Presl.** True commercial cinnamon, a moderately sized 16 m tall tree, native to tropical Asia, including India, Sri Lanka, and Myanmar. More in humid tropical forests with much variation in III flavor. Possibly indigenous to Western Ghats. Cultivated for bark used as spice, in perfumery and Ayurvedic medicines for its hypoglycemic, digestive, antispasmodic, and antiseptic properties. Bark oil is considered to have a strong germicidal and fungicidal activity.
24. ***Cinnamomum wightii*** **Meissn.** Up to 8 m tall tree with smooth, gray, pustular bark. Endemic, as understory tree in high elevation to montane evergreen forests of the Western Ghats—Anamalai, Palani, and Nilgiris. Bark yield oil with hepatoprotective activity. Vulnerable.

In addition, *Cinnamomum caudatum* Nees., a tree found in evergreen forests of Eastern Himalayas and Karnataka (Coorg), is reported, whereas *Cinnamomum curvifolium* (Lam.) Nees., an evergreen shrub to tree, native to East Asia, including Eastern Himalayas, has been reported with leafy branchlets containing ca. 35% of an essential oil, with main constituent safrole (from 80–95%). However, needs collection.

25. ***Curcuma aeruginosa*** **Roxb. (Zingiberaceae).** An herb having aromatic rhizomes with pink tips and grayish-blue or blue center. Resembles *C. caesia*. Native to tropical Asia, including Northeast India and Kerala. Medicinally important, yielding essential oil, and possessing antioxidant, antibacterial, and antituberculosis properties. Cultivated.
26. ***Curcuma amada*** **Roxb.** Mango ginger, a rhizomatous aromatic herb, closely related to *C. longa*, except for raw mango flavor. Native to Indian Subcontinent, with greater variability in northeast forest tracts, Assam, and West Bengal. Dried rhizome used as spice, while extract and essential oil used in Ayurveda for cytotoxicity, platelet aggregation inhibitory, and antioxidant activities. Cultivated.

27. ***Curcuma amarissima* Rosc.** A 1 m tall herb with rhizome, yellow inside, blush green outside, large, and bitter. Distributed from China to Northeastern India. Known for medicinal properties.
28. ***Curcuma angustifolia* Roxb.** East Indian arrowroot (*Tikhur*), a fast growing rhizomatous annual Ayurvedic plant. Native of tropical Asia, including sub-Himalayan tract, from Kumaon Hills eastwards to North Bengal and Peninsular India. Presence of curcumin and a phytosterol is reported in the rhizome, while essential oil exhibited antimicrobial, antifungal activities, etc. Cultivated.
29. ***Curcuma aromatica* Salisb.** Wild turmeric, a cosmetic perennial herb. Native of tropical Asia. Occur in degraded forests and plantations of Eastern Himalayas, Chota Nagpur, Koraput, and the Western Ghats. Ornamental, medicinally used for antioxidant, antimicrobial, anti-inflammatory activities. Close to *C. australasica*. Cultivated.
30. ***Curcuma aurantiaca* Van Zijp.** A medium-sized plant with small rhizomes and tubers, cream-colored inside. Native to Thailand and Java and recorded from moist deciduous, evergreen forest, and wet patches of Kerala, Tamil Nadu, India. Cultivated as an ornamental plant.
31. ***Curcuma caesia* Roxb.; syn. *Curcuma malabarica* Velay., Amalraj and Mural.** Black turmeric, *Kala-haldi*, is a perennial herb with bluish-black rhizome. Distributed wild in Eastern Himalayas, Central India, Karnataka, and Kerala. Important owing to medicinal properties of rhizomes as stimulants, antidiarrheal, diuretic, antiemetic, wound cleaner, and skin disorder. Cultivated, endangered.
32. ***Curcuma comosa* Roxb.** Native of much of Asia, including northeast region of India. Used as a food ingredient and medicinally in traditional folk system for treatment of uterine inflammation.
33. ***Curcuma coriacea* Mangaly & M. Sabu.** Recently described, rhizomatous perennial herb with very small nonaromatic rhizome, branched, and yellowish-white inside. Endemic to the grasslands of South Western Ghats. Endangered, listed in the IUCN Red List of Threatened Species (Romand-Monnier 2013).
34. ***Curcuma decipiens* Dalzell.** Herbs, with small rhizome (3–5 × 2 cm), ovoid conical and pale yellow white. Endemic to exposed laterite soil and rocky hill slopes of Peninsular India, including Kerala. Used in Ayurveda for skin diseases, anti-inflammatory, antibacterial, antifungal, antidiabetic activities, etc. Rare.
35. ***Curcuma ecalcarata* Sivar. & Balach.** Herbs with small rhizome (3–5 × 2–2.5 cm), conical in shape, yellow inside. Endemic to moist deciduous and semi-evergreen forests and the plains of South Western Ghats. A new natural source of pinocembrin and piperitenone.
36. ***Curcuma ferruginea* Roxb.** Bronze plume ginger is an attractive species with leaves ferruginous red. It blooms in spring and is distributed from Indochina (East India) region to Gangetic Plains.
37. ***Curcuma haritha* Mangaly & Sabu.** An herb, closely related to *C. aromatica*, differing in plant morphology and yellowish-gray, nonaromatic rhizome. Endemic to South Western Ghats in high-altitude grasslands and coastal region. Being evaluated pharmacognostically.

38. ***Curcuma inodora* Blatt.; syn. *C. purpurea* Blatt.** An attractive, perennial, scentless herb with showiest flowers. Distributed in tropical moist deciduous forests of southern Peninsular India. Cultivated as an ornamental and used in traditional medicine for muscular pain, psychosomatic disorders, and constipation. Essential oil is obtained from fresh rhizomes and leaves.
39. ***Curcuma karnatakensis* Amalraj, Velay., & Mural.** A small herb, endemic to Karnataka. Believed to grow only in Hirehalli, Uttar Kannada. Discovered in 1990.
40. ***Curcuma kudagensis* Velay., V.S. Pillai, & Amalraj.** A new ornamental species record from Karnataka around 1990. Synonym of *C. thalakaveriensis* Velay., Amalraj, & Mural., distributed in South Western Ghats
41. ***Curcuma latifolia* Roxb.** An herb with large rhizome, pendulous tuber conical (somewhat pear-like), pale yellow inside, and toxic in nature. Distributed in Thailand, Northeast India (Mizoram, West Bengal), and Western India. Cultivated in Thailand. Allied to *C. zedoaria* and *C. aromatica.*
42. ***Curcuma longa* Linn.; syn. *C. domestica* Valeton.** Turmeric (*Haldi*) is a rhizomatous, perennial herb native to Indian Subcontinent. Grows wild as undergrowth in forests with significant variability (Chota Nagpur). Widely cultivated in tropics, as one of the most commonly used spices for coloring and flavor; natural dye, and source of essential oil and *curcumin*. Used in traditional medicine, *Sidha*.
43. ***Curcuma leucorhiza* Roxb.** A stemless perennial plant growing ca. 2 m tall. Distributed in East India and Gangetic Plains. Called East Indian arrowroot, because its rhizome is the source of edible starch. Essential oil is extracted and being investigated for pharmaceutical properties.
44. ***Curcuma montana* Roxb.** An herb with short light yellow rhizome. Endemic to grasslands of Peninsular India and Eastern Himalayas and Northeast India.
45. ***Curcuma neilgherrensis* Wight.** A dwarf herb distributed in grasslands of Western Ghats. A folk medicine plant used in the management of diabetes mellitus. Also, known for antioxidant and anticancer properties.
46. ***Curcuma oligantha* Trimen.** A perennial herb with ornamental flowers. Distributed in semi-evergreen forests of South India, Sri Lanka, and Myanmar. Variability is found in flower color, white, classified as var. *oligantha* and yellow as var. *lutea*. Pharmaceutical properties are being investigated.
47. ***Curcuma peethapushpa* Sasidharan & Sivarajan.** A new species (Sasidharan and Sivarajan 1988), needing further investigations. Endemic to Peechi in Kerala.
48. ***Curcuma pseudomontana* J. Graham.** Hill turmeric, a rhizomatous perennial herb with small rootstock/tubers. Endemic to moist deciduous and semi-evergreen forests' hill slopes of Peninsular India. Used in local/tribal medicine and as a source of starch. Reported for anticancer and antiproliferative/cytotoxic activities. Listed vulnerable in the IUCN Red List of Threatened Species (Romand-Monnier and Contu 2013).

49. ***Curcuma rubescens* Roxb.** Ruby Ginger, an herbaceous soft-wooded, unusual perennial ginger with red stem and very attractive tall 20–30 cm waxy inflorescences, deep pink at the top. Distributed in middle Gangetic Plains (Bihar).
50. ***Curcuma sylvatica* Val.** An herb of medicinal value. Distributed in northeastern region, Kerala, and Karnataka. Recorded with highest concentration (320 mg/100 g) of biologically active peptide turmeric (DARE/ICAR 2010).
51. ***Curcuma vamana* M. Sabu & Mangaly.** A new herb species with small rhizome (3 × 0.5 cm), conical in shape, and orange within. Distributed in evergreen, semi-evergreen, and moist deciduous forests of South Western Ghats.
52. ***Curcuma zedoaria* Rosc.** White turmeric (*Kachur*), a perennial herb with aromatic rhizome having thin brown skin, bright orange color, and hard interior. Native to India and Indonesia. Semi-protected in Eastern Himalaya and Northeast. Cooked fresh or used for preparing pickles. The essential oil produced from the dried roots used in perfumery and soap fabrication, as well as an ingredient of tonics. Medicinally used for anti-inflammatory, antivenom, digestion improving, and gastrointestinal stimulant properties. An ecotype described as *C. raktakanta* Mangaly & Sabu. is distributed in coconut grooves on alluvial soil of South Western Ghats, with medium size (3–4 × 2.5–4 cm), conical, grayish to yellow inner, whitish toward the periphery, and aromatic rhizome. It produces antidiabetic and antihyperlipidemic effect by enhancing insulin release.
53. ***Murraya elongata* A. DC. ex Hook.f. (Rutaceae).** A shrub endemic to northeastern region of India and Myanmar. Murrangatin, a new monomeric coumarin, is isolated from its leaves. Related to curry leaf, *M. koenigii*.
54. ***Murraya exotica* L.; syn. *M. paniculata* (L.) Jacq.** *Kamini*, an evergreen shrub or small tree with glossy, aromatic foliage and white bell-shaped flowers with a heady scent. Native to southern China, Taiwan, and India. Leave extract has medicinal properties, used in folklore medicine. Related to *M. koenigii*. Cultivated as ornamental/hedge plant.
55. ***Myristica andamanica* Hook.f. (Myristicaceae).** A handsome tree endemic to evergreen hill forest, rainforest, or mixed forest of Andaman and Nicobar, India. Common adulterant for nutmeg, *M. fragrans*. Tannins, triterpenes, and alkaloids are the chief bioactive components present in its leaves, contributing to early tissue approximation, and increased tensile strength of the wound to heal.
56. ***Myristica beddomei* King.** Wild nutmeg (*Jangali Jaiphal*), evergreen dioecious tree, is found in evergreen and semi-evergreen forests of South India (Nilgiris and Anamalai hills). The fruit pericarp is a rich source of nutritional compounds and hence may be used as a functional food. Needs further investigation.
57. ***Myristica dactyloides* Gaertn.** Bitter nutmeg or mace/nutmeg (*Jatiphala*) is ca. 20 m tall tree. It is endemic to mid and high elevation zones of forests of Sri Lanka and India's Western Ghats. Harvested for its fruits. Vulnerable (Haridasan et al. 2015).
58. ***Myristica malabarica* Lam.** *Bambay-jayphal*, dioecious 25 m tall tree. Endemic to humid tropical forests of the Western Ghats. It is harvested from the wild for

its seeds, which are a source of oil. Its false nutmeg, therefore, an adulterant for nutmeg. Produce good rootstock for grafting the true nutmeg, possess hepato-protective activity. Vulnerable due to habitat loss (CAMP 1998).

59. ***Pimpinella urceolata* Wall. ex Banerjee.** Wild anise, erect to ascending perennial or biennial, 45–100 cm tall herb found Eastern Himalayas (Bhutan, Sikkim, and Darjeeling). Related to aromatic herb, Anise, *P. anisum*.
60. ***Piper argyrophyllum* Miq. (Piperaceae).** A climbing shrub found in evergreen and semi-evergreen forests of Western and Eastern Ghats. Twenty-three compounds, a novel neolignan, nine known neolignans, and 13 known alkaloids, were isolated from a methanol extract of stems (Singh et al. 1996). Rare.
61. ***Piper attenuatum* Buch. -Ham. ex Miq.** A climbing shrub with thin stems and papery, oval leaves, native to low and medium elevation forests of Himalayas in Northeast India and Eastern Ghats. Fruit is known for medicinal properties with piperine, piperlonguminine, and other active constituents and has shown highest percentage of crotepoxide.
62. ***Piper barberi* Gamble.** A slender climbing shrub with fruits known for medicinal properties. Endemic to evergreen forests of South Western Ghats (Pullathupara, Kollam district, Sholayar in Thrissur district, Triveni and Agasthamalai). Critically endangered, nearly extinct (Saha et al. 2015).
63. ***Piper falconeri* C. DC.** Endemic to Tripura and adjacent northeastern hills. An insecticidal amide reported from leaves and stems of this plant (Parmar et al. 1993).
64. ***Piper galeatum* C. DC.** Stout woody climbing shrub, endemic to evergreen and semi-evergreen and forests of South Western Ghats. Source of bold fruits.
65. ***Piper hamiltonii* C. DC.** *Jangali Paan* (*Awani Pan*), a perennial dioecious shrub common to Sikkim-Terai, North of West Bengal and Khasi hills. Recently reported from Western Ghats and Peninsular India (Geethakumary et al. 2014). Carminative and diuretic. Source of resistance to betelvine blackfly (*Aleurocanthus rugosa*).
66. ***Piper hapnium* Buch. -Ham. ex Hook.f.** Climbing shrub endemic to moist deciduous forests of South Western Ghats, including low-lying of hills of Courtallam, Tamil Nadu. Rare.
67. ***Piper hookeri* Miq.** Climbing shrub, endemic to moist deciduous to evergreen forests of North Western Ghats. Closely related to *P. hymenophyllum*.
68. ***Piper hymenophyllum* Miq.** Wild betel, a slender pubescent climber endemic to evergreen and semi-evergreen forests of Peninsular India. Cholinesterase inhibitors obtained from the aerial parts. Essential oil was active against both bacteria and yeast (Ratnam et al. 2015).
69. ***Piper longum* L.** Long pepper (*pippali*) straggling shrub, sometimes ascending and climbing, native to evergreen forests of Indian Subcontinent (northeast and south). The fruit consists of many minuscule fruits embedded in the surface of a flower spike, contain piperine. Cultivated for fruits, which are dried and used as spice. Ayurvedic plant with properties for weight loss, clearing of skin, slowing aging, purifying blood, improving digestion, etc. Close relative of black pepper, *P. nigrum*.

70. ***Piper mullesua* Buch. -Ham. ex D. Don; syn. *P. brachystachyum* Wall. ex Hook. f.** *Pahari peepal*, slender woody climber, described from evergreen and shola forests of Peninsular and Northeast India. Medicinal plant for treatment of migraine with insomnia, colic pain, cough, colds, rheumatism, gout dyspepsia, and bronchitis and recently found to possess insecticidal properties. Endangered due to overexploitation
71. ***Piper nigrum* L.** Black pepper, a flowering vine with single-seeded drupe fruit, native to evergreen forests of Western Ghats, South India; wild forms are mostly dioecious with genetic diversity for fruit color and other traits. Accession INGR 8100 has been registered for novel proliferating spikes. Cultivated for its fruits, which are dried and used as spice since antiquity for its flavor and medicinal value. Its extracts exhibit anti-inflammatory, antioxidant, and anticancer activities.
72. ***Piper peepuloides* Roxb.** Wild pepper is an evergreen climber found in tropical and subtropical evergreen forests of Northeast India (Himalayas). Fruits are used as spice and extract have insecticide and larvicide activity. Produce Ayurvedic medicine, like *P. mullesua.*
73. ***Piper pseudonigrum* Velay. & Amalraj.** It is an epiphytic, robust evergreen climbing shrub, described from the evergreen forests of South Western Ghats (Velayudhan and Amalraj 1992). The seeds are harvested from the wild for local use as a spice. Closely related to *P. nigrum*, as a component of primary gene pool.
74. ***Piper schmidtii* Hook.** *Nilgiri pepper*, a climbing shrub with red or yellow berries, endemic to shola forests of South Western Ghats with greater variability in evergreen forests of Nilgiris Hills. Locally used as spice. Also, found in Assam. The neolignan schmiditin, lignin galgravin and friedelin, beta-sitosterol and its beta-O-glucoside have been isolated from it. The extract exhibited antiamoebic activity.
75. ***Piper silentvalleyensis* P. N. Ravindran, M.K. Nair, & R. Ashokan.** Slender extensively branched climbers, a new and the only species with bisexual flower and minute fruit reported from evergreen forests of South Western Ghats. It could not be collected from its type locality and appear rare. Phylogenetically it is close to *P. mullesua* and distant from *P. longum* (Ravindran and Babu 1994).
76. ***Piper sugandhi* Ravindran, K.N. Babu, & V.G. Naik. and *P. sugandhi* var. *brevipinnis*.** A stout woody climber, described from Sugandhagiri project area of the Western Ghats. Closely related *P. nigrum, P. pseudonigrum*, and *P. trichostachyon* as part of primary gene pool of black pepper (Babu et al. 1993). Allied to *P. galeatum.*
77. ***Piper sylvaticum* Roxb.** Mountain long pepper, herbaceous, dioecious climbing shrub with cylindric spikes. Distributed in subtropical and *terai* forests of Assam, Bengal, and Gangetic Plains. Phytochemical investigation of the roots and stems resulted in the isolation of piperine, piperlonguminine, N-isobutyldeca-trans-2-trans-4-dienamide, and β-sitosterol. Genetically close to Long pepper.
78. ***Piper thomsonii* (C.DC.) Hook. f.** Long pepper, an herbaceous climber with variation in leaf size, distributed in Eastern Himalayas, northeastern region, and

Gangetic Plains. An accession INGR 8009 has been registered for its unique trait of sex change from male to bisexual plant. Genetically close to *P. sylvaticum*.

79. ***Piper trichostachyon* (Miq.) DC.** Stout woody climber with 10 cm long spikes, endemic to evergreen moist deciduous forests of Peninsular India and Western Ghats. Related to *P. galeatum* and *P. nigrum* (Babu et al. 1993). Medicinal, a new alkaloid designated piperstachine (VII) has been isolated from stem.
80. ***Piper wallichii* (Miq.) Hand. -Mazz.** A perennial plant, monoecious, creeping on ground and climbing on trees in China, Himalayas, and Northeast India. Medicinally used for treatment of rheumatoid arthritis and inflammatory ailments and known for antibacterial, antifungal, anti-inflammatory, and antiplatelet aggregation bioactivities. Fifteen known compounds including four triterpenoids (1–4), one sterol (5), one diketopiperazine alkaloid (6), and nine phenolics (7–15) were isolated from its stem (Shi et al. 2015).
81. ***Piper wightii* Miq.** Stout glabrous climber with reddish berries, endemic to evergreen and shola forests of South Western Ghats (Nilgiris Hill top/high elevation). Medicinal, methanolic fruit extract resulted in the isolation of 12 compounds (Boll et al. 1996), all new for this species. Close to *P. hymenophyllum*, referred as synonym.

In addition to the above *Piper* species, there are number species, which have been reported/recorded from either of the two centers of origin/diversity in the Indian Subcontinent (Fig. 11.2a, b). However, they are known either by herbarium, such as *P. laxivenum* C. DC. and *P. tenuirameum* C. DC. (medicinal value), from the Eastern Himalayas, or single collection, such as *P. cornilimbum* C.DC., the wild pepper, from Eastern Himalayas, or recorded as endemic and rare from a specific region such as *P. makruense* C. DC., endemic to Manipur and adjacent hills; *P. meeboldii* C. DC., endemic to Tripura and adjacent North Eastern hills; *P. muneyporensis* C.DC. and *P. nagense* C.DC., endemic to Manipur and adjacent Northeast Hills; *P. mungpooanum* C.DC., endemic to Eastern Himalayas; *P. oldhamii* C.DC. and *P. ovatistigmum* C.DC., endemic to Eastern Himalayas and Khasia and Jaintia Hills; *P. phalagense* C.DC., endemic to Sikkim Himalayas and Manipur hills; *P. pykarahense* C.DC., a climber described from evergreen forests of the Western Ghats and Tamil Nadu and listed in 1997 IUCN Red List of Threatened Plants (Walter and Gillett 1998); and *P. sikkimense* C.DC. and *P. sonadense* C.DC., endemic to Eastern Himalayas (Sikkim) (Deb 1958; Nayar 1996; Jamir and Pandey 2003). All these require further exploration for recollection, characterization and evaluation for assessment of their distinctiveness and potential genetic value. Biosystematic studies are required for tracing the phylogenetic relationships to facilitate conservation and use.

82. ***Trachyspermum khasianum* H. Wolff.; syn. *Carum khasianum* C.B.Cl. (Apiaceae).** Occur wild in Assam, Khasia-Jaintia Hills, and the Peninsular India. Medicinal. Related to Persian cumin, *Carum carvi*.

83. ***Vanilla andamanica* Rolfe (Orchidaceae).** This is climbing orchid with creamy white fragrant flowers identified by RA Rolfe in 1918. It is endemic to Andaman and Nicobar Islands. It is the wild relative of commercial *Vanilla planifolia*, used as the source of "vanilla" essence for flavoring ice creams, cakes, etc. Genetically distant and reported vulnerable (IUCN 1996).
84. ***Vanilla walkeriae* Wight.** Vanilla orchid is a scandent, non-pseudo-bulbous, leafless epiphyte, with fragrant pods, climbing by aerial roots. Native to evergreen forests South India and Sri Lanka. Also, medicinal to treat fever in cattle. Rare, yet to be collected for evaluation and use.
85. ***Vanilla wightiana* Lindl. ex Hook.f.** Leafless climbing epiphyte, rooting from nodes. Endemic to moist deciduous and semi-evergreen forests of the South Western Ghats and Sri Lanka. Medicinal Plants. Reported endangered from Andhra Pradesh (Rao et al. 2000).
86. ***Zingiber anamalayanum (Zingiberaceae).*** A new species described from the South Western Ghats (Sujanapal and Sasidharan 2010). It needs further investigations.
87. ***Zingiber capitatum* Roxb.** Wild ginger, it is an erect perennial herb. Rhizome emits long fleshy tuber-bearing roots. Found in tropical Himalayas, from Kumaon to Sikkim and Khasi hills, Northeast India, and Chota Nagpur Plateau as forest undergrowth. Traditional medicinal plant, antioxidant, and antimicrobial. A variant has been described as var. *elatum* or Z. *elatum* Roxb.
88. ***Zingiber cassumunar* Roxb.; syn. *Z. montanum* (J. König) Link ex A. Dietr.** Native to Thailand, Indonesia, and sub-Himalayan tract to southern peninsular region of India as forest undergrowth. Widely used for joint and muscular pain by Thai massage therapists. Contains the essential oils. Exhibit strong fungitoxic action against *Rhizoctonia solani*, the damping-off pathogen. Related to galangal.
89. ***Zingiber cernuum* Dalz.** Curved-stem ginger, a perennial herb, endemic to moist shady places of semi-evergreen forests of Western Ghats.
90. ***Zingiber chrysanthum* Roscoe.** Golden flowered ginger, naturally occur in tropical Himalayan forests, Sikkim, and Northeast India. Rhizomes are used medicinally. Cultivated in Himalayan Gardens.
91. ***Zingiber clarkei* King ex Baker.** It is an unusual ginger species with 2 m tall leafy stem habitat to shady situations in the subtropical forests of Bhutan and Northeastern India from 600 to 1500 m. Needs further investigations.
92. ***Zingiber ligulatum* Roxb.** A rhizomatous, perennial herb with sessile leaves, reported to occur in tropical region of East India and Myanmar. Used as traditional medicine. Very close to *Z. roseum*.
93. ***Zingiber neesanum* (Grah.) Ramamoorthy; syn. *Zingiber macrostachyum* Dalz.** Perennial rhizomatous herbs with shoots covered by sheathing leaf bases, distributed in shady habitats as forest undergrowth in South India (Western Ghats, Tamil Nadu) (Pushpakaran and Gopalan 2014) and Myanmar. Essential oil isolated from the rhizomes. Rare.

94. ***Zingiber nimmonii* (J. Graham) Dalzell.** An herb with small, purplish lilac inside, strongly aromatic rhizome. Endemic to moist deciduous forests and plains of Western Ghats. Medicinal, volatile oil isolated from the rhizomes is caryophyllene-rich.
95. ***Zingiber officinale* Rosc.** Ginger, a tropical rhizomatous herbaceous perennial plant. Probably originated in tropical Asia (India), probably wild in tropical and subtropical evergreen forests of southern Peninsular India, as undergrowth with significant variability. Rhizome is used as spice (flavoring), yields essential oils, and has many medicinal uses. Widely cultivated in tropics.
96. ***Zingiber purpureum* Roscoe.; syn. *S. montanum* (J. Koenig) Link ex A. Dietr.** Bengal ginger, an herb with branched, tuberous, aromatic rhizomes. Native of tropical Asia, from Northeast India to Indonesia. Possess medicinal and insect repellent activities. Essential oil extracted. Member of secondary gene pool of ginger (Jiang et al. 2006) and *zerumbet* ginger (Theilade 1999).
97. ***Zingiber roseum* (Roxb.) Roscoe.** An herb with thick, fleshy, white to pale yellow within, rhizome. Found in moist deciduous forests of India and Sri Lanka. Used in arthritis. Analysis of the rhizome essential oil revealed the presence of over 44 compounds of which 36 comprising 94.9% of the oil (Prakash et al. 2011)
98. ***Zingiber rubens* Roxb.** Bitter ginger, a small ornamental shrub ca. 2 m tall, native to the undergrowth of deciduous forests in western Assam, Khasia Hills, and the eastern Himalayas. Essential oil is extracted from roots. Source of cold hardiness.
99. ***Zingiber spectabile* Griff.** Beehive ginger, an herbaceous, perennial plant producing clumps of erect, unbranched, ornamental leafy stems with flowers looking like a beehive. Native to tropical Asia, including Northeast India. Rare. Member of secondary gene pool of ginger (Bua-in and Paisooksantivatana 2010).
100. ***Zingiber squarrosum* Roxb.** Native to Indian Subcontinent from Arunachal Pradesh to Myanmar to Andaman and Nicobar Islands.
101. ***Zingiber wightianum* Thwaites.; syn. *Z. squarrosum* Wight.** *Malayinchi*, an herb with thick, fleshy rhizome, distributed in evergreen forests of south Peninsular India and Sri Lanka. Medicinal.
102. ***Zingiber zerumbet* (L.) Sm.** Bitter ginger, a perennial vigorous herb, probably originated in India and spread eastward through Polynesia. Wild in sub-Himalayan tract to Gangetic Plains to Peninsular India. Well known for food flavoring, ornamental and rhizome extracts are employed in Ayurveda to cure diseases. Cultivated. Sources of disease resistance in ginger (Kavitha and Thomas 2008).

Like *Piper* species in *Zingiber* too, there are few species documented in the literature based on herbarium specimen, such as *Z. cylindricum* Thwaites., from South India, or reported for their medicinal properties, such as *Z. intermedium*, or by Botanical Survey of India, such as *Z. marginatum* Roxb., from remote areas of

Eastern Himalayas, like Sikkim and hills of Manipur, and Meghalaya, without details on their availability in situ or in collections, whereas species such as *Zingiber spectabile* Griff. is recorded in literature native to tropical Asia, but not exactly in India. These taxa require further exploration for collection, characterization, biosystematic evaluation for tracing the phylogenetic relationships and taxonomic status, and evaluation for assessment of their distinctiveness and genetic potential to facilitate conservation and use.

11.3 Perspective

In case of important spices, such as *Piper*, *Zingiber*, etc., a large number species have been described by arthodox taxonomists, raising the numbers without much representative basic information on essential features and potential genetic diversity. They are first needed to be evaluated bisystematically for their specific status, which will help in ascertaining their taxonomic entity and tracing the phylogenetic relationship between the wild relatives and with the cultivated species to visualize and select appropriate breeding strategies for introgression of desirable genes from these species into cultivated species. Their detailed characterization and biosystematic evaluation would help facilitate their scientific classification based on genetic divergence and cross-compatibility and promote targeted conservation and the use in crop improvement. Similar exercise is required to deal with the wild relatives belonging to cinnamon, *Cinnamomum*, and Turmeric *Curcuma*.

The principal taxa of most wild relatives and their diverse populations found in different ecologies need their evaluation for desirable traits, particularly, against the yield-reducing factors of cultivated species, such as biotic and abiotic stresses and principal biochemical components to assess genetic diversity and their potential value as genetic resource, and facilitate their rightful use in breeding program.

Most of the spices and condiments are also important medicinal plants and so are their wild relatives, commonly used as folk medicine. Thus, documentation of the associated traditional knowledge held by the tribal communities and rural physicians practicing in the indigenous systems of medicine is very important, needing consideration.

Recognizing the outcrossing breeding nature and vegetative mode of propagation of majority spices, there is a need for in situ conservation of wild and weedy populations of both cultivated species and their wild relatives in their natural habitat, particularly those of *Zingiber* and *Piper* species, which is the need of the hour for conservation of available genetic diversity and to allow this population to respond to changing climate. For example, to facilitate in situ conservation in the Western Ghats, there is an urgent need to establish the gene sanctuary in the hotspots of *Curcuma*, *Cinnamomum*, *Piper*, and *Zingiber* species diversity, such as Silent Valley. On similar grounds, in vitro conservation can also be visualized for virus indexing in natural populations and maintenance of virus-free genetic resource.

Characterization of wild genetic resources using molecular techniques, phytochemical and pharmaceutical evaluation, analysis, and identification of principal chemical component/molecule along with pharmacological properties is another area needing attention.

References

Babu KN, Naik VG, Ravindran PN (1993) Two new taxa of *Piper* (Piperaceae) from Kerala India with a note on their origin and inter-relationship. J Spices Aroma Crops 2:26–33

Baker JG (1886) Scitamineae In: Hooker JD (ed) The Flora of British India 6:198-264

Boll PM, Prasad AK, Tyagi OD, Wengel J, Olsen CE, Kumar N, Bisht KS, Parmar VS (1996) Neolignans, cyclohexanes and alkaloids from *Piper wightii*. Recueil des Travaux Chimiques des Pays-Bas 115(1):9–12

Bua-in S, Paisooksantivatana Y (2010) Study of clonally propagated cassumunar ginger (*Zingiber montanum* (Koenig) Link ex Dietr.) and its relation of wild *Zingiber* species from Thailand as revealed by RAPD markers. Genet Resour Crop Evol 57:405–414

CAMP Workshops on Medicinal Plants, India (January 1997) (1998) *Myristica malabarica*. The IUCN Red List of Threatened Species 1998: e.T31219A9615327

Cappelle C, Morchen M, Hilbert JL, Rambaud C (2007) Regeneration and molecular characterization of a male sterile interspecific somatic hybrid between *Cichorium intybus* and *C. endivia*. Plant Sci 172:596–603

DARE/ICAR Annual Report 2009–2010 (2010) Genetic Resources. ICAR, New Delhi, p 20

Deb DB (1958) Endemism and outside influence on the flora of Manipur. J Bombay Nat Hist Soc 55:313–317

Gamble JS (1925) Flora of the presidency of madras. Reprint, Botanical Survey of India, Calcutta, pp 1916–1935

Geethakumary MP, Sivadas D, Pandurangan AG (2014) *Piper Hamiltonii* C. DC. (Piperaceae) –a new distributional record for Peninsular India. Int J Plant Animal Environ Sci 4(4):18–20

Haridasan K, Ravikumar K, Saha D, Ved D (2015) *Myristica dactyloides*. The IUCN red list of threatened species 2015: e.T33526A50131225

IUCN (1996) IUCN/SSC Orchid Specialist Group. Publisher IUCN, ISBN 2-8317-0325-5, p 92

Jamir SA, Pandey HN (2003) Vascular plant diversity in the sacred groves of Jaintia Hills in northeast India. Biodivers Conserv 12:1497–1510

Jiang H, Xie Z, Koo HJ, McLaughlin SP, Timmermann BN, Gang DR (2006) Metabolic profiling and phylogenetic analysis of medicinal *Zingiber* species: tools for authentication of ginger (*Zingiber officinale* Rosc.) Phytochemistry 67:1673-1685

Kavitha PG, Thomas G (2008) Population genetic structure of the clonal plant *Zingiber zerumbet* (L.) Smith (Zingiberaceae), a wild relative of cultivated ginger, and its response to *Pithyum aphanidermatum*. Euphytica 160:89–100

Kostermans AJGH (1983) The South Indian species of *Cinnamomum*. Schaeffer (Lauraceae). Bull Botanical Surv India (BBSI) 25:90–133

Nayar MP (1996) Hot spots of endemic plants of India, Nepal and Bhutan. Tropical Botanic Garden and Research Institute, Thiruvananthapuram

Parmar VS, Sinha R, Shakil NA, Tyagi OD, Boll PM (1993) An insecticidal amide from *Piper falconeri*. Indian J Chem Sect B 32(3):392–393

Prakash O, Pant AK, Mathela CS (2011) Linalool-rich essential oil from the rhizomes of *Zingiber roseum* Rosc. J Essent Oil Res 18(6):638–639

Pushpakaran B, Gopalan R (2014) *Zingiber neesanum* (J. Graham) Ramamoorthy in CJ Saldanha and DH Nicolson (Zingiberaceae) – a new record for Tamil Nadu India. ZOO's PRINT 29(3):23–24

Rao YS, Mary K, Mathew KJ, Madhusoodanan MR, Sudharshan VKK, Potty SN (2000) Natural fruit set in *Vanilla wightiana* Lindl., an endangered species from Andhra Pradesh, India. J Spices Aromatic Crops 9:77

Ratnam KV, Md. Bhakshu L, Venkata Raju RR (2015) Phytochemical composition and *in vitro* antimicrobial activity of essential oil of *Piper hymenophyllum* Miq.: a rare wild betel. Int J Pharmacognosy Phytochem Res 7(1):68–71

Ravindran PN, Babu KN (1994) Chemotaxonomy of South Indian *Pieper*. J Spices Aromatic Crops 3(1):6–13

Ravindran PN, Nirmal Babu K, Sasikumar B, Krishnamoorthy KS (2000) Botany and crop improvement of black pepper. In: Ravindran PN (ed) Black pepper, *Piper nigrum*. Harwood Academic Publishers, Amsterdam, pp 23–142

Romand-Monnier F (2013) *Curcuma coriacea*. The IUCN red list of threatened species 2013: e.T44393421A44507719. http://dx.doi.org/10.2305/IUCN.UK.2013-1.RLTS.T44393421A44507719.en

Romand-Monnier F, Contu S (2013) *Curcuma pseudomontana*. The IUCN red list of threatened species 2013: e.T22486190A44506743. http://dx.doi.org/10.2305/IUCN.UK.2013-1.RLTS.T22486190A44506743.en

Sabu M (1991) A taxonomic and phylogenetic study of South Indian Zingiberaceae. Ph D Thesis. University of Calicut, Kerala

Saha D, Ved D, Ravikumar K, Haridasan K (2015) *Piper barberi*. The IUCN red list of threatened species 2015: e.T50126635A50131415

Sasidharan N, Sivarajan VV (1988) *Curcuma peethapushpa*, a new species of Zingiberaceae from India. Notes from the Royal Botanic Garden, Edinburgh 45(3):425–427

Shi Y, Yang L, Zhao J, Shi Y, Qu Y, Zhu H, Wang D, Yang C, Li X, Xu M, Zhang Y (2015) Chemical constituents from *Piper wallichii*. Nat Prod Res 29(14):1372–1375

Singh SK, Prasad AK, Olsen CE, Jha A, Jain SC, Parmar VS, Wengel J (1996) Neolignans and alkaloids from *Piper argyrophyllum*. Phytochemistry 43:1355–1360

Sujanapal P, Sasidharan N (2010) *Zingiber anamalayanum* sp. Nov. (Zingiberaceae) from India. Nord J Bot 28:289

Theilade I (1999) A synopsis of the genus *Zingiber* (Zingiberaceae) in Thailand. Nord J Bot 19:389–410

Thomas VP, Sanoj E, Sabu M, Prasanth AV (2009) On the identity and occurrence of *Amomum fulviceps* (Zingiberaceae) in India. Rheedea 19:13–17

Thomas VP, Dan M, Sabu M, Jabbar MA (2010) *Amomum andamanicum* (Zingiberaceae): a new species from the Andaman Islands, India. Blumea 55:295–299. doi:10.3767/000651910X550954

Thomas VP, Sabu M, Prabhu Kumar KM (2012a) *Amomum nilgiricum* (Zingiberaceae) a new species from Western Ghats, India. PhytoKeys 8:99–104. http://dx.doi.org/10.3897/phytokeys.8.2152

Thomas VP, Sabu M, Salim PM (2012b) On the identity and typification of *Amomum cannicarpum* (Zingiberaceae): a narrowly endemic and endangered plant from Western Ghats. South India Kew Bull 67:109–113

Velayudhan KC, Amalraj VA (1992) *Piper pseudonigrum*—a new species from Western Ghats. J Econ Taxon Bot 16(1):247

Velayudhan KC, Muralidharan VK, Amalraj VA, Gautam PL, Mandal S, Dinesh K (1999) *Curcuma* genetic resources. National Bureau of Plant Genetic Resources, New Delhi

Walter KS, Gillett HJ (eds) (1998) 1997 IUCN Red List of Threatened Plants. Compiled by the World Conservation Monitoring Centre. IUCN – The World Conservation Union, Gland, Switzerland and Cambridge, UK. lxiv + p 862

WCMC (World Conservation Monitoring Centre) (1998) *Cinnamomum heyneanum*. The IUCN red list of threatened species 1998: e.T38781A10144462. http://dx.doi.org/10.2305/IUCN.UK.1998.RLTS.T38781A10144462.en

12 Commercial Crops

12.1 Introduction

The major commercial crops grown/indigenous to India are cocoa (*Theobroma cacao* L.), coconut (*Cocos nucifera* L.), coffee (*Coffea arabica* L.), rubber [*Hevea brasiliensis* (Willd. ex A. Juss.) Müll. Arg.], sugarcane (*Saccharum officinarum* L.), tea (*Camellia sinensis* (L.) Kuntze), and tobacco (*Nicotiana tabacum* L.). *Cocos* is a monotypic genus, while *Theobroma cacao* and *Hevea brasiliensis* are of South American origin. The Northeast India is the major area of variability of cultivated tea (Wheatherstone 1992). *Camellia assamica* (Masters) Wight (Assam type) and *C. assamica* ssp. *lasiocalyx* (Planch ex. Watt.) Wight, which have gone through outcrossing with a non-tea-producing species *C. irrawadiensis* Barua, have generated significant amount of natural variability in primary gene pool. Besides, several other taxa, *Camellia caudata* Wallich, *C. irrawadiensis, C. taliensis* Melchior, *C. kissi* Wallich, and *C. drupifera* Lour, and other related genera, such as *Eurya*, *Pyrenaria*, *Schima*, and *Gordonia*, are also found in the forests of Northeast India (Singh 1999). *Eurya japonica,* sometimes placed in Theaceae and used as substitute of tea in China, is also available in these areas and the Western Ghats. In coffee, wild-related species, *Coffea bengalensis* and *C. fragrans,* are found in the northeast region and the Southern India, and *C. jenkinsii* and *C. khasiana* are endemic to the Khasi and Jaintia hills in the northeast region. *Coffea travancorensis,* related to *C. fragrans* and *C. crassifolia,* is localized in the Western Ghats.

The wild relatives of sugarcane that are of immediate interest from genetic resource point of view are those related to the taxa of the "*Saccharum* complex" consisting of three cultivated species *S. barberi*, *S. sinense*, and *S. officinarum*, which are genetically very close to each other (Mukherjee 1957; Daniels and Roach 1987). This complex will include *Saccharum* spp., *Erianthus*, *Sclerostachya*, *Narenga*, and *Miscanthus*. All the genera and species included in this complex are found in the Indian center of origin (Fig. 12.1), offering large species diversity for *Saccharum* germplasm. India is the center of origin *S. spontaneum* with maximum diversity found in the sub-Himalayan region, from Dehradun to Sundarbans, including Bengal,

A.K. Singh, *Wild Relatives of Cultivated Plants in India*,
DOI 10.1007/978-981-10-5116-6_12

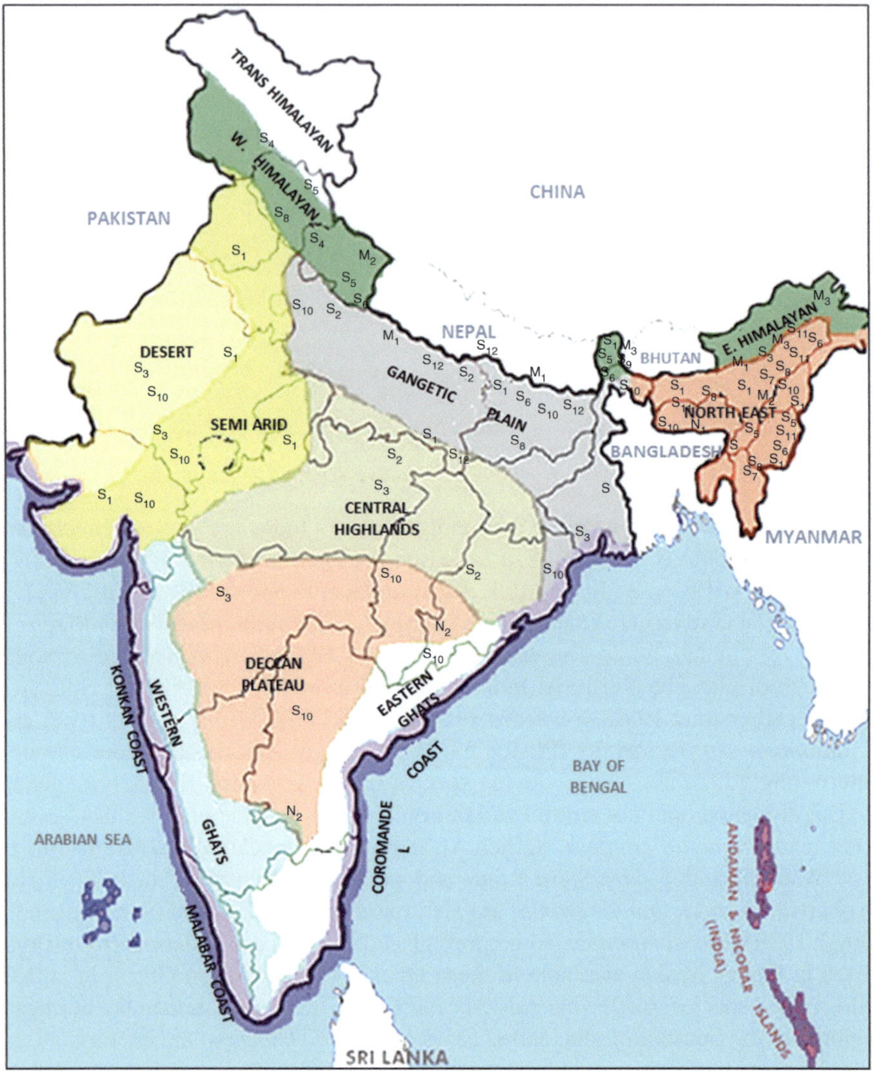

Fig. 12.1 Biogeographical regions associated with the distribution of wild *Saccharum* species and related genera, *Miscanthus* and *Narenga*
M1 = *Miscanthus fuscus*; M2 = *M. nepalensis*; M3 = *M. taylorii*; N1 = *Narenga fallax*; N2 = *N. porphyrocoma*; S = *Sclerostachya fusca*; S1 = *Saccharum arundinaceum*; S2 = *S. barberi*; S3 = *S. bengalense*; S4 = *S. filifolium*; S5 = *S. longisetosum*; S6 = *S. procerum*; S7 = *S. ravennae*; S8 = *S. rufipilum*; S9 = *S. sikkimense*; S10 = *S. spontaneum*; S11 = *S. wardii*; S12 = *S. williamsii*

Bihar, Orrisa, and northeastern region bordering with China and Myanmar, along with *S. arundinaceum*. *Saccharum bengalense*, and *S. ravennae*, are more confined to northern plains (Fig. 12.1). Among others, *S. rufipilum* occurs in hills of northern and northeastern region, while *S. filifolium* is confined to the Western Himalayas

(1500–2500 m). Species with concentration in northeastern region include *S. longisetosum* in Eastern Himalayas and *S. procerum* in Manipur and westward to Sikkim, while *S. sikkimense* confined to Sikkim Himalayas. *Saccharum williamsii* from Nepal could possibly be available in the border areas in Sikkim, North Bengal, and adjoining hills (Fig. 12.1). Some of the above species are treated under *Erianthus* as *E. filifolius* (=*S. filifolium*), *E. ravennae* (=S. *ravennae*), and *E. longisetosus* (=S. *longisetosum*). *Erianthus* is discontinuously distributed throughout India with maximum diversity in the *S. spontaneum* regions. *Ripidium* is also treated under *Erianthus* with *E. ravennae* = *Ripidium ravennae*. *Sclerostachya* and *Narenga* occur in the *terai* region of Northeast India. In genus *Narenga*, *N. porphyrocoma* is distributed mainly in the peninsular tract and *N. fallax* in Khasi and Naga hills of the northeast region (Fig. 12.1). *Miscanthus* species are confined to northeast region with *M. nepalensis* distributed from Nepal to northeast region, *M. taylorii* Bor.; syn. *M. nudipes* (Griseb.) Hack., is confined to Sikkim, while *M. wardii* to Assam valley and Lohit in Arunachal Pradesh (Fig. 12.1). These genera and their species grow together in many places, with overlapping flowering time, providing opportunity for gene flow and produce hybrid swarms. Tobacco, which is of South American origin, is represented by a wild species, *Nicotiana plumbaginifolia*. Dates (*Phoenix* L.) can also be considered as a commercial crop, represented by eight wild species distributed in Northeast Himalayan region and coastal and island ecologies. *Phoenix dactylifera* L. (commercial), *P. sylvestris*, and *P. canariensis* Chabaud are cultivated. *Phoenix sylvestris* occur throughout India in drier tracts, while others reported to occur in different parts are *P. acaulis* Roxb. ex Buch-Ham, *P. humilis* Royle, *P. robusta* (Bac.) Becc., *P. reclinata* Jacq., *P. zeylanica*, *P. paludosa* Roxb., and P. *rupicola* T. Anders (Vashishtha et al. 2005).

12.2 Wild Species Found in India

1. ***Areca triandra* Roxb. ex Buch-Ham.; syn. *A. nagensis* Griff. (Arecaceae).** A 1.5–4 m tall palm, native to tropical Asia including Eastern Himalayas, Nagaland, and island eco-region (Nicobar). Rare, found in forest shades along streams in Naga and Mikir hills of Assam and Nicobar. It is of high ornamental value and worthy of domestication. Used as masticatory and stimulant. Relative of areca nut, *A. catechu*.
2. ***Camellia caudata* Wall. (Theaceae).** Evergreen shrub or tree with elliptic leaf blade, native to East Asia, including Nepal to Eastern Himalayas, to northeast region of India.
3. ***Camellia irrawadiensis* (Hu) P.K. Barua.** A tall evergreen shrub, distributed in China, Myanmar, and Eastern Himalayas. Cross-compatible with cultivated *C. sinensis* L. and *C. assamica* to facilitate genetic introgression. Part of secondary gene pool of tea. Some cultivated populations of China hybrids were supposed to have this species or its close species as possible progenitors, while the other being *C. sinensis*.

4. ***Camellia kissi* Wall.** An evergreen shrub to small tree, native to East Asia including Sikkim, Assam, and Northeast Hill states, extending to Deccan Peninsula and Andaman Islands. Leaves are used as substitute for tea, and seed oil is used medicinally.
5. ***Camellia oleifera* Abel; syn. *C. drupifera* Lour.** Oil-seed camellia is an evergreen shrub growing up to 4 m. Native of China, extending to Myanmar and Northeast Hill region of India. It is an important source of edible seed oil. Close to *C. sinensis*, often confused, may be part of primary gene pool.
6. ***Camellia siangensis* T.K. Paul & M.P. Nayar.** Tree growing up to 5 m, endemic to Siang, Arunachal Pradesh. Kundu (2005) states that this species is found in the Western Ghats. Confused with *C. sinensis* may be part of primary gene pool. Rare, listed in The IUCN Red List of Threatened Species (Rivers and Wheeler 2015).
7. ***Camellia taliensis* (W.W. Sm.) Melch.** Native of China, extending to tropical Indochina, India, and Myanmar region. It is a hybrid between *Camellia irrawadiensis* Barua and *C. sinensis*. Recently domesticated in China and released as white tea. It is one of the most important endangered wild relatives of tea and harbors valuable gene resources for improvement (Liu et al. 2012).
8. ***Coffea bengalensis* Roxb. ex Schult. (Rubiaceae).** Bengal coffee, branched shrub with small beans, inferior to normal coffee. Native to tropical Himalayas and Northeast India, ascending up to 1000 m with varietal differentiation. Cultivated in Central and South India.
9. ***Coffea crassifolia* Gamble; syn. *Nostolachma crassifolia* (Gamble) Deb & Lahiri.** Reported from Western Ghats in the Anamalai and Travancore hills at an altitude of 1300 m. Previously thought extinct, but listed indeterminate in 1997 IUCN Red List of Threatened Plant (Walter and Gillett 1998).
10. ***Coffea fragrans* Wall.; syn. *Psilanthus fragrans* Wall. ex Hook.f.** A shrub distributed in northeastern region and in the Western Ghats. Morphologically like C. *bengalensis* in flower, and to *C. travancorica* and *C. wightiana* in fruit traits.
11. ***Coffea jenkinsii* Hook.f.; syn. *Nostolachma jenkinsii* (Hook.f.) Deb & Lahiri.** Confined to northeastern region, Khasi hills. Fruit ellipsoid.
12. ***Coffea khasiana* Hook. f.; syn. *Nostolachma khasiana* (Korth.) Deb & Lahiri.** A tree with 2.5 cm diameter, smooth, ventrally concave fruit. Found in Eastern Himalayas, Northeast India, particularly Khasi and Jaintia hills. An Indian relative of the coffee. *Coffea khasiana* and *C. jenkinsii* are placed under the genus *Nostolachma*.
13. ***Coffea travancorensis* Wight. & Arn.; syn. *Psilanthus travancorensis* (Wight & Arn.) J.-F. Leroy.** Bushy undershrub, native to evergreen forests of Western Ghats (Travancore, Nilgiri) and Sri Lanka; phylogenetically close to *C. fragrans.*
14. ***Coffea wightiana* Wight & Arn.; syn. *Psilanthus wightianus* (Wight & Arn.) J.-F. Leroy.** A deciduous shrub with much broader than long deep furrow fruit. Native to Western Ghats, Konkan Plains, and arid places from Coorg to Travancore.

15. ***Miscanthus fuscus*** **(Roxb.) Benth.; syn.** ***Saccharum fuscus*** **Roxb. (Poaceae).** A grass with erect culms, leaf sheaths longer than adjacent culm internode, glabrous on surface, outer margin hairy. Distributed in Upper Gangetic Plains and Northeast India and Indochina region. Phylogenetically close to *Saccharum*.
16. ***Miscanthus nepalensis*** **(Trin.) Hack.** Tufted perennial with short rhizomes; culms ca. 1.5 m high with fairy flower plumes. Native to East Asia, including all along Himalayas and Northeast Hill region of India. Potential ornamental plant.
17. ***Miscanthus taylorii*** **Bor.; syn.** ***M. nudipes*** **(Griseb.) Hack.** Pink fairy grass, a plant tufted, erect culms, 25–120 cm tall, distributed on mountain slopes in south China, Sikkim, Assam, and Nepal (Fig. 12.1). The presence of the *Miscanthus* spp. complex and their phylogenetic closeness to *Saccharum* spp. has been used to support North India being center of origin of sugarcane. Another species *M. wardii* Bor., endemic to northeast region and Assam, India, is very close to *M. taylorii* (*M. nudipes*), even considered synonym. May be part of the same species complex.
18. ***Narenga fallax*** **(Balansa) Bor. syn.** ***Saccharum fallax*** **Balansa.** Giant herb, possess characteristic bearded nodes, and a deep yellow-colored to deep copper-colored inflorescence. Native to Northeast India, particularly in cleared forest sites of Khasi and Naga hills.
19. ***Narenga porphyrocoma*** **(Hance) Bor.; syn.** ***Saccharum narenga*** **(Nees ex Steud.) Hack.** Perennial, culm solid, 1 cm in diameter with tall reed, up to 2 m high having solid internodes and bearded nodes. Found in *terai* and sub-Himalayan grasslands of Northeast and Peninsular India. Can be distinguished from *N. fallax* by shorter spikelet length. Source of temperature tolerance and resistance to red rot disease. Belong to secondary gene pool of sugarcane (Ram et al. 2001).
20. ***Nicotiana plumbaginifolia*** **Viv. (Solanaceae).** An annual weedy herb, 0.3-1 m tall, stems slender with leaves bearing similarity to those of the genus *Plumbago*. Native of Americas, naturalized to plains in dry localities of Maharashtra (Pune, Satara) and Tamil Nadu. Gene resource as member of tertiary gene pool of tobacco (Goodspeed 1954; Lewis and Nicholson 2007)
21. ***Phoenix acaulis Roxb.*** Dwarf date palm, a plant with trunk underground or, at most, grow to a few inches in height. Native to scrubland and savannas and in pine forests of northern India, Bhutan, and Nepal in altitudes between 350 and 1500 m. Member of tertiary gene pool of date (Pintaud et al. 2013).
22. ***Phoenix loureiroi*** **Kunth; syn.** ***P. humilis*** **Royle ex Becc. (Arecaceae).** A stout shrub, up to 2 m with cluttered leaf with an oblong drupe, green, orange to black fruit. Native to East Asia, including exposed slopes of scrub jungles and deciduous forests of Indian Subcontinent. Two varieties are found, var. *pedunculata* (Griff.) Govaerts (Kashmir) and var. *loureiroi*. Member of secondary gene pool of date (Pintaud et al. 2013).
23. ***Phoenix paludosa*** **Roxb.** Mangrove date palm, an unbranched, perennial tree that grows in clusters, ca. 5 m high, usually forming dense thickets. Native to coastal regions of Indian Subcontinent and the Far East. Cultivated ornamental.

Member of secondary gene pool of date (Pintaud et al. 2013). Source for poor soil tolerance.

24. ***Phoenix pusilla* Gaertn.** A 5 m tall species with usually single stem, but clumps do occur naturally. Endemic to coastal lowlands, ridges, and hills of Southern India and Sri Lanka. Cultivated as ornamental in landscape. Ayurvedic medicinal plant. Member of secondary gene pool of date (Sudhersan et al. 2008).
25. ***Phoenix rupicola* T. And.** Indian date, an 8 m tall and 20 cm wide tree, leaves are pinnately arranged and fruit is an oblong, yellow to orange drupe. Native to the mountainous forests of Eastern Himalayas, Northeast India. Naturalized in the Andaman Islands. Member of primary gene pool of date (Pintaud et al. 2013). Under threat, because of increasing settlement (Johnson 1998). Ornamental cultivated.
26. ***Saccharum arundinaceum* Retz. (Poaceae).** Hardy sugarcane, a perennial grass native to East Asia, including Indian Subcontinent. Habitat to grasslands along streams and field boarders in Sikkim, Assam, and Northeast Hill. Adaptable to dry habitats, on sandy soils. Member secondary gene pool and source of drought and disease resistance (Piperidis et al. 2000). Used as food, fiber, material for thatching, basket making, etc. Cultivated as hedge around plantations and for medicinal purpose.
27. ***Saccharum barberi* Jeswiet.** A grass with short canes, medium to slender thick, with high fiber and medium sucrose content. Originated in Northern India. At one time, it was the main cultivated species, now restricted to Gangetic Plains. Use in breeding is limited because of its shy flowering and low fertility that is causing genetic erosion (non-availability of natural populations). Part of primary gene pool (Flach and Rumawas 1996).
28. ***Saccharum bengalense* Retz.** *Munj* or *baruwa grass*, tall caespitose perennial, culms ca. 4 m high, distributed from West Asia to northwest plains of India, extending east and southward to dry sandy soils and open habitats. Part of secondary gene pool (Cai et al. 2005). Ornamental, used in folklore medicine too.
29. ***Saccharum filifolium* Steud.; syn. *Erianthus filifolius* (Nees ex Steud.) Hack.** Stout perennial with woody base, culms up to 2 m high. Distributed from West Asia to Northwest Himalayas in subtemperate forest edges and bordering riverbanks in the Indian Subcontinent. Member of tertiary gene pool of sugarcane (Moore et al. 2014).
30. ***Saccharum longisetosum* Narayan ex Bor.** Perennial, culms 1–3 m tall, 0.5–1 cm in diam with many nods. Native of East Asia, including Indian Subcontinent, in hilly sides of Uttarakhand, Eastern Himalaya (Sikkim) and Northeast Hill region. Variety *hookeri* is widely distributed along riverbanks, forest edges, and cleared sites. Member of secondary gene pool (Aitken et al. 2007).
31. ***Saccharum procerum* Roxb.** Perennial, forming large clumps. Culms are very robust, up to 7 m tall. Native to East Asia, including Indian Subcontinent. Abundant on the sandy soil of Middle Gangetic Plains, Sikkim, and northeast region (Manipur) in cleared forest. Used for forage, fiber and is an attractive

biofuel. Based on phylogenetic affinities with *S. arundinaceum*, it is part of tertiary gene pool (Cordeiro et al. 2003) of sugarcane.

32. ***Saccharum ravennae* (Linn.) Murray**. Ravenna grass, a tall bunchgrass, with bamboo-like stem producing large attractive flowers, 3 m above ground. Native to Africa, Europe, Asia, including Indian Subcontinent, in Northeast hill region, and *terai* region along riverbanks, with much variability. An invasive species, ornamental, and used for soil stabilization. Part of tertiary gene pool (Hodkinson et al. 2002).
33. ***Saccharum rufipilum* Steud.; syn. *Erianthus fulvus* Nees ex Hack.** Stout perennial with hard woody base; culms up to 3.5 m tall. Distributed in northwest Himalayas, eastwards up to Assam and Northeast Hill region; along sandy riverbanks, cleared sites/open habitats. Also, found in Dehra Dun and Siwaliks.
34. ***Saccharum sikkimense* (Hook. f) Narayan ex Bor.** A perennial caespitose with erect 30–60 cm long culms. Distributed in tropical India and Eastern Himalayas (Sikkim), along forest edges, cleared sites, and riverbanks. Also, occurs in medium/high elevations, up to 1800 m.
35. ***Saccharum spontaneum* L.** *Kans* is a perennial grass, ca. 3 m tall. Native to Africa and Asia, including Indian Subcontinent in waterlogged habitat, particularly disturbed sites and sandy riverbanks and open lowland of north/northwestern plains, *terai* region, and further down south. A paleotropic invasive weed creating habitat suited for rhinoceros. Source of vigor, cold tolerance, hardiness, and disease resistance in sugarcane breeding program, as part of secondary gene pool (Shukla 1996).
36. ***Saccharum wardii* (Bor) Bor ex Cope.; *Erianthus wardii* (Bor) Bor.** A grass found in Eastern Himalayas, Assam, and Northeast India. Recorded in 1983 (Pal et al. 1983). Member of tertiary gene pool of sugarcane (Nair et al. 1999).
37. ***Saccharum williamsii* (Bor) Bor ex Cope.** A tall grass, native in *terai* regions of Uttar Pradesh, Bihar, and Nepal of the Indian Subcontinent. Part of tertiary gene pool of sugarcane (Nair et al. 1999).
38. ***Sclerostachya fusca* (Roxb.) A. Camus (Poaceae).** A tall grass, native to India and Indochina region. Well adapted to waterlogged areas of sub-Himalaya, Upper Gangetic Plains, and Nepal. It crosses with sugarcane as part of tertiary gene pool (Sreenivasan and Sreenivasan 1984). Intergeneric hybrid has been raised to introgress waterlogging resistance, heavy tillering, and earliness. It is closely related to *Saccharum spontaneum* with similar habits.

12.3 Perspective

In most commercial crops, unexplored areas and the inadequately sampled region of related wild species need immediate attention with respect to collection, before genetic variability is permanently lost, because of diverse pressures, particularly in case of those that are under threat. Perennial nature of commercial crops suggests for regular exploration and collection after specific intervals of, say, about 25 years. This may help in collection of new genotypes evolved responding to climatic change as well as to new races/strains of pathogens and insect pest, facilitating continued use of evolving wild genetic resources. Biochemical and molecular characterization may be another priority area, for example, in case areca nut dwarf, tall, and suckering wild and weedy types may help facilitate their use.

In genus *Camellia*, the outcrossing nature has resulted in natural hybridization among species, even between the caffeine-producing taxa belonging to *C. sinensis* and the non-caffeine-producing species, such as *C. irrawadiensis* and other closely related species, generating high heterogeneity in the natural populations. Thereby opportunities exists for search for new variability (recombinations). This may even be happening in the present-day commercial seedling populations, reflecting possible scope/role of interspecific hybridization in genetic improvement of tea. This demands greater emphasis on documentation of available wild species germplasm with a comprehensive information on morphological, chemical, cytogenetical, and molecular characterization of the species and the genetic variability within, in forms of natural populations. This will provide improved profile and information on phylogenetic relationships among wild and the cultivated species. Documentation shall help promote conservation and greater use of these alternative genetic resources in tea improvement.

The evaluation of *Saccharum* complex consisting of five related genera and *Saccharum* spp. has helped in generating basic information on phylogeny, polymorphism, and levels of genetic diversity of the complex. This has resulted in publications of the number of catalogues, leading to use *Saccharum* species to the extent that today all commercial varieties grown around the world are the products of interspecific hybridization. However, it has limited to few species and clones. Further precise characterization involving molecular genetics techniques and evaluation of related wild species can further help in widening of genetic base and enhancement of productivity, particularly with breeding resistance/tolerance to diseases and pests, and wider adaptability.

References

Aitken K, Li J, Wang L, Cai C, Fan YH, Jackson P (2007) Characterization of intergeneric hybrids of *Erianthus rockii* and *Saccharum* using molecular markers. Genet Resour Crop Evol 54:1395–1405

Cai Q, Aitken KS, Fan YH, Piperidis G, Jackson P, McIntyre CL (2005) A preliminary assessment of the genetic relationship between *Erianthus rockii* and the "*Saccharum* complex" using microsatellite (SSR) and AFLP markers. Plant Sci (Elsevier) 169:976–984

Cordeiro GM, Pan YB, Henry RJ (2003) Sugarcane microsatellites for the assessment of genetic diversity in sugarcane germplasm. Plant Sci 165:181–189

Daniels J, Roach BT (1987) Taxonomy and evolution. In: Heinz DJ (ed) Sugarcane Improvement through breeding. Elsevier, Amsterdam, pp 7–84

Flach M, Rumawas F (1996) Plants yielding non-seed carbohydrates. Plant Res SEAs 9:145

Goodspeed TH (1954) The genus *Nicotiana*. Chron Bot 16:403–404

Hodkinson TR, Chase MW, Lledo´ MD, Salamin N, Renvoize SA (2002) Phylogenetic of *Miscanthus*, *Saccharum* and related genera (Saccharinae, Andropogoneae, Poaceae) based on DNA sequences from ITS nuclear ribosomal DNA and plastid trnL intron and trnL–F intergenic spacers. J Plant Res 115:381–392

Johnson D (1998) *Phoenix rupicola*. The IUCN red list of threatened species 1998: e. T38629A10140589

Kundu SR (2005) A synopsis of Theaceae in Indian subcontinent: its distribution and endemism. Acta Bot Hungar 47(3-4):369–377

Lewis RS, Nicholson JS (2007) Aspects of the evolution of *Nicotiana tabacum* L. and the status of the United States *Nicotiana* germplasm collection. Genet Resour Crop Evol 54:728

Liu Y, Yang SX, Ji PZ, Gao LZ (2012) Phylogeography of *Camellia taliensis* (Theaceae) inferred from chloroplast and nuclear DNA: insights into evolutionary history and conservation. BMC Evol Biol 12:92. doi:10.1186/1471-2148-12-92

Moore PH, Paterson AH, Tew T (2014) Chapter 1. sugarcane: the crop, the plant, and domestication of sugarcane: physiology, biochemistry, and functional biology. In: Moore PH, Botha F (eds) Sugarcane: physiology, biochemistry & functional biology. Wiley, Ames, pp 1–17

Mukherjee SK (1957) Origin and distribution of *Saccharum*. Bot Gaz 199:55–61

Nair NV, Nair S, Sreenivasan TV, Mohan M (1999) Analysis of genetic diversity and phylogeny in *Saccharum* and related genera using RAPD markers. Genet Resour Crop Evol 46:73–79

Pal DC, Uniyal BPK, Mudgal V (1983) *Saccharum wardii* (Bor) Bor ex Cope (Poaceae), a new record for India. Bull Bot Surv India (BBSI) 24:196–198

Pintaud JC, Ludeña B, Aberlenc-Bertossi F, Zehdi S, Gros-Balthazard M, Ivorra S, Terral JF, Newton C, Tengberg M, Abdoulkader S, Daher A, Nabil M, Saro Hernández I, González-Pérez MA, Sosa P, Santoni S, Bouguedoura N, Moussouni S, Si-Dehbi F (2013) Biogeography of the date palm (*Phoenix dactylifera* L., Arecaceae): insights on the origin and on the structure of modern diversity. ISHS Acta Hortic 994:19–36

Piperidis G, Christopher MJ, Carroll BJ, Berding N, D'Hont A (2000) Molecular contribution to selection of intergeneric hybrids between sugarcane and the wild species *Erianthus arundinaceus*. Genome 43:1033–1037

Ram B, Sreenivasan TV, Sahi BK, Singh N (2001) Introgression of low temperature tolerance and red rot resistance from *Erianthus* in sugarcane. Euphytica 122:145–153

Rivers MC, Wheeler L (2015) *Camellia sinensis*. The IUCN red list of threatened species 2015: e.T62058171A62058174.

Shukla U (1996) The grasses of North East India. Scientific Publishers, Jodhpur

Singh ID (1999) Plant improvement. In: Jain NK (ed) Global advances in tea science. Aravalli Books International Pvt Ltd, New Delhi, pp 427–448

Sreenivasan J, Sreenivasan TV (1984) *In vitro* propagation of a *Saccharum officinarum* (L.) and *Sclerostachya fusca* (Roxb.) A. camus hybrid. Theor Appl Genet 67:171–174

Sudhersan C, Al-Shayji Y, Manuel SJ (2008) Date palm crop improvement via T *x* D hybridization integrated with *in vitro* technique. Acta Hortic 829:219–224

Vashishtha BB, Saroj PL, Gunjeet K, Awasthi OP (2005) Arid fruits. In: Dhillon BS, Tyagi RK, Saxena S, Randhawa GJ (eds) Plant genetic resources: horticultural crops. Narosa Publishing House, New Delhi, pp 168–189

Walter KS, Gillett HJ (eds) (1998) 1997 IUCN Red List of Threatened Plants. Compiled by the World Conservation Monitoring Centre. IUCN – The World Conservation Union, Gland, Switzerland and Cambridge, UK. lxiv + p 862

Wheatherstone J (1992) Historical introduction. In: Wilson KC, Clifford MN (eds) Tea: cultivation to consumption. Chapman and Hall, London, pp 1–23

13 Medicinal and Aromatic Plants

13.1 Introduction

In India, around 9000 species have been described as of ethno-medico importance, identified by the local communities and the Indian Systems of Medicine, such as Ayurveda, Siddha, Unani, and Homeopath. Sixty-seven species are mentioned in Rigveda, 81 in Yajurveda, and 290 in Atharvaveda. While the properties of 1100, 1270, and 1150 plants are described in *Charak Samhita* (700 BC), *Sushruta Samhita* (200 years BC), and *Astang Hridayam* (AD 700), respectively (Singh et al. 2016).

Regarding cultivated medicinal plant species, at present around 18 species are under commercial cultivation and approximately the same number having potential for commercial cultivation are being promoted with development of cultivation practices and have come under marginal cultivation. They include both native and introduced plant species. National Medicinal Plants Board (NMPB) has worked out a list of 960 medicinal plants that are under trade, of which 178 are consumed in higher volumes, including around 36 cultivated and rest harvested from the wild habitat of temperate and tropical forests, road sides, etc. (Ved and Goraya 2007). From wild relative genetic resource point of view, the native species out of these 36 or more cultivated species, wild species belonging to genera, such as *Aconitum*, *Andrographis*, *Asparagus*, *Berberis*, *Cassia*, *Chlorophytum*, *Cymbopogon*, *Hyoscyamus*, *Mentha*, *Ocimum*, *Saussurea*, *Terminalia*, *Withania*, etc., are of significance and would be part of present documentation. Besides these, many dual-purpose genus/species of fruit and vegetable, such as *Allium*, *Dioscorea* (steroid-yielding yams), *Phyllanthus*, and *Solanum*; spices and condiments such as *Curcuma*, *Zingiber*, and *Piper*; forage such as *Mucuna*; ornamentals such as *Rosa* (source of rose oil); etc., have been documented and discussed under the respective crop groups, based on their primary use.

The 10 + 1 phytogeographical regions of the country are rich homes of medicinal and aromatic plants (Singh 2017). However, the Himalayan region is the richest and has high degree of endemism. For example, genus *Saussurea* presents the highest diversity in alpine habitats of the Himalayas. A total of 62 *Saussurea* species

A.K. Singh, *Wild Relatives of Cultivated Plants in India*,
DOI 10.1007/978-981-10-5116-6_13

have been recorded from the region, of which 37 are native to Himalayan region, 8 are endemic, and 28 are near endemic. Twenty-seven of 28 near endemic species are of medicinal significance and used in indigenous treatment of various diseases. The genus *Saussurea* shows high habitat specificity, as 16 species were recorded to be restricted to one or two habitats only (Butola and Samant 2010), whereas *Hyoscyamus* spp. grow from Kashmir to Garhwal in the Northwestern Himalayas.

Chlorophytum originated in the tropical and subtropical region of Africa and is represented by 13 species, of these 8 are endemic to the Indian Subcontinent (Nair 1974), whereas in the arid region *Withania*, native of India, has another species *W. coagulans* Dunal., found wild in western Rajasthan area bordering with Pakistan, besides the cultivated *W. somnifera* Dunal. Monotypic *Tinospora cordifolia* Miers is distributed throughout India as a climber, with the plants climbing on Neem tree showing higher level of alkaloid contents (Srivastava et al. 1994).

Among medicinal tree, 20 species of *Terminalia* have been found distributed in tropical and subtropical states of India. *Terminalia arjuna*, *T. alata*, *T. paniculata*, *T. chebula*, *T. bellerica*, *T. travancorensis*, *T. procera*, *T. myriocarpa*, *T. manii*, *T. bialata*, *T. coriacea*, *T. crenulata*, *T. citrina*, *T. catappa*, and *T. pallida* yield indigenous drugs, tannins, gums, oils, wood (for matchboxes, splints, pulp), fodder, and certain organic compounds from their leaves, trunk, bark, or fruits. Terminalias have evolved inter- as well as intraspecific variation, because of predominant outcrossing nature (Parkinson 1936; Srivastav 1993), resulting in recombination and segregation, together with mutations, and natural selection. In some cases, intraspecific variation appears so great, to indicate involvement more than one species. *Terminalia arjuna, T. tomentosa, T. chebula, T. bellerica, and T. citrina* are examples of such species complexes (Hooker 1878; Bahadur and Gaur 1980; Parkinson 1936; Srivastav et al. 1992). This reflects the possibility of directed interspecific gene introgression in further genetic improvement of cultivated species. Medicinally cultivated *Terminalia* species are *T. arjuna*, *T. bellerica*, and *T. chebula*.

13.2 Wild Species Found in India

1. ***Aconitum deinorrhizum* Stapf.; syn. *A. heterophylloides* (Brühl) Stapf. (Ranunculaceae).** Annual, biennial, or perennial herb, distributed on alpine grassy slopes of Jammu and Kashmir and Himachal Pradesh. Poison is prepared from the root of this species. Cultivated, roots and leaves are used in rheumatism and headache. Related to Indian aconite or *Vatsnabh* (*A. ferox* Wall.).
2. ***Aconitum balfourii* Stapf.; syn. *A. lethale* Griff.** A perennial herb, widespread in Kumaon and Garhwal Himalayas on shady slopes between 3000 and 4200 m altitudes. Tuberous roots are rich sources of pseudoaconitine (0.4 to 0.5%) and aconite alkaloids. Possess antimicrobial activity. Rare, threatened due to overexploitation.
3. ***Aconitum falconeri* Stapf var. *falconeri* and var. *latilobum* Stapf.** Erect biennial herb, up to 1 m tall. Native to the subalpine and alpine zones of the Garhwal

Himalayas with varietal diversity. Variety *latilobum* is endemic to Bashahr, Himachal Pradesh. It has yielded two new norditerpenoid alkaloids, faleoconitine (1) and 3′-methoxyacoforestinine besides known compounds (Atta-Ur-Rahman et al. 2000).

4. ***Aconitum kashmiricum* Stapf ex Coventry; syn. *A. heterophyllum* var. *bracteatum* Stapf.** Erect herb 10–30 cm high with branched stem, and 2–3 cm long paired tubers. Endemic to alpine areas of Jammu and Kashmir. The dried root is analgesic, anti-inflammatory, antipyretic, antiperiodic, aphrodisiac, astringent, cholagogue, febrifuge, and tonic. It is used in the treatment of liver disorders, dyspepsia, diarrhea, indigestion, nausea, vomiting, throat pain, anorexia, piles, and coughs. Seeds are stimulant, aromatic, emmenagogue, stomachic, carminative, antipyretic, and hyperacid, useful in hiccup. Listed endangered in the 1997 IUCN Red List of Threatened Plants (Walter and Gillett 1998). Cultivated.
5. ***Aconitum laciniatum* Stapf.; syn. *A. ferox var. laciniata* Brühl.** An herb with tuberous roots, distributed in the subalpine and alpine Himalayas (Sikkim) between altitudes of 3330 and 4200 m. Used for treating chronic infections and inflammatory conditions. Closely related to *A. ferox*. Cultivated as medicinal plant in Nepal.
6. ***Aconitum palmatum* D. Don; syn. *A. bisma* Rapaics**. A perennial herb, native to tropics of Uttarakhand, Eastern Himalayas (Arunachal Pradesh, Sikkim), Nepal, Bhutan. Used in treatment of rheumatism and diarrhea. Two new diterpenoid alkaloids extracted.
7. ***Aconitum spicatum* Stapf; syn. *A. napellus* L.** A tuberous highly valued herb, yielding deadly poison. Native to East Asia, including India in alpine zone of the Himalayas of Sikkim and Chumbi, Singalila Range, Darjeeling, and Nepal. Known for antipyretic and analgesic activities.
8. ***Albizia kalkora* (Roxb.) Prain. (Fabaceae/Leguminosae).** Silk tree, an attractive shrub or small deciduous tree, 3–8 m tall, native to East Asia, including China and the Eastern Himalayas, Northeast Hill region and Central Highlands of India. Wild relative of *Shirish*, *A. lebbeck*.
9. ***Albizia orissensis* Sahni & Bennet.** A new tree species reported from Orissa in 1974. Rare, found in Northern Eastern Ghats (Koraput). Needs investigations.
10. ***Albizia procera* (Roxb.) Benth.** *Safed Shirish*, a fast-growing, semi-deciduous, drought-tolerant tree. Native to East Asia in moist deciduous and semi-evergreen hill and swamp forests, and lowland savanna woodlands from Northern India to Southeast Asia.
11. ***Albizia thompsonii* Brandis.** Rare tree wood, found in South Western and Eastern Ghats, and Upper Gangetic Plains.
12. ***Aloe barbadensis* Mill.; A vera (L.) Burm.f. (Xanthorrhoeaceae)** *Ghratkumari*, a stemless/short-stemmed succulent 0.5–1 m tall plant. Native to the Mediterranean region of southern Europe and North Africa. Introduced to India in *Vedic* times as evidenced by *Rig Veda*. Escaped from cultivation and naturalized in semi-wild state all over India with variability. A natural blood

purifier providing natural glow to skin. Used in joint pain, liver, vata, digestive problems, and more. Cultivated.

13. ***Alpinia malaccensis* (Burm. f.) Roscoe (Zingiberaceae).** A robust 2–4 m tall herb with leafy stem, strongly aromatic when bruised. Native to Indochina region of tropical Asia, including Northeast India. Rhizome yields essential oil. Potential ornamental plant.
14. ***Alpinia smithiae* Sabu & Mangaly.** Newly described endemic species from South India (Sabu and Mangaly 1991). Produces essential oils. Endangered.
15. ***Alpinia zerumbet* (Pers.) B. L. Burtt & R. M. Sm.; syn. *A. speciosa* (J.C. Wendl.) K. Suhum.** Shell ginger, a perennial, rhizomatous, evergreen 2.4–3.0 m tall herb with colorful funnel-shaped flowers. Native to East Asia, including Northeast India. Rhizomes used as substitute for ginger spice and in medicine. Cultivated.
16. ***Andrographis echioides* (L.) Nees (Acanthaceae).** False water willow, an annual erect herb with hairy stem, 10–50 cm high. Native to wastelands of plains of India and Sri Lanka. Used for hair falling and hair growth, and anti-inflammatory activity. Close to *A. paniculata*.
17. ***Andrographis nallamalayana* J. L. Ellis; syn. *A. beddomei* C.B. Clarke.** A prostrate to decumbent, 30–60 cm tall herb with thick root stock. Endemic to Eastern Ghats in Nallamala biosphere, Kurnool district. Properties like *Kalmegh*, *A. paniculata*; genetic resources. Rare and Endangered.
18. ***Andrographis serpyllifolia* (Rottler ex Vahl) Wight.** A trailing herb. Found in dry deciduous forests of Eastern Ghats and Karnataka. Used medicinally in cancer treatment. Genetic resource for *A. paniculata*.
19. ***Artemisia japonica* Thunb.; syn. *A. parviflora* Roxb. ex D. Don (Asteraceae-Anthemideae).** A hardy, herbaceous, nonaromatic, perennial 1 m tall herb. Native to East Asia, including the Indian Himalayas. Used in traditional medicine to treat fever as febrifuge, eczema and other skin diseases, and digestive and ethnoveterinary problems because of depurative properties. Extracted essential oil is antiviral and anthelmintic, whose composition has been analyzed (Rashmi et al. 2014). Also, cultivated.
20. ***Artemisia nilagirica* (CB Clarke) Pamp.** An aromatic, herbaceous perennial shrub. Distributed in open habitats of Nilgiris of the Western Ghats, Eastern Ghats, and Western Himalayas. Extracted essential oil has antimicrobial and antifungal activities. The composition has been analyzed. Cultivated.
21. ***Asparagus curillus* Buch. Ham. ex Roxb. (Asparagaceae).** *Shatawar*, a shrub distributed in tropical and temperate climate of the central Himalayas. Ripe fruits cause abortion; tuberous roots with honey are given for dysuria, diabetes, and dysentery. In Ayurveda, a novel androstane-type steroidal glycoside 1 is isolated from its roots.
22. ***Asparagus fysonii* Macbr**. Erect, tall, armed shrub with smooth stem, and globose, pulpy berry fruit. Endemic to grasslands of the South Western Ghats. Possess medicinal properties.
23. ***Asparagus racemosus* Willd**. *Shatavari*, woody perennial climber, 1–2 m tall and prefers to root in gravelly, rocky soils high up in piedmont plains, at

1300–1400 m elevation. Native to Africa and tropical Asia, including Indian Subcontinent. Ayurvedic practitioners use it as reproductive tonic, and for digestive and nervous disorders.

24. ***Asparagus rottleri* Baker**. An herb, endemic to Peninsular India. Tuber is used as coolant and possesses medicinal properties. Under threat, though data deficient.
25. ***Asparagus sarmentosus* L.** *Shatamuli*, climbing herb with linear to lanceolate leaves and berry fruit. Native to Indian Subcontinent in the subtropical North India. Used medicinally, fresh root demulcent, nourishing, and aphrodisiac.
26. ***Atropa acuminata* Royle ex Lindl. (Solanaceae).** Indian belladonna is a perennial branched herb ca. 1.5 m tall. Distributed from West Asia to Indian Subcontinent, in West Himalayan states. All parts of the plant contain the alkaloids atropine, hyoscyamine, and bellodonnine, which are used as sedative and antispasmodic, in convulsive disorders, and as an antidote for poisoning. Close relative of cultivated *A. belladonna*, even considered a synonym. Mentioned endangered.
27. ***Berberis chitria* Lindl. (Berberidaceae).** Perennial evergreen shrub ca. 4 m tall. Native throughout the Himalayas of the Indian Subcontinent. Genetic diversity is reflected with occurrence of botanical variety *occidentalis* in Kashmir. Source of traditional medicine *Daruharidra*. A new aporphine base characterized as O-methylcorydine-N-oxide together with berberine, palmatine, jatrorrhizine, and oxyacanthine is found. Close to *B. aristata*.
28. ***Berberis lycium* Royle.** An evergreen, ca. 3 m tall shrub. Native to the Himalayas and hills of Northwest India, Madhya Pradesh, and Nilgiri of Tamil Nadu in the Indian Subcontinent. Various parts used for food and in folk medicine. Have shown hypoglycemic, hyperlipidemic, hepatoprotective, anticarcinogenic, and antipyretic properties. Roots known as *Darhald* used for diaphoretic treatment.
29. ***Berberis micropetala* C.K. Schneid.** A shrub with internodes 2.5–3.5 cm long, spines, and estylose berries. Endemic to Nagaland.
30. ***Berberis sublevis* W.W. Smith.** An evergreen shrub of sturdy habit, 1.5–2 m high. Endemic to Manipur. Genetic diversity is reflected with the presence of variety *microcarpa* with marginal spinules in Meghalaya.
31. ***Berberis tinctoria* Leschen.** Nilgiri barberry, spiny or glabrous, deciduous, evergreen to semi-evergreen 3–5 m tall shrub, endemic to high hill shola forests of the Western Ghats and Nilgiris and Palani hills, having a lot of medicinal properties, including antioxidant and antibacterial activity of root.
32. ***Berberis wardii* C.K. Schneid.** A profusely leafy, thickly bushy shrub, with shortly stylose, purplish black berries. Endemic to open grasslands of Nagaland. Under threat.
33. ***Bryonopsis laciniosa* (L.) Naudin (Cucurbitaceae).** *Shivlingi*, an herbaceous, slender, extensive, monoecious climber. Wild, common all over India. A prominent Ayurvedic herb, traditionally used as an aphrodisiac and pro-fertility compound, antimicrobial too.

34. ***Cassia alata*** **L.; syn.** ***Senna alata*** **(L.) Roxb. (Fabaceae/Leguminosae).** An attractive shrub with flower buds in columns, giving an appearance of yellow candles. Invasive, native of tropical Americas, introduced and naturalized to Indian Subcontinent. Wild relative of Senna, *S. alexandrina* Mill. Cultivated as ornamental and for erosion control. Antifungal, used to treat fungal infection, ringworm, etc.
35. ***Cassia montana*** **Heyne ex Roth; syn.** ***Senna montana*** **(Roth) V. Singh.** Large shrubs to small tree, common in dry stony hills and along riverbanks of South Western Ghats and Peninsular India. Medicinal, stem bark extracts possess analgesic and anti-inflammatory effects.
36. ***Cassia obtusifolia*** **L; syn.** ***C. tora*** **Sensu auct.;** ***Senna obtusifolia*** **(L.) H.S. Irwin & Barneby.** A poisonous annual herb or undershrub, 1.5 m tall. Native to Americas, naturalized in the Indian Subcontinent as weed. Possess toxicological and hematological effects. Its leaves, seeds, and root are used in folk medicine. Seeds used to remove heat from the liver and to improve eyesight.
37. ***Cassia sophera*** **L.; syn.** ***Senna sophera*** **(L.) Roxb.** *Kasundi*, shrubby herb, found all over India in plains up to 900 m, wastelands, roadsides, and the forests. Leaves eaten as vegetable. Medicinally used in Ayurveda, Folk, Homeopathy, and Siddha systems. Root bark used homeopathic treatment of osteoarthritis, asthma, and allergic rhinitis.
38. ***Celastrus paniculatus*** **Willd. (Celastraceae)**. *Malkangani*, a woody deciduous vine. Native to East Asia and Pacific, including Indian Subcontinent. Seeds are used to improve cognition. Preliminary evidence indicates its potential intestinal relaxant. Used to increasing memory, intellect, and concentration.
39. ***Chlorophytum borivilianum*** **Santapau & R.R. Fern. (Liliaceae)** *Safed musli*, a tuberous herb with lanceolate leaves, found in tropical wet forests of Peninsular India, including North Western Ghats. Therapeutically applied as sex/general health tonic. The medicinal value is derived from saponins and alkaloids present in the plant and is the source of its alleged aphrodisiac properties. Under threat due to overexploitation. Cultivated.
40. ***Chlorophytum malabaricum*** **Baker.** Delicate tuberous herb with white flowers arise in an erect spike, endemic to evergreen and shola forests of Peninsular India, including South Western Ghats and Nilgiris. Two other closely allied species, *C. nepalense* (Lindley) Baker and *C. khasianum* Hooker f. ($2n = 56$), are distributed in Eastern India. Considered vulnerable.
41. ***Commiphora mukul*** **Hook. ex Stocks., syn.** ***C. wightii*** **(Arn.) Bhandari (Burseraceae).** *Gugul*, a small tree, up to 4 m high, with thorny branches and thin papery bark. It is a native of the Middle East and the west Indian plains. It is sought for its gummy resin, which is harvested from the plant's bark through the process of tapping, and called Gugulipid (GL). It is being used in Ayurveda and Unani medicines. Cultivated commercially in India and Pakistan.
42. ***Croton malabaricus*** **Bedd. (Euphorbiaceae).** Subcanopy tree up to 20 m tall, endemic to evergreen and semi-evergreen forests of South Western Ghats. Poisonous, tribal medicine for diabetes. Related to *Croton tiglium* L.

43. ***Croton scabiosus* Bedd.** An endemic tree of semi-evergreen or deciduous forests of southern Eastern Ghats (Andhra Pradesh). Represented by a small population. Cake useful as manure. Related to *Jamaal Gota, Croton tiglium* L.
44. ***Curculigo orchioides* Gaertn. (Hypoxidaceae).** *Kali Musali* or golden eye grass, a bulb plant with partly subterranean stem and basal sheath of leaf fused around stem. Native to East Asia, including Indian Subcontinent. Curculigoside A, B, C, and D and curculigine A and D are found, which can be used against β-amyloid aggregation. Considered endangered.
45. ***Cymbopogon jwarancusa* (Jones) Schult.; syn. *Andropogon jwarancusa* Jones, (Gramineae/Poaceae).** It is a perennial grass with culms erect or geniculately ascending, up to 1.5 m tall. Native to Africa and Asia, including the Himalayas from Kashmir to Arunachal Pradesh and other parts of India, with genetic differentiation of variety *assamensis* BK Gupta in Assam and var. *ldakhensis* Gupta in Ladakh., yields essential oils and used as folklore medicine. The accessions from Thar Desert offer multiple stress-tolerant accessions for this C4, aromatic grass. Related to lemongrass, *Cymbopogon flexuosus* (Nees ex Steud.) Wats. Cultivated as ornamental.
46. ***Cymbopogon khasianus* (Hack.) Stapf ex Bor.; syn. *Andropogon khasianus* Munro ex Hack.** Perennial grass with a stout rootstock. Culms slender to robust, up to 2 m tall, confined to Meghalaya. Varietal diversity expressed as variety *nagensis* Bor. Related to *Cymbopogon* spp. Rich source of essential oil and methyl eugenol. Cultivated.
47. ***Datura metel* L. (Solanaceae).** *Dhatura*, an annual herb growing up to 1 m. It is slightly furry, with dark violet shoots, oval to broad oval leaves, and pleasantly scented 6–8 flower inflorescence. Native of Americas, but neutralized to the warmer parts of the world, including India. Used for skin condition, anxiety disorders, and respiratory ailments. Globally cultivated for its chemical and ornamental properties.
48. ***Datura stramonium* L.** *Dhatūra*, a foul-smelling erect, freely branching annual herb, 1.2 m tall weed. Native of North Americas, invasive and naturalized in many regions of the world, including India since ancient times. It contains atropine, used for asthma. The leaves are smoked. Closely related to *D. metel*.
49. ***Eclipta alba* L. Hassk. (Asteraceae/Compositae)**. *Bhringraj*, an annual herb, commonly found in moist places as a weed all over the world, including India. In Ayurveda medicine, the leaf extract is considered a powerful liver tonic, rejuvenant, and especially good for the hair. Leaves are used in preparing food. Grows in rice fields, in India.
50. ***Holarrhena antidysenterica* (Linn.) Wall. (Apocynaceae).** *Kutaj*, a small evergreen shrub to tree, native to East Asia, including Indian Subcontinent, especially in the sub-Himalayan tract. Used in the treatment of constipation, colic, diarrhea, dysentery, and piles. Also, a potent antioxidant.
51. ***Hyoscyamus albus* L. (Solanaceae).** White henbane, annual/biennial herb with long white hairs, up to 0.9 m tall. Native to Africa, Europe, West Asia, and recently recorded in Hyderabad, India (Swamy et al. 2014). Source of atropine, hyoscine, and hyoscyamine. Close to cultivated henbane, *H. niger*.

52. ***Hyoscyamus niger* L.** Black henbane, robust, leafy, poisonous plant growing up to 1 m. Native of Africa, Asia, including the Northwest Himalayas of India. It has been used as a medicine since centuries and has been described in most Indian traditional medicine systems. Source of atropine, hyoscine, and hyoscyamine. Cultivated.
53. ***Hyoscyamus pusillus* L.** It is an annual 8–30 cm tall herb. Native of Africa and Asia, including the Northwest Himalayas, bordering with Pakistan.
54. ***Ocimum canum* Sims. (Lamiaceae).** Camphor basil, a woody herb with hirsute stem. Paleotropic, and widely distributed in India. Described in Ayurveda and *Charaka Samhita*. The essential oil from leaves contains about 36 compounds and is a rich source of camphor, eugenol, β-caryophyllene, α-pinene, limonene, myrcene, camphene, naphthalene, caryophyllene, and valencene. Cultivated.
55. ***Ocimum gratissimum* L.; syn. *Ocimum viride* Willd.** *Ram tulsi*, an aromatic, perennial shrub, 1–3 m tall. Native to Africa, Madagascar, and Asia including India, in moist and dry deciduous forests and plains. Well known in the Indian system of medicine. Essential oils contain eugenol and show antibacterial activity; used as food preservative too. Cultivated.
56. ***Ocimum kilimandscharicum* Guerke.** Perennial woody shrub, up to 1–2 m tall, called *Kapur Tulsi*, because of having strong camphor scent. The leaves are purple when young, blue basil. It is a sterile hybrid, propagated by cuttings. Native of tropical East Africa, introduced, and escaped in India to wild, generating significant variability. Well known in Indian traditional medicine system for essential oil.
57. ***Phyllanthus fraternus* Webster (Euphorbiaceae).** *Bhuinanvalah*, an annual weed, native to Indian Subcontinent, found throughout the hotter parts of India. Used in Ayurveda and Siddha for the treatment of jaundice. Hepatoprotective in acute viral hepatitis. The phytochemical analysis of plant extract found alkaloids, flavonoids, tannins, glycosides, saponin, carbohydrates, resins, and phenols.
58. ***Phyllanthus niruri* Hook.** It is an erect, slender, branched, annual weed. Native of Americas, now found growing wild throughout tropical and subtropical Asia too, including India in most moist locations. An Ayurveda plant used in the treatment of kidney stones, gall bladder ailments, hepatitis, flu, cold, tuberculosis, and viral infections. Wild relative of *Phyllanthus amarus* Schum. & Thonn.
59. ***Plectranthus amboinicus* (Lour.) Spreng.; syn. *Coleus aromaticus* Benth.; *C. amboinicus* Lour. (Labiatae/Lamiaceae).** *Patharchur*, a tender fleshy perennial plant. Origin obscure, distributed in South Western Ghats. Leaves are used for flavoring food and drinks and cooked. Known for therapeutic antimicrobial and anti-inflammatory activities. Essential oils extracted from aerial parts. Cultivated in garden or as potted plant.
60. ***Plectranthus barbatus* And.; syn *Coleus forskohlii* Briq.** A tropical perennial plant, distributed in subtropical Western and Central Himalayas up to 2500 m on dry barren hills, in Northeast and Deccan Peninsula. Interesting from a scientific and medicinal standpoint because it produces forskolin associated with weight loss. Related to cultivated *Plectranthus* species.

61. ***Plectranthus hadiensis* var. *tomentosus* (Benth. ex E. Mey.) Codd; syn. *Coleus zeylanicus* (Benth.) L.H. Cramer.** Perennial, succulent herb or branched subshrub with aromatic leaves, spreading ca. 50–75 cm around. Native of South Africa, but cultivated and naturalized in South India in savanna and dry forest. Related to *Plectranthus* spp. Medicinal, rich source of bioactive phytochemicals, especially terpenoids.
62. ***Pavonia odorata* Willd**. *Sugandha Bala*, a small shrub, with mucilaginous leaf and white flowers. Native to tropical Asia, including India in scrub jungles and dry deciduous forests of South Western Ghats. Known for its rhizome essential oil, a natural cooling agent with stomachic properties and Ayurvedic drink *Shadanga Paniya*. Has 25 biological activities and 9 ethnobotanical uses.
63. ***Rauvolfia beddomei* Hook. f.; syn. *R. hookeri* Sriniv. & Chitra. (Apocynaceae).** A large dichotomously branched shrub. Endemic to evergreen forests of South Western Ghats.
64. ***Rauvolfia canescens* L. (syn. *R. tetraphylla*).** It is a small shrub that reaches up to 2 m in height. It is native to Americas, naturalized to India, and many other countries. The roots yield the drug deserpidine, which is antihypertensive and a tranquilizer.
65. ***Rauvolfia densiflora* (Wall) Benth ex Hook.f.** *Barachandrika*, a large shrub of 3–6 m high with milky juice. Distributed in Northeastern Hills, Eastern and Western Ghats. Medicinally important, being source of a group of alkaloids used in the treatment of hypertension.
66. ***Rheum australe* D. Don; syn. *R. emodi* Wall. ex Meisn. (Polygonaceae).** Indian rhubarb, a perennial herb ca. 3 m tall with thick and long roots. Native to East Asia including the Himalayas in India. Known for positive and balancing effect upon the whole digestive system. Stalks are eaten.
67. ***Rheum nobile* Hook.f. & Thomson.** Sikkim rhubarb, a giant herbaceous slow-growing plant, native to the Himalayas, from northeastern Afghanistan to Myanmar in alpine zone. The root is astringent, carminative, depurative, diuretic, purgative, and yields tonic. Stem used in Tibetan medicine.
68. ***Saussurea clarkei* Hook.f. (Asteraceae).** Herb, 10–15 cm tall, stout. Pappus brown. Endemic to the West Himalayas in Jammu and Kashmir on open hill slopes, ca 4500 m. Medicinal, wild relative of snow lotus. Listed rare in 1997 IUCN Red List of Threatened Plants (Walter and Gillett 1998).
69. ***Saussurea costus* (Falc.) Lipsch.** *Kuth*, a robust perennial herb, endemic to parts of the Himalayas in Jammu and Kashmir (Chenab Valley, Gurez-Tilel in Kishanganga Valley, Suru Valley) and the Western Ghats. Root bitter, aromatic, and aphrodisiac; essential oil is extracted from roots used in traditional Ayurveda and perfumery, since ancient times. Critically endangered (Saha et al. 2015). Cultivated.
70. ***Saussurea fastuosa* (Decne.) Sch.Bip.; syn. *S. forrestii* Diels.** Perennial herb 0.3–1.25 m tall. Pappus pale brown. Distributed in the Himalayas, in *Acer-Aesculus* forest or in open places at 2500–3500 m of Himachal Pradesh, Uttarakhand, and Sikkim.

71. ***Saussurea gnaphalodes* (Royle) Sch.Bip.** *Kuth*, almost stemless, dwarf, woolly haired, clustered perennial herb, 2–7 cm tall. Pappus pale brown or dull blue. Native to East Asia, including moist open places of Jammu and Kashmir, Himachal Pradesh, and Uttarakhand in the Western Himalayas. Sensitive to climatic change
72. ***Saussurea gossypiphora* D. Don.** *Kastūrī kamal*, a perennial herbaceous plant, strikingly woolly, 30 cm tall with flower heads embedded among wool. Native to open, stony slopes of Himalaya's altitude between 4300 and 5600 m. Aerial parts are used for blood and liver disorders and analgesic in Ayurveda. Sixteen compounds have been isolated.
73. ***Saussurea involucrata* Kar et Kir.** Snow lotus, a perennial herb 15–50 cm tall. Grows in high mountains of Tian-Shan and A'er Tai areas of China, Tibet, and Indian Himalayas. Part of Tibetan medicine system. Known for a broad range of bioactivities, promoting blood circulation, for treatment of inflammation and pain-related diseases. Considered endangered due to overexploitation.
74. ***Saussurea nagensis* C.E.C. Fisch.** Stems woody below, terete, thinly cottony. Endemic to Northeast India, on grassy slopes, ca 2150 m in Nagaland.
75. ***Saussurea obvallata* Wall.** *Brahmakamal*, a perennial herb growing up to 0.3 m tall. Distributed in the Himalayas (Jammu and Kashmir, Himachal Pradesh, and Uttarakhand) at ca. 4500 m altitude, Northern Burma and Southwest China. Tibetan medicinal herb. State flower of Uttarakhand. Cited in ancient Hindu epics.
76. ***Terminalia elliptica* Willd.; syn. *T. alata* Herb. Madr. Ex Wall.; *T. tomentosa* Wight & Arn. (Combretaceae).** *Asna* or *Saaj*, a large deciduous tree ca. 30 m tall, 1.4 m in diameter, with15–20 mm thick bark. Native to tropical Asia, including hill slopes, stream banks of the Himalayas, and moist and dry deciduous forests of Peninsular India. Produces timber wood, leaves used as fodder and feed for silkworms (*Antheraea paphia*), which produce the tassar silk, and bark used as medicine. Roots yield antifungal component, including three new glycosides. Part of *Terminalia* complex as genetic resource.
77. ***Terminalia myriocarpa* Van Heurck & Müll.Arg.** Hollock/East Indian almond, a fast-growing tree. Native to East Asia, including Indian Subcontinent, in sub-Himalayan region. Have been used for planting in Assam. Active ellagitannins have been isolated. Feed larvae of the moth *Acrocercops terminaliae*. Cultivated.
78. ***Terminalia paniculata* Roth.** *Kindal* (in timber trade), tropical deciduous tree to 30 m high, with a large natural distribution in tropical Asia in semi-evergreen and moist deciduous forests of Peninsular India, the Western and Eastern Ghats. Yields timber wood and food for silkworms. Bark known for anti-inflammatory activity.
79. ***Valeriana wallichii* DC; *V. jatamansi* (Valerianaceae).** Indian valerian/Tagar-Ganthoda, an erect, perennial rhizomatous herb ca. 1.3 m tall. Native to East Asia, including temperate zone of the Himalayas up to an altitude of 3000 m. Ayurvedic anodyne, antispasmodic, aromatic, calmative, carminative, diuretic, expectorant, nervine, relaxant, sedative, stimulant, and tranquilizer. Genetic resource for True Valerian, *Valeriana officinalis*.

80. ***Vetiveria zizanioides* (L.) Nash; syn. *Chrysopogon zizanioides* (L.) Roberty (Andropogoneae).** Vetiver or *Khus*, an erect perennial, densely tufted, awnless grass. Native of tropical Asia, including India in wet, damp marshy areas of Brahmaputra Valley, Gangetic Plains, extending down to the plains of Punjab and Rajasthan. Indian variant is known for sweet, earthy, woody balsamic highly priced essential oil used in perfumary and folk medicine. Could be utilized for grazing while young, hedge and erosion control. Cultivated.
81. ***Withania coagulans* (Stocks) Dunal.** *Rishyagandha* or *Paneer bandh*, a branched shrub up to 1 m tall. Distributed from West Asia to Indian Subcontinent, including dry hot and stony places of Western Rajasthan. The fruit is emetic and diuretic, also coagulating. Effective in the management of uncomplicated cases of *Prameha* (Type II Diabetes Mellitus). Genetic resource for *W. somnifera*. Cultivated.

13.3 Perspective

Many of the medicinal plant species have several synonymous names, indicating confusion in their identity. There is an urgent need for supportive biosystematic studies to review the species/plant identities and tracing the phylogenetic relationship between the so-called species identities/populations to facilitate conservation and use, with the support of appropriate breeding technologies.

Population assessment of the native, endemic, and rare/endangered species using standard ecological methods has been suggested for the quantification of the existing stock of these species in their natural habitats. Immediate assessment of the status of wild populations of medicinal plant species, particularly of those that are economically important and are under threat due to overexploitation, needs appropriate management interventions for building up populations of such species and to work out scheme to facilitate conservation and sustainable use. Review and rationalization of current schemes for incentive-based promotion on commercial cultivation so that the species of conservation concern and facing acute shortage in supply could be accorded needed focus. High priority to be accorded for development of best cultivation practices and to in situ conservation for the species mainly harvested from wild to avoid loss and genetic erosion.

Recognizing the need for predictive and effective use of herbal medicine/extract with greater confidence and with reproducible results, there is an urgent need for greater emphasis of phytochemical investigations for the descriptive biochemical profile, active ingredients, and their molecular characterization. This shall help in discerning the genetic variability regarding biochemical profile between the species and populations and identify the potential species/populations. Also, the use of the generated information should help protect the products and process under appropriate intellectual property rights systems. Also, studies can be initiated on the diversified use of medicinal plant species, for example, the use of *Terminalia* spp. as host of silkworms to improve economic profile.

References

Atta-Ur-Rahman FN, Akhtar F, Choudhary MI, Khalid A (2000) New norditerpenoid alkaloids from *Aconitum falconeri*. J Nat Prod 63(10):1393–1395

Bahadur KN, Gaur RC (1980) A note on the *Terminalia tomentosa* complex. Indian J For 3(4):367–369

Butola JS, Samant SS (2010) *Saussurea* species in Indian Himalayan Region: diversity, distribution and indigenous uses. Int J Plant Biol 1:43–51. doi:10.4081/pb. 2010.e9

Hooker JD (1878) The flora of British India. M/S Bishen Singh Mahendra Pal Singh, Dehradun

Nair VN (1974) Cytological studies in *Chlorophytum bruchae* Ansari. Curr Sci 43:161

Parkinson CE (1936) Indian terminalias of section pentaptera. Indian For Rec 1(1):1–26

Rashmi TR, Francis MS, Murali S (2014) Essential oil composition of *Artemisia japonica* Thunb. from Kerala. J Pharmacognosy Phytochem 3(4):160–163

Sabu M, Mangaly JK (1991) *Alpinia smithiae* (Zingiberaceae): a new species from South India. Edinb J Bot 48(1):69–71

Saha D, Ved D, Ravikumar K, Haridasan K (2015) *Saussurea costus*. The iucn red list of threatened species 2015: e.T50126641A50131430

Singh AK (2017) Revisiting the status of cultivated plant species agrobiodiversity in India: an overview. Proc Indian Natn Sci Acad 83(1):151–174. doi:10.16943/ptinsa/2016/v82/48406

Singh J, Lavania UC, Singh S (2016) Chapter 3. Indian traditional and ethno medicines from antiquity to modern drug development. In: Singh RJ (ed) Genetics resources and chromosome engineering of medicinal plants, Volume 6 medicinal plants. CRC Press, Boca Raton, pp 53–86

Srivastav PK (1993) Pollination mechanisms in genus *Terminalia* Linn. Indian For 119(7):147–150

Srivastav PK, Beek S, Siddiqui AA, Brahmchari BN, Thangavelu K (1992) Genetic divergence in leaf characters of *T. arjuna* and *T. tomentosa*. Silvae Genetica

Srivastava VK, Singh BM, Gupta V, Gupta R (1994) Evaluation and domestication studies on some native medicinal plants. In: Purkayastha RP (ed) Economic plants and microbes. Today and Tomorrow Publishers, New Delhi, pp 51–55

Swamy J, Annamma PS, Chandra Mohan K, Rasingam L (2014) *Hyoscyamus albus* (Solanaceae): a new distributional record for India. Rheeda 25(1):54–56

Ved DK, Goraya GS (2007) Demand and supply of medicinal plants in India. NMPB, New Delhi & FRLHT, Bangalore

Walter KS, Gillett HJ (eds) (1998) 1997 IUCN Red List of Threatened Plants. Compiled by the World Conservation Monitoring Centre. IUCN – The World Conservation Union, Gland, Switzerland and Cambridge, UK. lxiv + p 862

Floriculture Crops 14

14.1 Introduction

In case of ornamentals, from the wild relative of cultivated plant species perspective, the taxa, which are native/indigenous to India, fall in the scope and are of concern. A survey of indigenous ornamental plants showed that about 43 species of *Rhododendron*, 1234 of orchids, 43 of *Primula*, 20 of *Lonicera*, 14 of *Aster*, 64 of *Begonia*, 241 of *Impatiens*, 73 of *Iris*, 43 of *Jasminum*, 24 of *Hedychium*, 7 of *Pandanus* (CI), 37 of *Ixora*, 14 of *Gardenia*, 12 of *Crinum*, 11 of *Lilium*, 26 of *Barleria*, 37 of *Ipomoea*, 14 of *Tabernaemontana*, 10 of *Thunbergia*, and 37 of *Bauhinia* occur in India (Khoshoo 1968). This may include some duplicates/synonyms requiring biosystematic investigation.

The temperate Northern Hemisphere, including the Himalayas, presents great diversity of ornamental plant species. For example, in primrose, almost half of the known *Primula* species are from the Himalayas; in honeysuckles (*Lonicera*), a handful of species, including *Lonicera hildebrandiana,* are from the foothills of the Himalayas. The genus *Rosa*, which occupies the top position among ornamentals, is distributed in the temperate regions of the Northern Hemisphere, including forests of the Himalayas, where 11 species are found growing wild. These are *Rosa brunonii*, *R. foetida*, *R. gigantea*, *R. involucrata*, *R. longicuspis*, *R. macrophylla*, *R. moschata*, *R. rubiginosa*, *R. sericea*, *R. webbiana*, and *R. wichuraiana*. Of these six, *Rosa brunonii, R. gigantea, R. involucrate, R. longicuspis, R. sericea*, and *R. webbiana* are endemic to the Himalayan region, four of which have contributed in the development of modern rose.

The genus *Lilium* also has some species of Indian origin from the Himalayas and Nilgiri Hills. The ones originated in the Himalayan region are *Lilium nepalense*, *L. polyphyllum* D. Don, and *L. wallichianum* Schult., whereas *L. nilgiriensis* is native of Nilgiri Hills, Tamil Nadu. There are several species tolerant to abiotic stresses. For example, *Lilium regale* EH Wilson, *L. sulphureum*, and *L. sargentiae* are tolerant to alkaline soils and high temperature, *L. dauricum* Ker Gawl and *L. distichum* Nak to cold and moderately high temperatures, and *L. formosanum* to high

A.K. Singh, *Wild Relatives of Cultivated Plants in India*,
DOI 10.1007/978-981-10-5116-6_14

temperature. These species have been used in crop improvement program for developing Asiatic hybrids. The genus *Iris*, which is cosmopolitan in distribution, is also found in Kashmir from the valley bottom to high alpine areas between 1600 and 4500 m, growing under diverse habitat. *Hedychium* J. Köenig (Zingiberaceae) species are represented with concentration (24 out of 65) in Northeast India.

For orchids, temperate Western and Eastern Himalayas, Northeast Hill region, and tropical Western Ghats, Eastern Ghats, and island eco-regions offer major habitat, harboring 1100 species belonging to 157 genera (Vij 1995). Around 300 species belonging to 75 genera, including some of the monotypic genera, such as *Didicea*, *Proteoceras*, and *Risleya*, are endemic to India, mostly confined to the Northeast Hills and Western Himalayan region (Vij 1995). Of these around 100 species are endangered and listed in the list of threatened plants of India. From genetic resources point of view, orchids are cross pollinated by insects and birds. Natural hybrids of orchids both intergeneric and interspecific occur wild. This suggests that the cultivated species can be hybridized within and between genera with compatible species generating a wide range of variability for selection of flower color and to induce genetic improvement, exploiting the interspecific species genetic diversity in desired direction, as per cultivation and market demand.

Chrysanthemum is a member of family Asteraceae. It is a native of Europe and Asia. There are about 160 species in the genus. The species native to India are *Chrysanthemum atkinsonii* C.B. Clarke, *C. indicum* L., and *C. leucanthemum* L. Similarly, many species represent *Aster*. *Begonia* of family Begoniaceae, distributed in tropical and subtropical regions of the Himalayas and Peninsular India, is bred for variation in foliage and flower color. Monoecious nature of their flowers makes it easy. A high diversity of *Impatiens* species is found in tropical South India and the Eastern Himalayas, because of outcrossing, facilitated by pollinators. The genus *Crinum* L., habitat to tropical and subtropical regions, has 13 species, 1 variety, and 1 form. Of these five are endemic (Lekhak and Yadav 2011), such as *Crinum brachynema* Herb., *C. eleonorae* Blatt. & McC. f., *C. eleonorae* f. *purpurea* Blatt. & McC., and *C. woodrowii* Baker. They are endemic to Mahabaleshwar, Maharashtra. While, *Ipomoea* L. species are distributed in tropical, subtropical, and temperate regions of the country.

Among shrub ornamentals, evergreen shrub of *Gardenia* is distributed in tropical and subtropical regions of Peninsular India and Southwest India. Natural stands of *Tabernaemontana* species are found in open forests of tropical India. Similarly, *Thunbergia* species, which are vigorous annual or perennial vines and shrubs, are native to the tropical regions of the country. They are frequent escapes from garden, and may become invasive and weedy. Most of the *Barleria* species are distributed in the Peninsular India and in the northeast region.

In addition, there are ornamental palms and trees indigenous to India. Around 97 species and 3 cultivars of wild and semi-wild palms belonging to 21 genera are distributed in northeast region, Western Ghats, and Andaman and Nicobar Islands (Basu 1995). There are several important genera found in Peninsular India, such as *Areca*, *Borassus*, *Bentinckia*, *Caryota*, *Corypha*, *Phoenix*, and *Pinga*. *Borassus flabellifer* is the most widespread in semi-wild form, along the coast in Andhra Pradesh

(Basu 1995). Moist tropical forest in the sub-Himalayan region has *Calamus*, *Phoenix*, *Trachycarpus*, and *Wallichia*. *Phoenix sylvestris* known as silver date palm grows wild in semi-arid regions, while *Nypa fruticans* Wurmb and *Phoenix paludosa* Roxb. in Sundarbans are ornamental date palms.

14.2 Wild Species Found in India

1. ***Aerides crispa* Lindl. (Orchidaceae).** It is an epiphytic orchid with purple stem, native to semievergreen and evergreen forests of Western India at an elevation of 800–1200 m. Inflorescence is 40–50 cm long with 20–25 fragrant flowers.
2. ***Aerides emericii* Reichb.f.** An epiphytic orchid, endemic to the Andaman and Nicobar Islands. It has broad leathery leaves, on axil of these blooms, arching to pendulous, 30 cm long, and many flowered inflorescences.
3. ***Aerides maculosa* Lindl.** Fox brush orchid, a dwarf stout herb with crystalline pink, spotted magenta, fragrant flowers, endemic to evergreen forests of Peninsular India. The epithet *maculosa* means "spotted"; refer to the colored spots on the petals.
4. ***Aerides ringens* (Lindl.) C.E.C. Fisch.** It is miniature, hot orchid growing on trees. Found in moist deciduous and semievergreen forests and grasslands of the Peninsular/Southern India and Andaman Islands. Recently reported from Gujarat.
5. ***Aerides rosea* Lodd. ex Lindl. & Paxton.** It is one of the most beautiful and compact-growing species, epiphytic to tree trunks in open forests of China, Eastern Himalayas, Assam, and Southeast Asian countries. Cultivated.
6. ***Androcorys gracilis* (King & Pantl.) Schltr. (Orchidaceae).** A tuberous tiny orchid 12–20 cm tall with a spike of many flowers arranged loosely. Native to the Eastern Himalayas and South China. Another five species recorded in the region are *A. angustilabris*, *A. josephi*, *A. kalimpongensis*, *A. monophylla*, and *A. pugioniformis*.
7. ***Anoectochilus brevilabris* Lindl.** Jewel orchids, miniature in size with the stunning patterns and coloration of foliage. Genetically diverse terrestrial species of the Eastern Himalayas, native to the mountains of Sikkim, Nepal, Assam, and Vietnam. A variant of jewel orchid, *Anoectochilus nicobaricus* Balakr. & Chakra., was found wild endemic in Great Nicobar Islands.
8. ***Aster* spp. (Asteraceae/Compositae).** Mostly native to North America and Southern Europe. In India represented by few wild species like *Aster ageratoides* Turcz., a perennial erect herb, 40–100 cm tall, found in Northeast India with significant genetic variability and *Aster indamellus* Grierson., an herb, native to alpine and subalpine open forests, grassy slopes, and rock crevices of Western Himalaya and Western Ghats. They live in enumerate geographic entities. Cultivated.
9. ***Barleria acuminata* Wight ex Nees; Syn. *B. tomentosa* (Acanthaceae).** Subshrub with subsessile pink flowers. Distributed in dry deciduous forests

along the forest edges, scrub jungles, and plains up to 600 m in Peninsular India. Medicinal too.

10. ***Barleria courtallica* Nees.** Throated barleria, a 2 m tall prickly shrub with blue flower having yellow or white tube and throat. Endemic to evergreen forests of Peninsular India and Western Ghats.
11. ***Barleria cristata* L.** *Vajra danti* is an erect, branched evergreen shrub growing 1–3 m, native to Asia, including the Himalayas, up to 2000 m and in understory of dry deciduous forests of Peninsular India and Southeast Asia. Cultivated with pruning into shapely shrubs. Medicinal, used in Ayurveda for treatment of variety of diseases including anemia, toothache, and inflammatory disorders.
12. ***Barleria cuspidata* Heyne ex Nees.** *Kate Koranti*, a perennial armed shrub, up to 1 m high with yellow flowers solitary or crowded at axils of upper node. Found in plains, dry localities, and scrub jungles, from tropical Africa, Sri Lanka, and Peninsular India to Malaysia. Anthelmintic activity observed.
13. ***Barleria grandiflora* Dalz.** *Dev Koranti* a perennial stout herb to shrub with solitary white flower in leaf axil. Endemic to dry deciduous forests of North Western Ghats. A potential ornamental. Used as folk medicine antioxidant.
14. ***Barleria prionitis* L.** Erect armed shrub with golden yellow-orange porcupine flower. Native to dry evergreen to thorn/deciduous forests in India, Sri Lanka, and tropical Africa. Well-known Ayurveda plant. Known for anti-inflammatory, antinociceptive, and antioxidant activities. Cultivated as ornamental and hedge.
15. ***Barleria stocksii* Anderson.** Herbs, unarmed, with sparse hairs, 60 cm tall. Flower white with 3 cm long corolla tube, basally cylindric, yellowish, lobes elliptic-oblong. Endemic to Andhra Pradesh and Karnataka, recollected after 140 years (Kambhar et al. 2014).
16. ***Bauhinia acuminata* L. (Fabaceae/Leguminosae).** Dwarf white orchid bauhinia, 2–3 m tall tree. Native to Southeast Asia. Cultivated all over tropical Asia. Found, escaped, and naturalized in West Bengal. Medicinal, coolant, flowers eaten, and wild relative of camel foot tree *B. variegata*.
17. ***Bauhinia racemosa* Lam.** *Katmauli*, a small deciduous crooked, ca. 12 m tall tree, with white or pale-yellow flower. Common in dry and moist deciduous forests and foothills of India and Sri Lanka. Inner bark used to make ropes. Young leaves made into delicious chutney. Leaves medicinal, used to treat liver ailments.
18. ***Bauhinia stipularis* Korth.; syn. *B. nicobarica* (Balakr. & Thoth.) Bennet.** Nicobar creeper is a large, tendriled climber with smooth branches. Mature flowers' petals get recurved at maturity. Native to Great Nicobar Islands. Young shoots and leaves are eaten.
19. ***Bauhinia tomentosa* L.** Yellow bell orchid tree, a medium to large shrub with attractive bright yellow flowers having black to maroon-colored centers. Native to Africa and Asia, including India. Pharmacological studies revealed antioxidant, antibacterial, antifungal, antihypoglycemic, etc., properties. Stimulates the immune system and scavenges free radical generation in vitro. Widely cultivated.

20. ***Bauhinia vahlii* Wight & Arn.** Camel's foot/*malu* is a vigorous climbing shrub with yellow or white flowers and grows to the tops of the trees. Native to dry deciduous, tropical, and subtropical forests of the Indian Subcontinent. It is considered an important traditional folk medicine and is being investigated.
21. ***Bauhinia variegata* L.** *Kachnar*, a moderate-sized, deciduous, ornamental tree with purple flowers. Native to East Asia, including the Indian Subcontinent in dry evergreen to dry deciduous forests. Flowers and seeds (fried) are eaten and used in many Indian recipes. Ayurvedic herb, with anticancer, antioxidant, hypolipidemic, antimicrobial, anti-inflammatory, nephroprotective, hepatoprotective, antiulcer, immunomodulating, molluscicidal, and wound healing properties. Widely cultivated as ornamental.
22. ***Begonia burkillii* Irmsch; syn. *B. aceroides* Irmsch (Begoniaceae).** A rare rhizomatous dioecious species with compact size and uniquely colored leaves. Light pink flowers emerge from under the leaves in winter. Restricted to Arunachal Pradesh, India. Brought into cultivation by Rehka Morris.
23. ***Begonia canarana* Miq.** A monoecious, perennial up to 15 cm tall herb with white flowers. Endemic to moist shady places among rock crevices in evergreen forests of South Western Ghats (Karnataka).
24. ***Begonia crenata* Dryand.** Small annual succulent herb with tuberous roots, red stem, monoecious, with pink male (two sepals and two petals) and female flower (five perianth segments). Endemic to banks of streams in the Western Ghats.
25. ***Begonia griffithiana* Warb.** This is tall begonia with medium-sized green leaves. Found in Arunachal Pradesh, India, and Bhutan. Potential ornamental.
26. ***Begonia integrifolia* Dalzell.** A small, tuberous herb with short, reddish, fleshy stem, and monoecious pinkish flowers. Habitat to evergreen forests of Peninsular India, extending to Indochina and Malay region.
27. ***Begonia malabarica* Lam.** A 2 m tall shrub with long stem, repeatedly dividing, with many toothed and serrate leaves and rose-colored flowers. Distributed in moist deciduous forests of Peninsular India and Sri Lanka. Medicinally important, phytochemical analyses of leave extract resulted in identification and isolation of friedelin, epi-friedelinol, β-sitosterol, luteolin, quercetin, and β-sitosterol–3-β-D-glucopyranoside. Studies support its usefulness for several diseases.
28. ***Begonia sikkimensis* A. DC.** About 30 cm tall tuberous herb with rhizomatous rootstock and attractive patterned leaves. Native to China and the forests of the Indian Subcontinent in Arunachal Pradesh, Sikkim, West Bengal, Bhutan, and Nepal.
29. ***Begonia xanthina* Hook.** A 40 cm tall herb with mottled leaves, changing color through growth, and yellow flowers. Native to shaded, moist, and rocky areas of the Eastern Himalayas to southwestern Yunnan in China. Cold tolerant.
30. ***Bulbophyllum acutiflorum* A. Rich. (Orchidaceae).** Epiphytic rhizomatous herb, with six to eight tongue-shaped lips, cream-colored, speckled with pale

brown flower inflorescence. Endemic to shola forests of Peninsular India. Marginally cultivated.

31. ***Bulbophyllum cornu-cervi* King.** A miniature-sized, warm growing epiphyte, with several flowered inflorescences having ovate, acute, and shorter than the ovary floral bracts. Native to Indochina region including the Eastern Himalayas and Sikkim. Genetic resource for cultigen orchid.
32. ***Bulbophyllum ebulbum* King & Pantl.** A unifoliate epiphyte with creeping rhizome and many 1 cm long flowered racemose inflorescence. Endemic to the Eastern Himalayas at an elevation around 500 m. Cultivated marginally.
33. ***Bulbophyllum fuscopurpureum* Wight.** Yellow-brown purple orchid. An herb with ovoid pseudobulb having an inflorescence with four or five yellowish-pink to darker flowers. Habitat to shola forests of South Western Ghats. Cultivated marginally.
34. ***Bulbophyllum gamblei* Hook.f.** A small, cool to cold growing epiphyte or lithophyte with umbellate inflorescence with three to six flowers, held amid the leaves. Found in the Eastern Himalayas, Sikkim, and Assam. Cultivated on limited scale.
35. ***Bulbophyllum obrienianum* Rolfe.** An orchid with spotted petal flowers. Native to the Eastern Himalayas (Arunachal Pradesh). Cultivated on limited scale.
36. ***Bulbophyllum parviflorum* E.C. Parish & Rchb.f.; syn. *B. thomsonii* Hook.f.** A miniature-sized, warm to hot growing epiphyte with many small flowered inflorescences arising on a mature pseudobulb. Found in the Eastern Himalayas, in montane forests extending to the Far East. Cultivated on limited scale.
37. ***Bulbophyllum protractum* Hook.f.** A miniature, hot to warm growing epiphyte with cylindrical pseudobulbs with three to five flowered inflorescences. Native to Assam, Sikkim, Myanmar, and Vietnam. Two new bibenzyl derivatives isolated. Cultivated marginally.
38. ***Bulbophyllum roxburghii* (Lindl.) Rchb.f.** A medium-sized, hot growing epiphyte, with pseudobulbs having 5–12 flowered inflorescences that exceeds the height of the leaves. Found in tropical valleys of the Eastern Himalayas, Sikkim, and Assam. Cultivated on limited scale.
39. ***Calanthe chloroleuca* Lindl. (Orchidaceae).** A medium-sized, warm to cold orchid, with conical-ovoid to cylindric pseudobulbs. Inflorescence carries few to many acute floral bracts and primrose-scented flowers. Habitat to mossy banks and dense forests of the Eastern Himalayas, Assam, and Nepal. Genetic resource to cultigen.
40. ***Calanthe herbacea* Lindl.** A medium- to large-sized, cool to cold growing epiphyte herb with 6 to 20 flowered, pubescent, ridged inflorescences with floral bracts. Found in the Eastern Himalayas, Assam, and the Far East. Extracted from the wild. Genetic resource to cultigen.
41. ***Calanthe mannii* Hook.f.** A small- to medium-sized, warm to cold growing herb with short, cylindric pseudobulbs, carrying 10–20 flowered, pubescent, long inflorescences with glabrous, small, lanceolate, acute floral bracts. Habitat

to dense rainforests and secondary mossy grasslands in foothills of the Himalayas, Assam, and the Far East. Extracted from the wild. Genetic resource.

42. ***Calanthe whiteana* King & Pantl.** A large-sized, cool herb with conic pseudobulbs, having 10 to 20 flowered inflorescences, with one to two tubular sheaths and persistent, reflexed, lanceolate, puberulent floral bracts. Native to the Eastern Himalayas, Assam, and the Far East. Extracted from the wild. Genetic resource.
43. ***Crinum amoenum* Roxb. ex Ker-Gawl (Amaryllidaceae).** Perennial plant with highly scented, striking spidery white flowers, having red-tinted reverses borne in cluster atop a slender leafless stem. Native to the Himalayas, where it grows in moist places or along rivers.
44. ***Crinum asiaticum* L. [var. *asiaticum*; var. *procera*].** Giant crinum lily, with largest poisonous bulb. Native to Africa, Asia, including India and Australia. Cultivated as ornamental in many warmer regions and used as folk medicine.
45. ***Crinum brachynema* Herb.** *Karanphul*, a bulbous herb with 5–20 fragrant flowers borne in an umbel held by spathe. It is restricted to lateritic plateaus covered with soil along the margins of semievergreen forests in Gujarat and Maharashtra. Critically endangered. It has great potential as an ornamental.
46. ***Crinum malabaricum* Lekhak & Yadav.** A new aquatic species, fully immersed herb with tunicated bulbs rooted into watercourse beds with white sweet-scented flowers (Lekhak and Yadav 2012). Endemic to seasonal stream of Periya Village, Kasaragod, Kerala.
47. ***Crinum solapurense* Gaikwad et al.** A new species, from a swamp bordering the Bhima River in Solapur, Maharashtra. It resembles *C. viviparum* var. *viviparum* and *C. lorifolium* (Gaikwad et al. 2014).
48. ***Crinum viviparum* (Lam.) R Ansari & VJ Nair**. River crinum, an aquatic erect, 0.5–1 m tall herb with large bulbs, bloom underwater with six petals and white flowers. Habitat to rocky, shallow rivers and streams in India and Sri Lanka. Cultivated.
49. ***Crinum woodrowii* Baker.** A rare and bulbous 30 cm tall plant with large six to seven fragrant white flowers in an umbel arising from top of the scape. Endemic to Satara, Mahabaleshwar, Maharashtra. Critically endangered, thought to be extinct but was rediscovered in 2004 after a gap of 100 years (Punekar et al. 2004).
50. ***Cymbidium aloifolium* (L.) Sw (Orchidaceae).** Autotrophic, epiphytic orchid with ovoid pseudobulbs, bearing medium-sized, yellow to cream yellow, with a broad, central maroon-brown stripe, slightly fragrant flowers borne on pendant inflorescence. Occur in China to the Eastern Himalayas, to Western Ghats, to Andaman Islands. Extracted from the wild. Highly threatened, genetic resource.
51. ***Cymbidium devonianum* Paxton; syn. *C. sikkimense* Hook.f.** It is a primary Indian hybrid species with 25–30 cm long broad and blade like leaves, pendulous scapes with 15 to 25 green bases overlaid with olive and a bit of burgundy-spotting small flowers. It has been widely used in hybridization program. Found in subtropical and subalpine forests of Northeast India. Extracted from the wild. Genetic resource

52. ***Cymbidium gammieanum* King & Pantl.** A primary natural orchid hybrid of cross, *C. elegans* × *C. erythraeum*. It is a medium-sized, cool growing herb with a short pseudo-stem having 15–20 flowered inflorescences and minute floral bracts. Found in the Eastern Himalayas (Sikkim). Extracted from the wild. Potential genetic resource.
53. ***Cymbidium mackinnonii* Duthie.; syn. *C. goeringii* (Rchb.f.).** A cool growing herb, with small, ovoid pseudobulbs having solitary to few fragrant flowered inflorescence. Native of northwestern India to the Far East. Extracted from the wild.
54. ***Cymbidium whiteae* King & Pantl.** An herb with a 20–30 cm long scape with 10–12 flowers. Distributed in Sikkim, the Himalayas (1500–2000 m), and Assam. Extracted from the wild. Endangered (Nayar and Sastri 1987). Cultivated in baskets.
55. ***Dendrobium chrysotoxum* Lindl. The** golden bow orchid, a small-sized herb with clustered, grooved, clavate, or fusiform pseudobulbs, with arching spikes of 4 cm waxy, bright orange fragrant flowers having fringed lips. Native of East Asia, including the Eastern Himalayas and Assam. Extracted from wild. Cultivated.
56. ***Dendrobium crepidatum* Lindl. & Paxton; syn. *D. lawianum* Lindl.** Epiphyte with 7–20 cm long stem and flowers having white petals with pale pink margins and lips white with a large pale-yellow spot in the middle, in short peduncles from the nodes of leafless stem. Native to tropical forests of Southeast Asia, the Eastern Himalayas, and Western Ghats.
57. ***Dendrobium herbaceum* Lindl.** Herb with 30–40 cm long stem and four to six white flowers in slender racemes on leafless stem. Endemic to moist deciduous, semievergreen, and shola forests of Peninsular India (Andaman). Extracted from the wild.
58. ***Dendrobium lituiflorum* Lindl.** An epiphyte with bent raceme bearing 1–5, long-lasting, fragrant flowers on short scape, with large bracts arising out of the nodes all along the dried cane on short racemes. Native to broad-leaved forests in China, the Eastern Himalayas, Assam, and the Far East, found on tree trunks. Extracted from the wild.
59. ***Dendrobium microbulbon* A. Rich.** Warm to cool growing small epiphyte with conical to ovoid, purplish-brown, clustered pseudobulbs. Three to ten white flowered inflorescences arise at the apex of an elongate leafless pseudobulb. Endemic to evergreen forests of Western Ghats. Extracted from the wild. Used in folk medicine.
60. ***Dendrobium ovatum* (L.) Kraenzl.** A miniature- to medium-sized, hot to cool growing epiphyte with subcorymbose inflorescence of several to many flowers. Endemic to moist deciduous forests and plains of Western Ghats. Ayurveda plant.
61. ***Dendrobium praecinctum* Rchb.f.; syn. *D. pauciflorum* King & Pantl.** Large-sized, warm to cool growing epiphyte with branched stem, apex of each branch carries one to five yellow/yellow white flowers. Native to East Asia, including the Eastern Himalayas and Assam.

62. ***Gardenia campanulata* Roxb. (Rubiaceae).** A large shrub or small tree, 4.5–6 m tall with fascicled male and solitary, sessile female flowers. Found in Bihar, Assam, Meghalaya, and Northeast India. The fruits are used as a cathartic and anthelmintic and as fish poison and larvicidal (Yusuf et al. 2009).
63. ***Gardenia gummifera* L.f.** *Kikemali*, a small tree, 3 m high, with solitary white turning to yellow flowers in loose cymes. Endemic to dry deciduous forests of Peninsular India. The yellowish resin produced is used in Ayurvedic medicines.
64. ***Gardenia jasminoides* J. Ellis.** Cape jasmine, a shrub with dark green shiny leaves and white fragrant flowers. Native to East Asia, including Northeast India. Widely cultivated ornamental species in warm temperate and subtropical gardens. Oldest cultivar: *veitchii*. Used as folk medicine.
65. ***Gardenia latifolia* Aiton.** *Papra*, a medium-sized, deciduous, long-lived tree with solitary white turning yellow flowers at the end of branches. Found in the dry deciduous forests all over India. Widely cultivated. Medicinally used in rheumatism, cuts, wounds, etc.
66. ***Gardenia resinifera* Roth.** *Dekamali*, a small tree, ca. 3 m tall with white fragrant axillary solitary flower. Found in dry deciduous forests of India, Bangladesh, and Myanmar. Cultivated for medicinal cumbi gum, which is excreted on the stem and buds of the plant and is antispasmodic, expectorant, carminative, and stimulant.
67. ***Hedychium coccineum* Buch. -Ham. ex Sm. (Zingiberaceae).** Orange ginger lily is a perennial bush growing to 1.5 m, with two close ranks of waxy leaves. The spike inflorescence has six rows of small orange fragrant flowers. Native to East Asia, including the Eastern Himalayas of India, Bhutan, and Nepal. Grows on forest edges and mountain grasslands. Cultivated, early bloomer.
68. ***Hedychium coronarium* J. Koenig.** Butterfly ginger, robust perennial plant with stalks topped with 6–12 enjoyable, fragrant, pure white flowers in long cluster. Native to East Asia, including the Indian Himalayas. Introduced to other parts of the world during slavery era. Considered invasive.
69. ***Hedychium dekianum* Rao & Verma**. Ginger lily, a perennial herb with spike inflorescence having white flowers. Found in Northeast India (Meghalaya).
70. ***Hedychium flavescens* Carey ex Roscoe.** Yellow ginger, a ca. 2.5 m tall perennial plant with strongly fragrant pale-yellow flowers. Native of the Eastern Himalayas and Nepal, invasive in nature. Introduced to other parts of the world. Cultivated. Parents of commercial hybrids highlighting the potential role of interspecific breeding in lilys' improvement.
71. ***Hedychium gardnerianum* Griff.** *Kahili ginger* is a 2.4 m tall perennial herb with long bright green leaves clasping the tall stem with a spike holding fragrant pale yellow/red flowers above foliage. Native to the Indian Himalayas. Introduced in tropics as an ornamental cultigen and has become an invasive weed. Cytotoxic against human small cell (lung cancer).
72. ***Hedychium marginatum* CB Clarke**. Yellow butterfly ginger, a perennial rhizome-forming herb with lemon-yellow flowers. Found in Manipur along

forest borders. A potential ornamental. Another different kind with red flowers is referred as *Hedychium greenii.*

73. ***Hedychium spicatum*** **Sm.** *Kapur kachari,* a small hardy ginger with green leaves and large orange/white flowers. Native to East Asia, including the Indian Himalayas. Cultivated, known as *Shati* in Ayurveda, used for treatment of cough, hiccough, fever, and asthma. Rhizome extract/essential oil is antibacterial and antifungal.
74. ***Impatiens acaulis*** **Arn. (Balsaminaceae).** A stemless perennial herb with beautiful white or pink flowers in racemes. Commonly found gregariously growing on moist rocks in grasslands of South India (Western Ghats) and Sri Lanka.
75. ***Impatiens denisonii*** **Bedd.** Scapigerous herb with raceme of 7–11 pink flowers. Endemic to wet rocks/tree trunks in forests of South Western Ghats (Nilgiris).
76. ***Impatiens glandulifera*** **Royle.** Himalayan balsam, a large annual 1 to 2 m tall herb with hooded shape pink flowers. Native to the Himalayas, grows densely as invasive species. Introduced and cultivated in the Northern Hemisphere, becoming problematic because of invasive nature.
77. ***Impatiens lawsonii*** **Hook.f.** Epiphytic annual, scapigerous, tuberous herb with four to six white or pink flowered straight scapes. Endemic on moist tree trunks of shola forests of South Western Ghats.
78. ***Impatiens levingei*** **Gamble ex Hook.f.** Small herb with small or no stem and few white flowers confined toward the apex of the scape. Endemic to moist rocky places in shola forests of Western Ghats.
79. ***Impatiens neo-barnesii*** **CEC Fisch**. Epiphytic acaulescent delicate pendulous herbs with scape carrying cream or nearly white flowers. Endemic to moss-covered tree trunks in South Western Ghats. Listed endangered in 1997 Red Data List (Walter and Gillett 1998).
80. ***Impatiens orchioides*** **Bedd.** Epiphytic, scapigerous herb with tuberous roots and reddish-brown or pale brown six to ten flowers borne in scape. Endemic to shola forests, Western Ghats. Listed vulnerable (Threatened Taxa of Kerala – Kerala State Biodiversity Board www.keralabiodiversity.org/index.php/...of.../109-threatened-flora).
81. ***Impatiens rufescens*** **Benth. ex Wight & Arn.** Terrestrial annual herb with large, axillary, solitary or in pairs, rose pink or purplish, ca 1.8 cm across flowers. Endemic to swampy and marshy areas of South Western Ghats.
82. ***Impatiens scapiflora*** **Hook.** Stemless herb with 4 cm across flowers toward apex. Endemic to wet rocks in evergreen forests/grasslands of Western Ghats. Confused with *I. acaulis*, which has the lower petal single lobe, whereas otherwise it is deeply bilobed.
83. ***Impatiens tenella*** **Heyne ex Hook. f.** A 20–60 cm tall herb with binate, axillary ca 1 cm across, dark pink, lilac to white flowers. Endemic to moist areas in grasslands of South Western Ghats. Listed endangered.

In addition, *Impatiens nilagirica* C. Fischer. has been reported endemic to Nilgiri Hills in the Western Ghats with very few records. It is listed in Red Data Book

(Nayar and Sastry 1988) and endangered in 1997 Red List of Threatened Plants (Walter and Gillett 1998).

84. ***Ipomoea digitata*** **L. (Convolvulaceae).** *Bilaikand*, an extensive perennial climber with beautiful pink/purple flowers. Native to Caribbean Islands, but naturalized in India with diversity from east to west coast. Grown as ornamental. Tuberous roots are used in Ayurveda for liver and spleen ailments and health promoting effects.
85. ***Ipomoea eriocarpa*** **R.Br.** *Buta*, creeper, stem hispid with small pale pink with a deeper throat flowers in axillary on sessile or peduncled cymes. Native to Africa, Asia, and Australia, including dry deciduous forests, scrub jungles, and plains of Karnataka and Kerala. Used as vegetable and fodder.
86. ***Ipomoea hederifolia*** **L.** Slender twining herb with deep red flowers in cyme. Native of tropical America. Naturalized to Tropical Asia, including Kerala, India, cultivated too.
87. ***Ipomoea nil*** **(L.) Roth.** Morning glory, a climbing annual herb with trifoliate leaves and flowers in various shades (blue, pink, or rose with white stripes or color edges). Native of tropical Americas. Naturalized in India and Sri Lanka and cultivated, commercial interspecific hybrid developed, reflecting the scope of interspecific breeding.
88. ***Ipomoea ochracea*** **(Lindl.) G. Don.** A woody perennial climber with sulfur-yellow flowers. Native of Africa, recorded in Western Ghats, India (Shimpale et al. 2012).
89. ***Iris crocea*** **Jacquem ex RC Foster (Iridaceae).** It is a bulbous perennial plant with golden yellow flowers. Native to Central Asia and grassy woods of Jammu and Kashmir, India. Cultivated, hybrids cultivar known, reflecting the scope of interspecific hybridization. Several isoflavonoids isolated from its rhizomes.
90. ***Iris decora*** **Wall.; syn.** ***I. nepalensis*** **D. Don.** This is rare dwarf plant with violet or pale blue flowers, self-incompatible, pollinated by insects, reflecting possible role of interspecific breeding. Native to China, India (Himachal Pradesh), Bhutan, and Nepal. Cultivated. Root is aperient, deobstruent, diuretic, and purgative.
91. ***Iris hookeriana*** **Foster.** Rhizomatous perennial herb with two or few blue purple flowers borne on short stalk. Native to tropical Asia on mountain slopes and alpine meadows of the Indian Western Himalayas. Cultivated. Medicinal too, anthelmintic.
92. ***Iris kashmiriana*** **Baker.** A rhizomatous perennial herb with straight, sword-shaped, glaucous leaves and tall, thick stem with two branches, holding 2–3 white, cream or pale blue, lilac, lavender, or blue purple flowers. Insect pollinated. Native to Western (Kashmir) and Central Himalayas of India. Cultivated, interspecific hybrids cultigen known. Antioxidant, used in traditional medicine.
93. ***Iris wattii*** **Baker ex Hook.f.** Bamboo iris, rhizomatous perennial herb with bamboo-like stem and lavender blue flowers. Native to grasslands and forest

margins of Indochina region. Cultivated, hybrid cultigen known, reflecting the potential of interspecific breeding.

Additionally, *Iris sikkimensis* Dykes is a rhizomatous perennial herb with pale or light green thin leaves, slender stem, and two or three lilac or purple flowers, with a white beard and orange tips. Sterile, like *I. hookeriana* and *I. kemaonensis*, thought to be the hybrid of two. Recorded in Sikkim. However, its presence has been questioned (Bahali 2005).

94. ***Ixora notoniana* Wall. ex D. Don. (Rubiaceae).** Large shrubs or small tree with terminal corymbose cyme inflorescence having subsessile, rose/red tinge flowers. Endemic to South Western Ghats along the margin of evergreen forests in medium and high elevations. Genetic resource for cultigen.

In addition, *Ixora goalparensis* Bremek., a 2 m tall shrub with white flowers, documented from Assam, while similar looking species, *I. longibracteata* Bremek., has been recently rediscovered from Jaldapara National Park, West Bengal, India (Karthigeyan and Arisdason 2015), whereas *Ixora rosella* Kurz., syn. *I. grandifolia* var. *rosella* (Kurz) Hook.f., known by herbarium, and *I. andamanica* and *I. nicobarica*, by type collection from Andaman and Nicobar, need further exploration for presence and distinctiveness.

95. ***Jasminum angustifolium* (L.) Willd. (Oleaceae).** *Banmallika* is a small climbing shrub with solitary or in three white star-like fragrant flowers. Display heterostyle promoting insect pollination. Native of India, Andaman Islands, and Sri Lanka forests. Used in sacred offerings, perfumery, and medicinal with anti-inflammatory and analgesic activities.
96. ***Jasminum arborescens* Roxb.** *Navamallika*, a large erect shrub with fragrant flowers in lax trichotomous cymes. Habitat to evergreen and semievergreen forests of the Northwest Himalayas, Peninsular India, and Sri Lanka. Genetic resource.
97. ***Jasminum attenuatum* Roxb. & G. Don.** A woody, stout, glabrous vine with white flowers. Native to China, India (Northeast Hills), Myanmar, and Thailand.
98. ***Jasminum auriculatum* Vahl.** *Juhi,* small climbing bushy plant with many beautiful and heavy gardenia-type scented flowers. Native to the Indian Subcontinent. Harvested from the wild and cultivated. Yields essential oils, used in perfumery. Used for sacred offerings.
99. ***Jasminum flexile* Vahl**. It is a climbing shrub with many white flowers borne in cymes up to 15 cm long. Genetic diversity is represented by var. *ovatum*, var. *travancorense*, and var. *flexile*. Native to evergreen, semievergreen, and shola forests and plains of Peninsular South India. Its absolute is known as *Taruni*.
100. ***Jasminum humile* L.** A 2–4 m tall evergreen shrub with clusters of usually six yellow, scented flowers. Native to Asia in dry valleys including the Indian

Western Himalayas and Bhutan. Grown as ornamental. Medicinal, astringent and tonic for the heart and bowels.

101. ***Jasminum lanceolarium* Roxb.** Robust, large climbing shrub with white very fragrant flowers on branched inflorescence. Native to East Asia, including China, Northeast India, and Indochina region. Used in traditional medicine, compounds like secoiridoid glucosides and flavanone glycosides isolated.
102. ***Jasminum multiflorum* (Burm.f.) Andr.** Star jasmine, an evergreen branched climbing shrub, with eight lobed white flowers in dense capitate cymes without fragrance. Native to moist deciduous to evergreen forests of the Indian Subcontinent and Indochina region. Cultivated as ornamental for offerings and folk medicine.
103. ***Jasminum parkeri* Dunn.** Dwarf jasmine, a domed evergreen shrub of 12 cm, with cute unscented yellow flowers. Endemic to the Northwest Himalayas, first collected from Chamba, Himachal Pradesh. Hardy. A potential ornamental, for rockeries.
104. ***Jasminum ritchiei* C.B. Clarke.** Climbing branches, nearly glabrous shrubs with three to nine white flowers on sub-paniculate cymes. Distributed in moist deciduous forests of Peninsular India and Sri Lanka. Extracted oil is used in toothache.
105. ***Jasminum rottlerianum* Wall ex DC.** *Vanamallika*, climber or reclining shrub, up to 2 m high with terminal subcapitate cymose cluster of 7–13 white flowers. Habitat to Western Peninsula, Konkan southward to Kerala. Ayurvedic, leaves, used in eczema.
106. ***Jasminum scandens* (Retz.) Vahl.** A scandent climbing shrub with drooping branches having white fragrant flowers. Habitat to moist deciduous forests of Western Ghats and Northeast India. Roots used in ringworm. Genetic resource.
107. ***Lilium mackliniae* Sealy (Liliaceae).** *Siroi lily* is a shade-loving lily, having bell-shaped flowers with pale bluish-pink petals, showing seven colors under the microscope. Endemic to Siroi Hill ranges of Manipur. State flower, with a postal stamp, rare, and endangered. Grows well in cultivation. Leaves used in skin disease.
108. ***Lilium wallichianum* Schult.f.** A bulbous plant with up to four trumpet-shaped white to creamy yellow flowers. Native to India and Myanmar, growing on slopes and grasslands. Variety *wallichianum* in the Himalayas and var. *neilgherrense* (Wight) H. Hara., in Nepal and South Western Ghats. Grown for fragrant flowers.

In addition, *Lilium roseum* Wall. ex Hook.f. has been documented by the Botanical Survey of India from West Himalayas in 1853 and 1892 but needs further exploration.

109. ***Lonicera hildebrandiana* Collett & Hemsl. (Caprifoliaceae).** Giant honeysuckle, an evergreen climber, with 12–14 cm long slender G-curving tubes that expand near the tip into wide flaring lobed flowers, starts white, turns

yellow, and finally dull orange. Self-incompatible, cross pollinated. Native of Southeast Asia, found in Indian foothills around Myanmar border.

110. ***Lonicera ligustrina*** **Wall.** An evergreen, semievergreen, or deciduous shrub ca. 1.5–2.5 m tall with bisexual, funnel-shaped yellowish-white flower. Native to evergreen forests of China, India (Nilgiri too), Bhutan, Nepal, and Myanmar with varietal diversity.
111. ***Lonicera tangutica*** **Maxim.** A deciduous spreading ca. 4 m tall shrub with bright yellowish-white tubular flower in peduncled pairs in leaf axils on stalk. Distributed in forests of Bhutan, China, India (Sikkim), and Nepal (1200–4000 m). Cultivated.
112. ***Pleione humilis*** **Lindle (Orchidaceae).** An epiphyte orchid, with ovoid-conical pseudobulbs carrying pure white single-flowered inflorescence, enveloped in green tubular sheath. Native of East Asia, including the Himalayas from Nepal to Sikkim. Growing on moss or smooth trunks of trees.
113. ***Polyalthia fragrans*** **(Dalz.) Bedd (Annonaceae).** A 15–20 m tall tree with fragrant leaves and flowers. Endemic, canopy tree in evergreen forests of the Western Ghats of India. Possess antioxidant and anti-inflammatory properties. Yields timber too.
114. ***Polyalthia meghalayensis*** **Prakash & Mehrotra.** Erect shrub described by Ved Prakash and Mehrotra (1990) from Meghalaya. Endemic to evergreen forests of Meghalaya, ca 800 m.
115. ***Polyalthia suberosa*** **(Roxb.) Thrwaites.** A 4–6 m tall tree branching from base with yellowish-green, mostly suffused with purple, solitary flowers. Found in evergreen forests of Indo-Malayan region, including the Western Ghats.
116. ***Primula denticulata*** **Sm. (Primulaceae).** Drumstick primrose, a beautiful Himalayan herb 8–32 cm tall, with heteromorphic, mauve to pinkish-blue flowers in compact spherical heads. Native to moist alpine regions of East Asia, from Afghanistan to the Himalayas (Kashmir to Assam) to China.
117. ***Primula macrophylla*** **D. Don.** A perennial 0.3 m tall herb with narrowly lanceolate leaves and purple, violet, or lilac flowers, often with a darker eye. Distributed from Afghanistan to southeast Tibet. Extremely variable with several botanical varieties, including dwarf var. *moorcroftiana* from Kashmir and NW India and var. *macrocarpa* from Nepal to Bhutan. Cultivated.
118. ***Primula sikkimensis*** **Hooker.** A perennial herb with leaves forming rosette and yellow flowers on a long pedicel, rarely cream-white, becoming dark green when dry. Native to the Himalayas, China, Nepal, Bhutan, India (Arunachal Pradesh, Sikkim), and Myanmar.
119. ***Primula stuartii*** **Wall.** A perennial 20–30 cm tall herb with one rarely two umbels of yellow flowers. Native to the Indian Subcontinent (Himachal Pradesh, Uttarakhand). Hybridizes with *P. macrophylla*.

Primula is a cultivated genus. Interspecific hybridization in cultivar development is the prime approach. Half of its species are from the Himalayas suggesting great promise.

120. ***Rhododendron arboreum* Sm. (Ericaceae).** It is an evergreen shrub/small tree with a showy bright red flower. Extremely variable in stature, hardiness, foliage, and flower color, leading to classification into subsp. *nilagiricum* (Zenker) Tagg., subsp. *cinnamomeum*, etc. Native to China, India, and Indochina region. It is the state tree/flower of Uttarakhand, Himachal, and Nagaland. Ayurvedic, used in increased serum cholesterol, triglycerides, atherosclerosis, etc. Cultivated and genetic resource as parent of most cultivated hybrids.
121. ***Rhododendron barbatum* Wall. ex G. Don.** Large shrub to small tree with tubular campanulate, fleshy, and deep to blood red flowers borne in densely umbellate inflorescence. Native to China, India (Eastern Himalayas), and Bhutan. Cultivated, gene resource for ornamental features, and possess medicinal potential.
122. ***Rhododendron campanulatum* D. Don.** A wild low-growing form, with shiny green foliage and bluish-purple pink metallic bloom. Native to the Indian Subcontinent. Cultivated ornamental. Known for medicinal use in body ache and throat pain.
123. ***Rhododendron formosum* Wall.** A shrub with bristled evergreen leaves and one to six open white trumpet flowers and flushes of apricot or pink in each inflorescence. Native to Northeast India. Genetically variable with varieties. Vulnerable.
124. ***Rhododendron grande* Wight.** Large 6–15 m tall tree with flowers from cream to yellow to pink, spotted with a blotch on the top petal. Native to mixed forests of the Eastern Himalayas of India, Nepal, Bhutan, southern Tibet, and China.
125. ***Rhododendron johnstoneanum* Watt ex Hutchinson.** Double diamond, a beautiful scented spreading evergreen shrub, with reddish brown peeling bark, dark green hairy leaves, and creamy yellow double flowers. Endemic to Manipur and Lushai Hills of Mizoram. Cultivated in the UK.
126. ***Rhododendron macabeanum* Watt ex Balf.f.** A large evergreen shrub/tree with glossy dark green leaves and bell-shaped, pale to deep yellow with a purple blotch, in dense rounded trusses flowers. Endemic to Manipur and Naga Hills. Cultivated.
127. ***Rhododendron subansiriense* Chamberlain.** A species endemic to India (Eastern Himalayas). Listed vulnerable in the IUCN Red List of Threatened Species (WCMC 1998). Cultivated rarely.
128. ***Rhododendron thomsonii* Hook. f.** A shrub 2–4 m tall, with leathery oblong-elliptic to ovate or orbicular to obovate leaves and red flowers. Native to northern India, Eastern Himalayas (Sikkim), Nepal, Bhutan, and China.
129. ***Rhododendron triflorum* Hook. f.** Evergreen shrubs/small tree with pale yellow flowers, with lobes tinged pink and brown spots inside. Native to the Indian Eastern Himalayas, Nepal, Bhutan, and the Far East. Intraspecific variation classified into variety *triflorum*, var. *bauhiniiflorum* (Watt ex Hutch.), endemic to Manipur. Cultivated.

In addition, there are species, such as *Rhododendron concinnoides* Hutch. & Kingdon, documented endemic to Arunachal Pradesh, introduced to the UK, and endangered, and *R. wattii* Cowan, with small population, endemic to Manipur, India, appearing natural hybrid of *R. macabeanum* x *R. arboretum* but questioned for presence, and listed Vulnerable in the IUCN Red List of Threatened Species (1998), needs further studies.

130. ***Rosa clinophylla*** **Thory (Rosaceae).** Erect shrub, 6 m tall with usually solitary white flowers having base of yellow shade on leafy branchlet. Reported from the Himalayas, India, Nepal, and Thailand. Cultivated in nurseries/gardens for sale.
131. ***Rosa longicuspis*** **Bertol.** A robust climber up to 6 m with flowers about 5 cm across, silky on the back, sweetly scented. Found in the Eastern Himalayas, India from Assam to Yunnan, and Sichuan, China, with varietal differentiation. Cultivated.
132. ***Rosa macrophylla*** **Lindl.** Himalayan rose, deciduous tall bush, with one to two terminal deep pink flower with essence. One of the most variable species. Native to Asia, including the Himalayas from Punjab to Sikkim. Grown in nurseries and gardens.
133. ***Rosa sericea*** **Lindle**. The silky rose, erect shrubs, 1–2 m tall with solitary, axillary flowers. Native to southwestern China extended to North India and the Himalayas (Sikkim), Bhutan, and Nepal. Cultivated as ornamental hedge.
134. ***Rosa webbiana*** **Wall.** Webb's Rose, common deciduous shrub with singly borne pink/red flowers with white center, scented. Fruits edible, possess high vitamin C content. Native to the Western Himalayas, from Pamir in Central Asia to Afghanistan, Kashmir, and Tibet.

Additionally, *Rosa moschata* Mill. Musk rose, suspected to have originated in the Western Himalayas, India, and *Rosa multiflora* Thunb., a native to eastern Asia, naturalized to the Eastern Himalayas, Assam, and Uttarakhand as invasive species, need further exploration.

135. ***Tabernaemontana alternifolia*** **L.; syn.** ***T. heyneana*** **Wall. (Apocynaceae).** A small tree with milky sap and white flowers borne in corymb-like cyme. Native to India in the Western Ghats. Source of latex/rubber. Ayurvedic plant, antimicrobial, used to treat skin infections.
136. ***Tabernaemontana coronaria*** **(Jacq.) Roxb.** It is a spreading, many-branched shrub with cymes of four to six salverform, waxy, pure white flowers. Native to parts of India (northeast), China, and Thailand. Medicinal, antioxidant. Cultivated.
137. ***Tabernaemontana divaricata*** **R.Br. ex Roem. & Schult.** An evergreen 1.5–1.8 m tall shrub, with characteristic milky latex and single "pinwheel" shape white unscented or double-scented flowers. Native and common garden plant in India, now throughout the tropics. Ayurvedic plant with antioxidant, cytotoxic properties.

138. ***Thunbergia fragrans* Roxb. (Acanthaceae).** A perennial climbing twiner with fragrant and attractive white flowers. Native to East Asia, including India and Sri Lanka, and invasive. Widely cultivated in the tropics and subtropics, medicinal too.
139. ***Thunbergia grandiflora* Roxb.** *Neel lata* is an evergreen vine with tuberous root and grow ca 20 m with blue to mauve flowers 8 cm across and 4 cm long tube, which is pale yellow inside. Native to East Asia, including India (northeast), Nepal, and Burma. Widely naturalized and invasive. Cultivated.
140. ***Thunbergia laurifolia* Lindl.** Laurel clock/blue trumpet vine, a fast-growing plant with trumpet-shaped, short broad tube, pale blue, white outside and yellowish inside flowers, in pendulous inflorescences. It is native to India (Starr et al. 2003) and Indochina, invasive. Cultivated as ornamental in the tropics. Medicinal, possess antioxidant, anti-inflammatory, and anticancer properties.
141. ***Thunbergia mysorensis* (Wight) Anderson.** Mysore trumpet vine or Indian clock vine is a woody-stemmed evergreen climber with hanging blossoms of flowers in a curved-shaped, yellow/maroon to brownish with outer red tip. Native to southern tropical India. Cultivated in tropical and subtropical gardens.
142. ***Vanda wightii* Reichb f. (Orchidaceae).** Epiphyte, with two to three flowers that are fragrant, 4–4.5 cm across, dirty brownish/greenish-yellow with light tessellations on sepals and petals, white at base and back. Distributed in deciduous forests of India (South Western Ghats) and Sri Lanka. Thought to be extinct (Singh 2001).

14.3 Perspective

The collection of ornamental plants in India mainly constitutes introductions made during medieval and British rules. Despite the richness of plant diversity and potential ornamental value of many indigenous plant species, only few regions, such as Cape, have been explored to some extent, but Bay Islands, Northeast Hills, including the alpine zones of the Himalayas are almost untouched and require thorough exploration, particularly for collection of wild species genetic diversity. Among the explored taxa, there has been confusion regarding the identity and taxonomic status of many, resulting in a number of synonym. There are only few organizations/institutes identified as the centers for biosystematic studies and/or devoted to research on this aspect in India, which needs attention from conventional plant science departments.

Despite huge ornamental and possible medicinal significance of many and high remuneration value as commercial crops, the ornamental plant genetic resources have not been a priority area in the national research and agribusiness systems. Hence, there is very limited knowledge available for many taxa and their members, for example, in case of genera *Gardenia*, *Hedychium*, etc. Therefore, there is a need for greater scientific attention, business entrepreneurship, and adoption of modern technique of protective cultivation in floriculture.

Many of the wild relatives of ornamental species are themselves potential ornamental and can be brought into regular cultivation through domestication and development of agronomic practices for their productive cultivation. Therefore, there is need for ornamental evaluation of the wild relatives of cultivated ornamental crop species for their direct exploitation, beyond the use as genetic resource. It may require research input for production of quality planting material, such as seeds, bulbs, corms, rhizomes, tubers, etc., and use of advance the biotechnological approaches, such as tissue cultured plants, synthetic seeds, etc.

The use of wild relatives for generation of desired diversity and broadening the genetic base through interspecific hybridization is most promising in case of ornamentals. This is because of their breeding behavior, which is often supported by cross pollination through insect pollinators that help facilitate generation of new genetic recombinants/variation in nature, whereas the vegetative propagation helps in fixing these variations and large-scale asexual multiplication for cultivation. The gametic sterility that may result in hybrid cultivars because of interspecific genomic incompatibility is of least consequence as flower, foliage, and canopy shape, size, and color are of the main economic consequences and not the seed yield. The number of artificial intergeneric hybrid cultivated on commercial scale in case of crinums, orchids, lilies, etc., corroborates this.

References

Bahali DD (2005) Does *Iris sikkimensis* Dykes occur in India? Curr Sci 89(10):1665–1666

Basu SK (1995) Genetic resources of palms. In: Chadha KL, Bhattacharjee SK (eds) Advances in horticulture Vol 12. Ornamental plants. Malhotra Publishing House, New Delhi, pp 203–222

Gaikwad SP, Garad KU, Gore RD (2014) *Crinum solapurense* (Amaryllidaceae), a new species from Maharashtra, India. Kew Bull 69:9095

Kambhar SV, Harihar NS, Katrahalli KS (2014) Recollection of endemic species *Barleria stocksii* T. Anderson (Acanthaceae) in Karnataka State, India, after 140 years. Check List 10:419

Karthigeyan K, Arisdason W (2015) *Ixora longibracteata* Bremek. (Rubiaceae), an addition to flora of India, with notes on its status and distribution. Taiwania 60(2):95–98

Khoshoo TN (1968) Genetic improvement of ornamentals in India-prospects, experimental approaches, and suggestions. Indian J Genet 28a:87–98

Lekhak MM, Yadav SR (2011) Karyotype studies in two critically endangered and endemic *Crinum* species (Amaryllidaceae) from Northern-Western Ghats of India. Nucleus 54(1):25–30

Lekhak MM, Yadav SR (2012) *Crinum malabaricum* (Amaryllidaceae), a remarkable new aquatic species from Kerala, India and lectotypification of *Crinum thaianum*. Kew Bull 67:521–526

Nayar MP, Sastri ARK (eds) (1987) Red data book of Indian plants, vol 1. Botanical Survey of India, Calcutta, India, pp 1–371

Nayar MP, Sastri ARK (eds) (1988) Red data book of Indian plants, vol 2. Botanical Survey of India, Calcutta, India, pp 1–273

Prakash V, Mehrotra BN (1990) A new species of *Polyalthia* (Annonaceae) from Meghalaya, India. Nord J Bot 10(1):45–47

Punekar SA, Kavade SP, Datar MN, Lakshminarasimhan P, Rao PSN (2004) *Crinum woodrowii* Baker (Amaryllidaceae), hitherto assumed to be extinct, rediscovered after a century from Mahabaleshwar, India. Curr Sci 87:1049–1051

Shimpale VB, Kshirsagar PR, Pawar NV (2012) *Ipomoea ochracea* (Convolvulaceae) – a new record for India. Rheedea 22(2):99–102

Singh DK (2001) Orchid diversity in India; an overview. In: Pathak P, Seghal RN, Shekhar N, Sharma M, Sood A (eds) Orchids: science and commerce. FRI, Dehradun, pp 35–65

Starr F, Starr K, Loope L (2003) *Thunbergia laurifolia* Blue trumpet vine Acanthaceae. http://www.hear.org/starr/hiplants/reports/pdf/thunbergia_laurifolia.pdf

Vij SP (1995) Genetic resources of orchids. In: Chadha KL, Bhattacharjee SK (eds) Advances in horticulture vol 12, Ornamental plants. Malhotra Publishing House, New Delhi, pp 153–181

Walter KS, Gillett HJ (eds) (1998) 1997 IUCN Red List of Threatened Plants. Compiled by the World Conservation Monitoring Centre. IUCN – The World Conservation Union, Gland, Switzerland and Cambridge, UK. lxiv + p 862

WCMC (World Conservation Monitoring Centre) (1998) *Rhododendron subansiriense*. The IUCN red list of threatened species 1998: e.T31243A9618851. http://dx.doi.org/10.2305/IUCN.UK.1998.RLTS.T31243A9618851.en

Yusuf M, Begum J, Hoque MN, Chowdhury JU (2009) Medicinal plants of Bangladesh (revised and enlarged). Bangladesh Council of Scientific and Industrial Research Laboratories, Chittagong, p 794

Agroforestry 15

15.1 Introduction

Agroforestry involves mixing of trees or other woody perennials in crop/animal production system to benefit from the resultant ecological, economical interactions and material gains. It includes trees, shrubs, bamboos, etc. In India, it is believed to have been initiated during *Vedic* era. *Khejri* or *sami* (*Prosopis cineraria*), *aswattha* (*Ficus religiosa*), *palasa* (*Butea monosperma*), and *varana* (*Crataeva roxburghii*) have been mentioned in ancient literature of *Rig Veda*, *Atharva Veda*, and other ancient scriptures. Traditionally, agroforestry systems vary from simple forms of shifting cultivation to established complex home gardens: from systems involving sparse stands of trees on farm lands, like plantation of *Prosopis cineraria* in the arid zone of Western India, to high-density complex multistoried homestead gardens in humid tropical lands of the Western Ghats. Indigenous wild species related to *khejri* (*Prosopis cineraria*), one of the earliest agroforestry trees, are few, but many more have naturalized from the Americas as weed in India.

The tree species that are generally used in agroforestry are either nitrogen fixer or efficient recycler of the available nutrients. Examples are *Acacia*, *Albizia*, *Alnus*, *Erythrina*, *Faidherbia*, *Ficus*, *Gliricidia sepium* (introduced), *Inga*, *Parkinsonia*, *Pilhecellobiuin*, *Prosopis*, *Robinia*, and *Sesbania* and the multipurpose trees, such as *Areca catechu*, *Azadirachta indica*, *Butea moonosperma*, *Dalbergia sissoo*, *Tectona grandis*, *Artocarpus* spp., *Casuarina equisetifolia*, *Mangifera indica*, *Phoenix dactylifera*, *Pterocarpus santalinus* (red sandal), *Ceiba pentandra*, *Leucaena leucocephala*, *Grevillea robusta*, *Bambusa arundinacea*, *Erythrina variegata*, *Syzygium aromaticum*, and *Ziziphus mauritiana*. Besides these, some of the species associated with agroecologies in this regard include willows (*Salix* spp.), poplars (*Populus* spp.), *Morus alba*, and *Acer* spp. in the temperate regions of Himalayas (Fig. 15.1) and coconut (*Cocos nucifera*) and pandanus (*Pandanus tectorius*) commonly seen near the canals and backwaters of the coastal region of Malabar. Palmyrah palm (*Borassus flabellifer*) is a common multipurpose palm grown in coastal Andhra Pradesh.

A.K. Singh, *Wild Relatives of Cultivated Plants in India*,
DOI 10.1007/978-981-10-5116-6_15

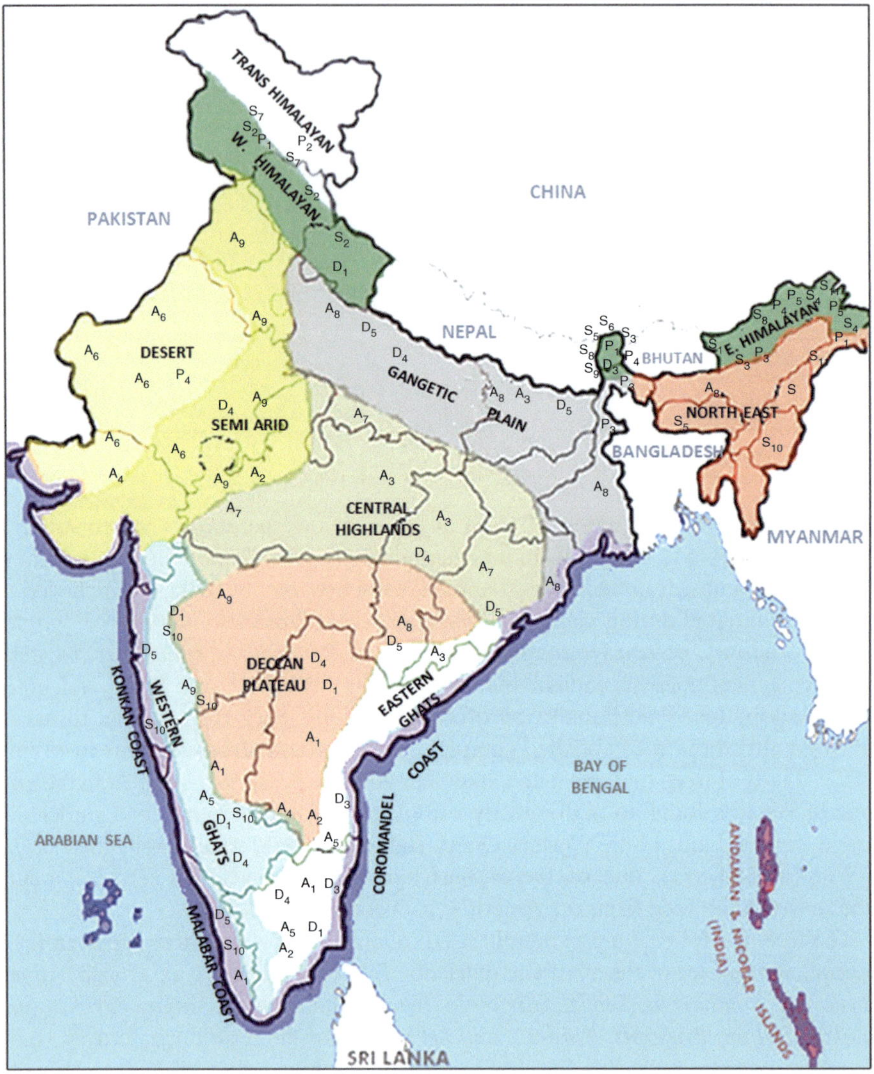

Fig. 15.1 Biogeographical regions associated with the distribution of wild relative species of some common agroforestry tree species, *Acacia*, *Dalbergia*, *Populus*, and *Salix*
A1 = *Acacia chundra*; A2 = *A. concinna*; A3 = *A. donaldii*; A4 = *A. eburnea*; A5 = *A. ferruginea*; A6 = *A. jacquemontii*; A7 = *A. leucophloea*; A8 = *A. pennata*; A9 = *A. senegal*; D1 = *Dalbergia congesta*; D2 = *D. coromandeliana*; D3 = *D. paniculata*; D4 = *D. latifolia*; D5 = *D. volubilis*; P1 = *Populus ciliata*; P2 = *P. euphratica*; P3 = *P. gamblei*; P4 = *P. jacquemontii*; P5 = *P. rotundifolia*; S1 = *Salix daltoniana*; S2 = *S. denticulata*; S3 = *S. longiflora*; S4 = *S. obscura*; S5 = *S. psilostigma*; S6 = *S. radinostachya*; S7 = *S. sclerophylla*; S8 = *S. sikkimensis*; S9 = *S. stomatophora*; S10 = *S. tetrasperma*; S11 = *S. wallichiana*

Nearly half of the agroforestry species are introductions to India and do not have any representation of their wild relatives in India. Whereas, the wild relatives of multipurpose indigenous agroforestry species, such as *Areca catechu*, *Artocarpus* spp., *Ficus* spp., *Mangifera indica*, *Ziziphus mauritiana*, etc., have been documented and discussed under the respective crop groups of their primary use. In this chapter, we discuss the wild relatives of indigenous agroforestry species, such as the temperate species belonging to some genera, such as *Acer*, *Morus*, *Pinus*, *Populus*, *Salix*, etc., found in the Himalayan region (Fig. 15.1), and the *Acacia*, *Dalbergia*, *Madhuca* spp., *Tectona grandis*, etc., that are naturally distributed in subtropics of the Himalayan region and tropics of Peninsular India, including both the Western and the Eastern Ghats (Fig. 15.1).

15.2 Wild Species Found in India

1. ***Acacia chundra* (Rottler) Willd. (Fabaceae/Leguminosae)**. A deciduous 8 m tall tree; bark rusty brown, rough, and peeling off in thin flakes; yellowish-white flowers in axillary of 1–3 clustered spikes. It is found in deciduous forests and plains of Peninsular India, Sri Lanka, and Myanmar. Wood is used to make agricultural implements. Heart wood yields katha and cutch.
2. ***Acacia concinna* (Willd.) DC**. *Shikakai*, a climbing shrub native to Asia, including the warm plains of Central and South India. The tree is food for the larvae of the butterfly. Used in traditional hair care, as fruit and bark contains saponins, a foaming agent used in shampoo. Leaves used for chutney and pods for health problems like skin disorders, itching, psoriasis, etc.
3. ***Acacia donaldii* Haines**. Small tree endemic to the sandstone, shale, and quartzite in Bihar, MP, and northeastern Ghats (Koraput region). Listed as data deficient in the IUCN Red List of Threatened Species (Chadburn 2012a).
4. ***Acacia eburnea* (L.f.) Willd.; syn. *A. campbellii* Arn**. *Pahari Kikar*, a large shrub or a small tree, with rough dark-grey bark and stalked pod. Found throughout India in dry regions. Under threat.
5. ***Acacia ferruginea* DC**. *Safed-khair*, a medium-sized 20 m tall tree with dark brown rough bark, cracking into pieces, and flat pod fruit. Native to the dry deciduous forests of Peninsular India and Sri Lanka. Figures in epic *Mahabharata*, sacred.
6. ***Acacia jacquemontii* Benth**. *Baonḷī*, an erect shrub, 1.8 m tall, with multiple shoots. It is very hardy and inhabitant to dry sandy soils, with high temperature, and excessive drought-prone areas. Native to Thar Desert and arid and semi-arid regions of India. Suitable for erosion control, afforestation, and social and agroforestry. Tribes use bark, roots, and gum to cure various diseases. Limited cultivation.
7. ***Acacia leucophloea* (Roxb.) Willd**. *Ronjh*, a 12-m tall thorny deciduous tree with crooked stem and light-yellow to cream color flowers in pedunculate glomerules, aggregated in terminal or axillary panicles. Native to Indo-Malayan

region including plains and dry deciduous forests of India. Multipurpose tree, including medicinal. Known agroforestry plant.

8. ***Acacia pennata* (L.) Willd**. *Agla bel*, *Biswal*, a woody climber with smooth stem and heads of cream flowers on axillary or terminal racemes. Paleotropic, including the roads and moist deciduous forest border of India, ascending to 1700 m in the Himalayas. Medicinal.
9. ***Acacia senegal* (L.) Willd**. A 5–12 m tall tree, native of sub-Saharan African desert, naturalized to arid and semi-arid regions of India from the West Coast to Andaman. Gum arabic is the main product. India produces some 90% of the global market. Also, used as forage, food (dried seed), rope, wood, and medicine for its astringent properties. Known multipurpose agroforestry tree.
10. ***Acer caesium* Wall. ex Brandis (Aceraceae)**. Asian maple, a 25 m tall tree with gray bark. Leaves are noncompound, with five shallow lobes and dentate margin. Native to Himalayas from Afghanistan to Indian Subcontinent to Western China. Indian type classified as ssp. *caesium*, which possesses seed dormancy. Cultivated.
11. ***Acer oblongum* Wall. ex DC**. Himalayan maple, a medium-sized evergreen to semi-deciduous 15–22 m tall tree. Distributed in Central, East, and Southeast Asia, including the humid climate of the Himalayan forests (600–2000 m) of India.
12. ***Albizia amara* (Roxb.) Boiv (Fabaceae/Leguminosae)**. *Krishna Siris* is a mid-sized, deciduous much-branched tree with thin bark whitish-yellow flowers. Native of Africa and dry deciduous forests of the Indian Subcontinent and Sri Lanka. The genetic variability is expressed in the form subspecies *amara* and subsp. *sericocephala*. Powdered leaves used as a soap substitute.
13. ***Albizia lebbeck* (L.) Willd**. *Siris*, a large deciduous, up to 15 m tall tree with yellowish-brown, rough fissured bark and spreading crown and umbellate clusters of greenish-white fragrant solitary flower in leaf axil. Probably native of tropical Asia, including deciduous forests and plains of India. Neutralized to tropics. Multipurpose tree, with ornamental, shade, forage, fuel, value, etc. Known agroforestry tree, cultivated.
14. ***Albizia procera* (Roxb.) Benth**. White siris, a deciduous 30 m tall tree with an open canopy, greenish-yellow, smooth bark, and whitish globose heads of flowers. Native to East Asian moist deciduous forests and plains, from North India to Southeast Asia to Australia. Potential wood and charcoal. Known agroforestry tree.
15. ***Albizia thompsonii* Brandis**. A deciduous tree endemic to tropical dry deciduous forests and open scrub forests of the Eastern Ghats. Variable, with variety *thompsoni* and var. *galbana* Haines. Listed near threatened in the IUCN Red List of Threatened Species (Chadburn 2012b).
16. ***Alnus nepalensis* D. Don (Betulaceae)**. Nepalese alder, a large deciduous tree with silver-gray bark. Native to East Asia, found all over Himalayas from Pakistan to southwest China. It helps erosion control, root nodules fix nitrogen, and wood is used for making boxes, firewood, etc. For these benefits, alder is

used in the traditional agroforestry system of Northeast Hill region of India, particularly Nagaland.

17. ***Alnus nitida* (Spach) Endl.** Western Himalayan alder, a deciduous 20 m tall tree. Native to Indian Subcontinent, from Pakistan to Indian Western Himalayan states and Nepal. Cultivated along streams or roadside for wood and bark, used in dyeing and tanning, and local medicine.
18. ***Anogeissus latifolia* (Roxb. ex DC.) Wall. ex Bedd. (Combretaceae).** *Dhaura*, a small- to medium-sized tree native to the dry deciduous forests of Himalayas, India, Nepal, Myanmar, and Sri Lanka. Source of wood gum/resin, leaves contain gallotannins, used for tanning and medicine. Known agroforestry tree.
19. ***Anogeissus sericea* Brandis.** An important hardwood tree and native to Indian Subcontinent. Two cytological variants, variety *sericea* important for wood, fuel, and fodder and var. *nummularia* King ex Duthie, a multipurpose hardwood tree, are both rare and endemic to the forest of Rajasthan. Known agroforestry tree.
20. ***Caesalpinia bonduc* (L.) Roxb; syn. *C. crista* L. (Caesalpiniaceae).** *Kantkarej*, a vine-like shrub, 6 m tall, scrambles over other vegetation. Stems armed with recurved prickles and yellow flower. Paleotropic, including tropical and subtropical India. Ayurvedic herb, seeds, leaves, and bark are used for antiperiodic, antipyretic, antispasmodic, anthelmintic, antivirus, antifungal, and anti-inflammatory properties.
21. ***Caesalpinia decapetala* (Roth) Alston.** A robust, thorny, evergreen 2–4 m tall shrub or climber, forming dense thickets. Stem is hairy and thorny with yellow flowers. Native to East Asia, including India. Invasive, serious weed in Africa. Cultivated in tropics and subtropics as hedge. Medicinal for relieving pain.
22. ***Dalbergia congesta* Graham ex Wight & Arn. (Fabaceae/Leguminosae).** Shrub or climber, endemic to semi-evergreen forests of South Western Ghats (Nilgiri Hills), now found all over India. Medicinal and endangered (Chadburn 2012c).
23. ***Dalbergia coromandeliana* Prain.** An erect shrub, endemic to scrub forests along the Cormandel Coast of Tamil Nadu, India. Medicinal and has shown antibacterial activity.
24. ***Dalbergia lanceolaria* subsp. *paniculata* (Roxb.) Thoth.; syn. *D. paniculata* Roxb.** A medium-sized deciduous tree and native to dry slopes of tropical Asia, including the Indian Subcontinent. Used as folk medicine and insecticide.
25. ***Dalbergia latifolia* Roxb.** Indian rosewood/*shisham* is a premier 25 m tall timber tree with grey bark, peeling in long fibers. Native to tropical Asia, including low-elevation tropical monsoon forests of India. It produces a hard, durable, heavy wood, which is valued timber and fuelwood. Medicinal. Popular in agroforestry.
26. ***Dalbergia volubilis* Roxb.** A woody climbing shrub. Native to moist deciduous forests of India, Sri Lanka, and Myanmar. Exhibit anti-inflammatory and anti-arthritic activities. Leaves juice applied to aphthae. An ecotype classified as var. *assamica* Thoth. Another species documented is *Dalbergia wattii* C.B. Clarke, reported endemic to Manipur and Nagaland. Wood is used for fuel.

27. ***Madhuca longifolia* (L.) J.F. Macbr.; syn. *M. indica* (Sapotaceae)**. A fast-growing ca. 20 m tall tropical tree, native to Central and North Indian plains and forests of the Indian Subcontinent, and Myanmar. Infraspecific variation classified into variety *latifolia* and var. *longifolia*. Multipurpose tree and yields food, fuel, and oil. Flowers are the base for beverage, *mahua*. Wood, medicinal, and *Antheraea paphia*, which produces *Tassar* Silk is fed on its leaves. Part of agroforestry with tribes.

Madhuca diplostemon (C.B. Clarke) P. Royen, a wild relative of *mahua* (*M. longifolia*) recorded from Deccan Peninsula, not found in recent surveys, is probably extinct.

28. ***Populus ciliata* Wall. ex Royle**. Himalayan poplar, a deciduous tree with tall clean straight trunk, growing up to 20 m tall. Native to East Asia, including the Western Himalayan states of India, Nepal, Bhutan, Sikkim, and Myanmar. Used for erosion control, forage, fuel, timber, and medicine. Known agroforestry tree.
29. ***Populus euphratica* Oliv. (Salicaceae)**. *Bahan*, a medium-sized to large-sized deciduous tree with rarely a straight stem, often bushy. Native to Africa, temperate Asia, and the Indian Subcontinent. Known as a desert poplar and is drought and salinity tolerant. Interspecific hybridization attempted in Iran. Grown in agroforestry for wood.
30. ***Populus gamblei* Dude**. A tree of about 15 m high. It is the southernmost poplar tree in the Northern Hemisphere, endemic to India (from Darjeeling hills, North Bengal, Arunachal Pradesh, and Sikkim). Reported wild from Nagaland too (Naithani et al. 2005).
31. ***Populus jacquemontii* Dode var. *glauca* (Haines) Kimura**. A beautiful poplar tree with bronze-purple leaves, maturing to blue-green on the upper surfaces and serrate margin with bisexual flower. Endemic to Eastern Himalayas and Nepal.
32. ***Populus rotundifolia* Griff**. A shrub or tree 2–10 m tall; bark grayish-white and smooth. Leaves broadly ovate or orbicular. Distributed on mountain slopes in China to the blue pine forests of Eastern Himalayas, Bhutan, etc.
33. ***Prosopis farcta* (Sol. ex Russell) J.F. Macbr. (Fabaceae/Leguminosae)**. A woody perennial dwarf shrub/tree weed with branches downy, slender, whitish, armed with pointed prickles, and small yellow flowers. Native to North Africa, temperate Asia, and the Indian Subcontinent. Used as green forage, pods yield dye, and seeds medicinal.
34. ***Prosopis cineraria* (L.) Druce; syn. *P. spicigera* L.** *Jand/shami/khejri*, sacred evergreen tree with bluish-green foliage. Native to Arabian Peninsula, West Asia, and India. Wood used to kindle the sacred fire, for charcoal, and in construction and agricultural implements making. Browsed by livestock; bark used for treating muscular and joint pains, leprosy, piles, and worm infestations; leaves and fruits for curing nervous disorders. Common agroforestry tree in western plains of India.

In addition, *Prosopis juliflora* (SW.) DC. is introduced from the Americas and Caribbean Islands and has become an invasive weed in India, used for firewood and to make barriers.

35. ***Salix daltoniana* Andersson (Salicaceae)**. Shrub or small tree of 1–5 m tall growing densely along the hill slopes and roadsides in temperate mountains of Central and Eastern Himalayas between 2743 and 4400 m.
36. ***Salix denticulata* Andersson; syn. *S. elegans* Wall. ex Andersson**. An elegant willow, a small tree/shrub, with twigs pubescent becoming hairless. Endemic to Himalayas from Afghanistan to India, Nepal, and China. A new flavonoid glycoside (1) has been isolated from the aerial parts. Closely related to white willow, *S. alba*.
37. ***Salix longiflora* Wall. ex Andersson**. Shrub or small tree; 2–6 m tall; branchlets dull, blackish, and glabrous. Leaves elliptic with the entire margins. Habitat among shrubs on mountain slopes and in valleys of Eastern Himalayas (Bhutan, Nepal, Sikkim).
38. ***Salix obscura* Andersson**. Tree 1–6 m tall; branchlets glabrous or sparsely pubescent and blackish. Leaves entirely oblanceolate, revolute, and brownish. Distributed along streams in blue pine forests of Eastern Himalayas. Rediscovered after 121 years (MoEF 2014).
39. ***Salix psilostigma* Andersson**. Shrubs/small tree with brown branchlets, pilose when young, with glabrescent leaves. Native to mountain slopes of Sikkim, Meghalaya, Bhutan, and Nepal.
40. ***Salix radinostachya* C.K. Schneid**. An ornamental shrubby willow, having deep-red stems in winter and bright-green large lanceolate leaves. Found in Eastern Himalayas (Sikkim), Bhutan, and China.
41. ***Salix sclerophylla* Andersson**. An erect shrub 2 m tall. Branchlets with many nodes, tortuous, and dull purplish-red in color. Found on the mountain slopes of Western Himalayas (Kashmir, Nepal). Meets requirements of fodder, fuel, and small timber in higher reaches of the cold desert of Ladakh.
42. ***Salix sikkimensis* Andersson**. Shrub 1–2 m tall; branches stout, angular, blackish, and finely pubescent. Leaves elliptic or oblanceolate. Found on streamsides in Eastern Himalayas (Bhutan, Nepal, Sikkim, and Arunachal Pradesh).
43. ***Salix stomatophora* Floderus**. New species, like *S. wallichiana* with leaves 3–4 × 1–1.5 cm, densely pubescent along the midrib but otherwise glabrescent. Distributed in Eastern Himalayas (Sikkim), Bhutan, etc.
44. ***Salix tetrasperma* Roxb**. Indian willow, a medium-sized dioecious tree, ca. 25 m tall, bark 10–12 mm thick, pale brown, rough, vertically fissured, blaze red; young branches silky pubescent. It is habitat to wet and swampy places; shedding leaves at the end of the monsoon. Native to the Indian Subcontinent (Manipur in northeast, Maharashtra, Karnataka, Kerala). Planted along water channels to control erosion.
45. ***Salix wallichiana* Andersson; syn. *S. disperma* Roxb. ex D. Don**. A deciduous fast-growing tree ca. 2–10 m tall and branchlets pubescent. Leaves elliptic,

acuminate, or acute and base cuneate or rounded; margins serrulate. Native to China, Eastern Himalayas, Bhutan, etc. Frost resistant. Medicinal herb.

Salix species are cross-fertile, and hybridization does occur, both in nature and cultivation. Therefore, there is a scope of interspecific breeding in genetic enhancement/improvement of cultivated species by introgressing desirable traits from wild relatives.

46. ***Salvadora persica* L. (Salvadoraceae)**. *Pilu* or miswak is a shrub to small evergreen tree with a crooked trunk. It has a pleasant fragrance and warm and pungent taste. Native to dry arid regions of subtropical to tropical Africa, extending to West Asia, Middle East, and India. Sticks are natural toothbrush supporting dental hygiene. Roots and stem are source of numerous active compounds, such as chlorhexidine gluconate of medicinal value, and oil. Grazed by livestock. Harvested from the wild and cultivated in agroforestry. Another species found in India is *bada pilu*, *Salvadora oleoides* growing 6–9 m tall.

15.3 Perspective

India has a long historical tradition of growing trees on farms and around homes. It has been proved beyond doubt that agroforestry is more profitable to farmers than agriculture or forestry for a certain area, like arid zones. Therefore, there is a need for an intensification of agroforestry systems. However, despite multiple benefits from agroforestry systems, majority of the farmers have been hesitant to adopt these systems on large scale primarily because of certain apprehensions about the tree components, such as long rotation, reduction in the gross area, and complicated legal procedures involved in tree farming, falling, trade, and market fluctuations. This situation demands an awareness drive regarding the merits of agroforestry.

Major genetic gain in forestry species, such as *Eucalyptus*, *Populus*, and *Salix*, has been achieved through interspecific hybridization. Therefore, wild species to be involved in the genetic improvement program needs to be evaluated for traits such as soil improvement, for marketable products, etc. as per potential economic value of cultivated species in the system. Also, it is for the compatibility with present and possible future crops and freedom from any negative physical or chemical effects on the soil and noncompetition with the crop to be associated.

As the interspecific breeding is a nonconventional method of tree improvement, to facilitate successful interspecific hybridization, much work is needed in the exciting area of biosystematics and biotechnology. They may help overcome some of the basic hybridization problems, and the use of parasexual means in genetic improvement of target species and in large scale multiplication, facilitating cultivation.

As genetic improvement of domesticated species requires introgression of genes from wild-related species/populations, the wild resource may require a domestication strategy to reduce pressure on wild resources and facilitate conservation with sustainable use. Therefore, domestication/conservation programs of wild relatives should either be by maintaining them in situ or ex situ stands of natural varieties/strains under protective management.

The benefit of agroforestry would include improvement of biodiversity, especially by the abundance of "edge effects" of the agroforestry systems permitting a synergistic improved integrated crop protection with their association with trees, chosen to stimulate the hyperparasite (parasites of parasites) population of crops. It is a promising way forward for further use of associated species, besides the conventional benefit of soil fertility and economic material gains, for example, the use of *Acacia* spp. as a host of lac insect, *Kerria lacca*. Further, in recent years innovations in the promotion and use of medicinal plants in agroforestry have shown potential resource toward the generation of income for local communities and provide for long-term benefit in forest management and forestry systems.

References

Chadburn H (2012a) *Acacia donaldii*. The IUCN red list of threatened species 2012: e.T19891442A20125493

Chadburn H (2012b) *Albizia thompsonii*. The IUCN red list of threatened species 2012: e.T33644A20121746

Chadburn H (2012c) *Dalbergia congesta*. The IUCN red list of threatened species 2012: e.T19892181A20055198

MoEF (Ministry of Environment and Forests) (2014) Annual Report 2013-2014, Ministry of Environment, Forests and Climate Change Government of India, New Delhi, p 4

Naithani HB, Deorani SC, Yaden TA (2005) *Populus gamblei* Dode – a tree new to Nagaland, India. Indian Forester 131(9):1212–1216

16 Cottage Industry Crops and Others

16.1 Introduction

There are plant species, which provide livelihood support to farmers or traditional artisans and craftsmen. These people have inherited their skill as an art form from their previous generations, which have been facilitated by the local availability of raw plant material required for such craft and/or cottage industries. It has helped in strengthening livelihood support of locals with additional income. Sericulture, i.e., rearing silkworms for silk production, is such an activity, involving cultivation of plant such as mulberry *Morus* spp. to host the silkworms. *Morus* is represented by four species, i.e., *M. indica* L., *M. alba* L., *M. laevigata* Wall., and *M. serrata* Roxb. Of these *M. serrata* and *M. laevigata* are natural to Himalaya.

Bamboo, cane, pandan, etc., are other plant species providing raw material for craft. As per the National Bamboo Mission (Sarma 2008), India has 128 species belonging to 18 genera. Of these 87 species are naturally occurring. The mission has identified 13 species belonging to seven genera of economic importance to support both traditional and industrial use, whereas another report based on the growth habits and utility have identified 21 species belonging to the genera *Bambusa* (9) and *Dendrocalamus* (7) and one species each of *Melocanna*, *Ochlandra*, *Pseudoxytenathera*, *Schizostachyum*, and *Thyrsostachys* for cultivation. They together with their ancestral wild species formulate the genetic resources. More than 50% of bamboo species occur in Eastern India. Other areas rich in bamboo diversity are the Western Ghats, Bastar, and Andaman and Nicobar Islands. *Bambusa* and *Dendrocalamus* are also found in other tropical regions of the country.

In case of cane, where long slender stem is used by furniture and handicraft industries, a total of 18 species are used. Of these, nine species, *Calamus brandisii* Becc., *C. delessertianus* Becc., *C. gamblei* Becc., *C. hookerianus* Becc., *C. pseudotenuis* Becc. ex. Becc. & Hook f., *C. rotang* L., *C. thwaitesii* Becc. & Hook f., *C. travancoricus* Bedd. ex. Becc. & Hook. f., and *C. vattayila* Renuka., are native to South India. Seven species, *C. acanthospathus* Griff., *C. erectus* Roxb., *C. flagellum* Griff., *C. gracilis* Roxb., *C. guruba* (Buch. -Ham.) ex Mart., *C. leptospadix* Griff.,

A.K. Singh, *Wild Relatives of Cultivated Plants in India*,
DOI 10.1007/978-981-10-5116-6_16

and *C. tenuis* Roxb., are native to Northeast India, while two species, *Calamus andamanicus* Kurz. and *C. pseudorivalis* Becc., belong to the Andaman and Nicobar Islands (KFRI 2004).

Though pandan is basically planted as ornamental plants on the seaward slopes and crests of frontal sand dunes to prevent wind erosion, its leaves are used in handicraft. The craftsmen generally collect the pandan leaves both from wild and cultivated plants. In addition, *kewra* is an industrial flavoring product distilled from the flowers *Pandanus odoratissimus*, which is used to flavor drinks and desserts in Indian cuisine. Other cultivated *Pandanus* species, *P. odorifer*, *P. amaryllifolius*, *P. sanderi*, and *P. dubius*, are all exotic. Indigenous *Pandanus* species are distributed throughout the tropical and subtropical islands of Indian Ocean. Based on a study of these, *P. palakkadensis* and *P. mangalorensis* are critically endangered; *P. unguifer* and *P. martinianus* are endangered; *P. diversus* (Silchar) and *Benstonea thwaitesii* are vulnerable; *P. emarginatus* is data deficient; and others are of least concern (Zanan and Nadaf 2013).

In addition, there are plant species, which may not fall in a specific category but are cultivated to harvest raw material for production of specific product for commercial use, for example, *Indigofera* spp. for dye; *Sapindus* spp. for base material of cosmetics; *Curcuma* species for minor products like arrowroot (starch); *Jatropha* spp. for industrial oil or biofuel; host plants like *Butea monosperma* (palasa), *Schleichera oleosa* (kusum) for rearing of useful insect such as *Kerria lacca*, etc. Besides, many as multipurpose tree, like *Bombax ceiba*, are cultivated. The wild relative genetic diversity of these is distributed in areas suited to their habitat.

16.2 Wild Species Found in India

1. ***Bambusa balcooa* Roxb. (Poaceae/Gramineae)**. *Barak*, a clumping bamboo without nodal roots. Native to Indochina and the Indian Subcontinent (Bangladesh and Nepal too). Used as source of food and for house construction, paper pulp, fiber, woven mats, etc. It is one of the commercial species, cultivated under homestead gardens.
2. ***Bambusa bambos* (L.) Voss**. Indian thorny bamboo, a tall, bright-green-colored, spiny, clumping bamboo, native to East Asia, including India. Cultivated for erosion control and used as food, forage, and material for paper, pulp, ply, craft, and fiber.
3. ***Bambusa nutans* Wall. ex Munro**. It is an evergreen or deciduous, clump-forming bamboo with thick-walled culms, 6–15 m tall, with cane diameter of 5–10 cm. Distributed wild on hill slopes and uplands in the well-drained soil of East Asia, including India. Used as source of poles and fiber for paper industry. Cultivated.
4. ***Bambusa pallida* Munro; syn. *B. pseudopallida* R.B. Majumdar**. Perennial with erect culms 15–20 m tall, woody without nodal roots. Native to East Asia, including India (Arunachal Pradesh, Meghalaya, Sikkim, West Bengal),

Indochina, and China. Source of cane and fiber for making baskets and mats. Marginally cultivated.

5. ***Bambusa polymorpha* Munro**. *Jama betwa*/Bengal bamboo, a large handsome tufted tall bamboo 25–30 m tall. Native to East Asia, including Northeast India, Myanmar, and Thailand. Shoots are edible and source of canes for house construction, woven matting, baskets, furniture, handicrafts, and pulp.
6. ***Bambusa tulda* Roxb**. *Deobans*, with culms 14 m tall, apically drooping. Native to the Indian Subcontinent, Indochina, and China. Most useful and source of food, shade/shelter, and cane/fiber. Extensively used in paper industry, furniture, making baskets, and concrete reinforcement. Cultivated. Very close to *B. nutans*.
7. ***Benstonea foetida* Roxb. (Pandanaceae)**. Dioecious, shrub to medium-sized tree, with a broad canopy, spirally arranged leaves, heavy fruit, and moderate growth rate. Distributed in the evergreen forest of the Western Ghats and East Coast of West Bengal. Another species, *Benstonea thwaitesii* (Martelli) Callm. & Buerki, is distributed in deep forests of the Western Ghats.
8. ***Bombax insigne* Wall. (Bombacaceae)**. Deciduous tree, ca. 25 m tall with showy, scarlet, pink, creamy, or white flowers. Native to rocky areas of evergreen, semi-evergreen, and moist deciduous forests of India and Myanmar. Variability classified into the variety *polystemon* Prain. and var. *andamanica* Prain. from Andaman. Related to *Semal* (*B. ceiba*). Variety *polystemon* reported critically endangered in the IUCN Red List of Threatened Species (WCMC 1998).
9. ***Calamus* spp. rattan palms: *Calamus tenuis* Roxb. (Arecaceae)**. *Jati*, a climber, growing in clumps with very large slender, scandent stem. Distributed in the tropical Himalayas of Assam, Bangladesh, and Myanmar. A commercially exploited rattan species, used in furniture making and craft, mostly from wild, occasionally cultivated. Other two species exploited commercially are *Calamus leptospadix* (Tita) and *C. floribundus* (Lejai). The other wild species of significance are *C. guruba* (Sundi), *C. flagellum* (Raidang), etc.
10. ***Dendrocalamus asper* (Poaceae/Gramineae)**. Dense-clumping species, 15–20 m tall, and 8–12 cm in diameter. Culms are grayish green, becoming dull brown on drying. Native to Southeast Asia, common in Northeast India and Sri Lanka. Young shoots consumed as a vegetable, and cane is used for building material in heavy construction.
11. ***Dendrocalamus brandisii* (Munro) Kurz.; syn. *Bambusa brandisii* Munro**.A giant clumping bamboo with smooth, straight, thornless culm ca. 25–30 m tall and 13–30 cm in diameter. Native of Northeast India (Jiribam, Manipur), Myanmar, Thailand, and Andaman. Used as vegetable and cane in construction, basketry, and handicrafts. Cultivated.
12. ***Dendrocalamus calostachyus* (Kurz) Kurz**. Indian Bamboo, a tufted large clustering bamboo with culms 20–30 m tall. Distributed in Northeast India and Myanmar. Extensively cultivated.
13. ***Dendrocalamus giganteus* Munro**. Giant bamboo, a giant tropical and subtropical, dense-clumping species, native to Southeast Asia, and found in India

too. Cultivated along river banks and in home gardens for erosion control, ornamental, vegetable, and cane and used in construction and handicraft.

14. ***Dendrocalamus hamiltonii* Gamble**. A species with culms growing up to 15–18 m high. Variation classified into variety *edulis*, which has palatable shoots. Native to East Asia, including the Indian Subcontinent. Yields vegetable, cane, and fiber suitable for weaving material.
15. ***Dendrocalamus longispathus* (Kurz) Kurz**. It is an evergreen, clumping bamboo with culms 10–20 m tall, green with white blooms when young, becomes grayish green when dry. Native to the Indian Subcontinent (Mizoram) and Indochina.
16. ***Dendrocalamus strictus* (Roxb.) Nees**. *Male bamboo*, a tropical/subtropical clumping bamboo 8–16 m tall, with often solid culm, branching from the base. Two varieties are known, *strictus* and *sericeus*. Native to the Indian Subcontinent and Southeast Asia. Cultivated; shoots are edible; cane used for rafters, baskets, sticks, etc., fiber for paper pulp, and leaves in thatching.
17. ***Freycinetia sumatrana* Hemsley (Pandanaceae)**. A climber with thorny leaves growing on large trees. Native of South Asia, including deep forests of Andaman and Nicobar Islands along with another species, *Freycinetia rigidifolia* Hemsley.
18. ***Girardinia palmata* (Forsk.) Gaudich.; syn. *G. diversifolia* (Urticaceae)**. *Himalayan nettle*, perennial herbaceous shrub. Grows wild in Indian states of Himalayas and Nepal. Bark yields fiber used in clothing, bags to floor mats, and rope.
19. ***Indigofera arrecta* A. Rich. (Fabaceae/Leguminosae)**. Bengal indigo, an erect, woody shrub of 3 m high. Native of Africa and Arabian Peninsula, naturalized in India along roads/forests. Used for erosion control and as soil improver and source of blue indigo. Cultivated in agroforestry. Closely related to *I. suffruticosa* and *I. tinctoria* with intermediate form, possibly of hybrid (Orwa et al. 2009).
20. ***Indigofera cassioides* Rottler ex DC**. Erect shrubs, 1.5 m tall. Native to grasslands and hill slopes of East Asia, including India and Indochina. Variants found from Lahul to Kumaon Himalayas, classified as var. *mussooriensis* SN Biswas; syn. *I. cedrorum* Dunn. Root decoction and powder used as medicine; yields dye too.
21. ***Indigofera coerulea* Roxb**. Well-known shrub, native to tropical Africa and India (Gujarat). A rare variant, var. *monosperma* (Santapau) Santapau found in North Gujarat (Kathiawar), whereas var. *caerulea* in Tamil Nadu with upper side of the leaflets glabrous and var. *occidentalis* with upper side hairy. Used as folk medicine for jaundice, epilepsy, night blindness, etc., besides being source of dye.
22. ***Indigofera constricta* (Thw.) Trimen**. Erect shrub with scarlet flower. Distributed in semi-evergreen forests of Peninsular India (Bundelkhand) and Sri Lanka. Rare.

23. ***Indigofera deccanensis* Sanjappa**. Perennial shrub. Endemic to slopes of hills and banks of rivers in Peninsular India (Bundelkhand, Northwest Deccan Plateau). Used for green manure.
24. ***Indigofera dosua* Buch. -Ham. ex D. Don**. A shrub, 5–20 cm tall with pink flowers. Distributed on valley slopes in Eastern Himalayas (India), Indochina, and Malay region. The variant, variety *simlensis* (Ali) Sanj., is endemic to Lahul to Kumaon Himalayas, northwestern Himalayas. Dye yielding.
25. ***Indigofera gangetica* Sanjappa**. Perennial shrub. Found in Himachal Pradesh, Uttarakhand, and upper Gangetic Plains. Used as manure and fodder.
26. ***Indigofera glandulosa* Wendl**. A subshrub, up to 75 cm tall, branchlets pubescent with pink flowers. Found in scrub jungles, fallow, and exposed barren lands of Peninsular India. Variety *sykesii* Griffth ex Baker found in northern Dcccan. Forage; seeds have three times protein of wheat.
27. ***Indigofera trita* L.f**. Asian indigo is spreading to erect; annual or short-lived perennial herb or shrub. Native to tropics, including India as weed. Variable, variety *trita*, is an erect undershrub, distributed in the wastelands and scrub forests of West Asia, Indo-Malesia to Australia. Others are ssp. *subulata* and ssp. *marginata* (Schinz) Polhill, which are shrubs endemic to South Western Ghats.
28. ***Indigofera tinctoria* L**. Suffrutescent herbs, ca. 1.5 m tall. Native to Africa; Asia, including the Indian Subcontinent; and Australia. Habitat to degraded forest, scrub jungles, and plains. Soil improver, yield tannin/dye, and used in treatment of fever, liver and spleen disorders, rheumatoid arthritis, gout, gray hairs, etc. Cultivated.
29. ***Indigofera trifoliata* L**. Prostrate herbs with radiating branches. Distributed in degraded forests and scrub jungles of Indo-Malesia, China, and Australia. Infraspecific taxon, subspecies *trifoliata* L.; syn. *I. barberi* Gamble, recorded from Tamil Nadu and Andhra Pradesh. Rare.
30. ***Jatropha* gossypifolia L. (Euphorbiaceae).** *Ratanjot*, a shrub, purplish green with glandular hair on younger parts. Native of South America, naturalized in Peninsular India as weed. Medicinal and gene source for *J. curcas*, a potential biofuel.
31. ***Jatropha tanjorensis* J.L. Ellis & Saroja**. Up to 3 m tall shrub, distributed in the plains of Peninsular India. Phenotypically intermediates between *J. curcas* L. and *J. gossypifolia* L. Cytological and biochemical results suggest it to be an interspecific hybrid between the two (Prabhakaran and Sujatha 1999). Therefore, it is a pre-bred genetic resource demonstrating a possibility of interspecific gene introgression.
32. ***Jatropha villosa* Wight var. *ramnadensis* Raman**. Undershrub, endemic to dry deciduous forests of Peninsular India, evolved as new ecotypes, from naturalized *J. villosa* in southern Eastern Ghats, Ramanathapuram, and Tamil Nadu. Genetic resource for *J. curcas*.
33. ***Melocanna baccifera* (Roxb.) Kurz; syn. *M. bambusoides* Trin. (Poaceae/ Gramineae).** Muli bamboo, a tall, small-culmed bamboo with greenish young culms and straw-colored old culms. It is native to the Indian Subcontinent and

Indochina region. Young shoot used as vegetable, while cane for basketry and construction. It is one of the two bamboos belonging to the genus.

34. ***Millettia pinnata* (L.) Panigrahi.; syn. *Pongamia pinnata* (L.) Pierre; *P. glabra* Vent. (Leguminosae/Fabaceae)**. *Karanj* or *Pongamia*, a large canopy tree, ca. 15–25 m tall, with small clusters of white, purple, and pink flowers. Native to East and South Asia, including India, Southeast Asia, and Australia, with wide variation. Cultivated in landscaping, known for tree–seed–oil, a source of industrial oil/biofuel, fuel, medicine to cure infectious diseases and host for *lac* insect. Drought resistant.
35. ***Morinda* spp. (Rubiaceae)**. Cultivated *Morinda citrifolia*, Noni is a multipurpose tree with wild relatives such as *M. angustifolia* Roxb., a medium-sized tree, ca. 6 m tall, found in deciduous forests of India, Sri Lanka, Myanmar, and China. *Morinda persicifolia* Buch. -Ham. is found in tropical Himalayas; *M. reticulata* Gamble (syn. *M. coreia* Buch. -Ham) and *M. citrifolia* L. in South Eastern Ghats and Western Ghats (Andhra Pradesh, Tamil Nadu, Karnataka, Kerala) and Andaman and Nicobar Islands (Gamble 1925).
36. ***Morus alba* L.; syn. *M. serrata* Roxb**. *Shahatut*, a deciduous shrub/small-sized 10–20 m tall monoecious tree with serrated margin leaves. Native to China, naturalized in India. Drought tolerant. Cultivated as ornamental in agroforestry and revegetator and for rearing silkworms. Timber used for furniture, fruit eaten, and medicinal.
37. ***Morus macroura* Miq.; syn. *M. laevigata* Wall**. Himalayan mulberry, a medium-sized deciduous, monoecious, long-life, hardy tree with long-fruited aggregate of berries. Native of deciduous/mixed forests of China and tropical Asia, including India in subtropical Himalayas. Resistant to drought, salinity, and frost. Source of wood and fiber. Cultivated for fruits in gardens.
38. ***Ochlandra* spp. (Poaceae/Gramineae)**. A genus of Indian bamboo with around 11 species endemic to the Western Ghats. *Ochlandra travancorica* (Bedd.) Gamble. Tall reed-like grasses with culms to 5 m tall, endemic to evergreen/semi-evergreen forests of South Western Ghats. Others are *O. beddomei*, a shrub with culms 10–12 m high, 3–4 cm in diameter, endemic to South Western Ghats.
39. ***Oxytenanthera nigrociliata* (Buse) Kurz (Poaceae/Gramineae)**. *Kaligoda*, a densely tufted evergreen bamboo with culms 10–18 m high and ca. 10 cm in diameter. Distributed in East Asia–southern China, India, Myanmar, Thailand, and Indonesia. Harvested from the wild, source of food, culms split, and used to make utensils and basketry.
40. ***Pandanus furcatus* Roxb.; syn. *P. nepalensis* H. St. John (Pandanaceae)**. Himalayan screw pine, arborescent, ca. 5 m tall and perched on stilt-like aerial roots. Native to East Asia, including the Indian Subcontinent. Ornamental leaves and fiber used for weaving mats, ropes, house-building materials, hats, and carry bags.
41. ***Pandanus kaida* Kurz.; syn. *P. unipapillatus* Dennst**. A large shrub, kept short in cultivation. Distributed in marshy areas and swamps of the Western Ghats.

Used for fencing and insect repellent and in medicine, while leaves for making mats, boxes, and hats.

42. ***Pandanus leram* Jones**. Nicobar breadfruit, an evergreen tree. Distributed in the interior of littoral forest, from the Andaman and Nicobar Islands to Indonesia. Local gathers its edible fruit from the wild, while use leaves and fiber in craft.
43. ***Pandanus mangalorensis* Nadaf & Zanan**. A new species with multiple branching, triangular inflorescence, and drupes. Endemic to Mangalore, Karnataka, South India. Critically endangered (Zanan and Nadaf 2012a)
44. ***Pandanus martinianus* Nadaf & Zanan**. A new species differing in dimension and shape of syncarp and arrangement of drupes. Endemic to the foothills of West Siang, Arunachal Pradesh, and Dhemaji, Assam, India. Endangered (Zanan and Nadaf 2012b, 2013).
45. ***Pandanus odorifer* (Forssk.) Kuntz.; syn. *P. fascicularis* Lam**. *Kewda* or fragrant screw pine is an aromatic small-branched screw pine. Distributed in mangrove forests and sea coasts of temperate and tropical Asia, including the Western Ghats, East Coast, and Andaman and Nicobar Islands. Yields flavoring agent, essential oil, fiber for craft, and medicine. Cultivated.
46. ***Pandanus palakkadensis* Nadaf, Zanan, & Wakte**. A new species, closely related to *P. canaranus* and *P. furcatus* with drupe having flat pileus and elevated shoulders, endemic to Palakkad, Kerala, India (Nadaf et al. 2011). Critically endangered (Zanan and Nadaf 2013).
47. ***Pandanus unguifer* Hook. f**. A clump-forming shrub with simple slender, decumbent stem. Leaves lorate, more abruptly acuminate, and gradually narrowed to base. Distributed in South Sikkim and Darjeeling, West Bengal. Endangered (Zanan and Nadaf 2013). Close to Himalayan screw pine, *P. furcatus*.

Besides, above species there are more wild *Pandanus* species, characterized with good fiber content facilitating craft, fragrant spadix, and edible fruit.

48. ***Pinus gerardiana* Wall. ex D. Don (Pinaceae)**. *Chilgoza pine*, a slow-growing 10–20 m tall tree with deep, wide, and open crowns, native to eastern Afghanistan and the Northwestern Himalayas, India. Known for edible nut, *Chilgoza*, cultivated as cash crop. Listed near threatened in the IUCN Red List of Threatened Species 2013 (Farjon 2013).
49. ***Pinus kesiya* Royle ex Gordon**. *Khasi pine*, a large tall (30–35 m) tree, with whorled branches and an oval crown. Most widely distributed in Asia from Khasi Hills, Meghalaya, India, to further east up to Vietnam. Yields gum/resin; wood is soft and light and used for diverse purposes, including boxes, paper pulp, craft, etc.
50. ***Pinus wallichiana* A. B. Jacks**. Himalayan pine, a coniferous evergreen tree with leaves in bundles of five needles. Native to valleys in Himalayas, Karakoram, and Hindu Kush mountains and Afghanistan to southwest China. Cultivated,

for wood, firewood, gum/resin, and source of turpentine. Popular is an ornamental tree for parks/gardens.

51. ***Rhus chinensis* Mill.** A deciduous ca. 6 m tall tree, native to East and Southeast Asia, including tropical and subtropical India. Possesses compounds with strong antiviral, antibacterial, anticancer, hepatoprotective, antidiarrheal, and antioxidant activities; cultivated ornamental; source of tannin, dye, etc. and galls used as medicine.
52. ***Rhus griffithii* Hook. f.** Trees, up to 10 m tall. Distributed in Eastern Himalayas (1300–2000 m), West Bengal, Sikkim, Arunachal Pradesh, Assam, and Meghalaya. Young leaves are used as vegetable. Other species found are *Rhus mysorensis* G. Don, a bushy small tree, ca. 8 m tall, common in foothills, scrub jungle to 900 m. Northwest India to the Peninsular India; *R. punjabensis* J. L. Stewart, small 5–15 m tall trees, distributed in NW India and variety *sinica* in Himalayas and East Asia.
53. ***Rubia edgeworthii* Hook. f.** A perennial scandent to climbing herb. Found on grassy slopes in Northwest Himalayas. Ornamental, related to cultivated madder *Rubia cordifolia*, whose rhizome and roots are source of a red pigment, textile dye. Other *Rubia* species found in India are *R. manjith* Roxb., *R. tibetica* Hook. f., *R. wallichiana* Decne., etc.
54. ***Sapindus trifoliatus* L.; syn. *S. laurifolius* Vahl**. *Ritha*, a tree up to 18 m high with ovoid drupe fruits 16–18 mm across, lobed, smooth, brownish-yellow, and rusty. Endemic to semi-evergreen and moist deciduous forests and plains of South Asia. Cultivated as ornamental. The drupes contain saponin, a natural surfactant used in soap and shampoo industry; medicinal and insecticidal. Other sister species, *S. mukorossi* Gaertn., occur in North India east to the Himalayas and *S. emarginatus* Vahl., a ca. 10 m tall tree, inhabitant to dry deciduous forests and margins of grasslands in South Asia (Western Ghats). They can be used for the same purpose.
55. ***Shorea talura* Roxb.; syn. *S. roxburghii* G. Don**. A tree up to 25 m tall with simple large leaves. Native to Indo-Malayan region, including southern Eastern Ghats and Cauvery Basin. Commercially used as a host of *lac* insect in Mysore and Madras. Another sister species is *Shorea tumbuggaia* Roxb., endemic to Andhra Pradesh and Tamil Nadu. Valued for timber and pharmaceutical properties. Threatened.
56. ***Schleichera oleosa* (Lour.)**. *Kusum* is a large nearly evergreen tree. Native to Southeast Asia and the Indian Subcontinent, in the foothills of Himalayas, western Deccan to Sri Lanka. Host for *Kerria lacca*; yields kusum oil from the seed used in hairdressing and medicine.

16.3 Perspective

Raw material of cultivated species of several taxa used in cottage industry, craft, etc., has reduced pressure on main sources, such as wood for furniture. However, their limited cultivation and predominant use from the wild have now started putting pressure on their populations leading to the depletion of these resources. There is a need that their wild relative's gene pool is conserved and evaluated for economic potential for use in improvement of cultivated species or for direct commercial exploitation. This would require greater research efforts on these neglected or marginally cultivated species and their wild relatives, for generation of further knowledge on them both from basic and applied point of view either for their direct exploitation or use in the improvement of cultivated species. This will help in overcoming the inherent challenges faced in the exploitation of wild relatives. For example, bamboo, being cross-pollinated species, exhibit greater variation in the wild, which can be profitably exploited and visualized for in situ conservation of populations in specific ecological niche, harboring specific genetic variability. Also, from conservation point of view, shifting cultivation in northeast and other anthropogenic development resulting from population explosion has been adversely effecting bamboo gene pools in the entire country, though more specifically in Northeast India. Vigorous flowering of few important bamboo species in the northeast is one of the natural threats for frequent disaster and natural mode of erosion. It is required that monopodial group of bamboo species spreading over a large area be declared as gene sanctuaries for the concerned group, whereas sympodial species distributed sporadically be conserved through community-based approach.

These efforts shall facilitate conservation and sustainable use of wild genetic resources of such taxa in genetic improvement of cultivated species on long-term basis or alternative sources of raw material for direct use. Also, it may lead to diversification in use of these groups of taxa in nontraditional manner and develop more innovative products, such as jewelry, blinds, natural fiber, etc.

References

Farjon A (2013) *Pinus gerardiana*. The IUCN red list of threatened species 2013: e.T34189A2850009

Gamble JS (1925) Flora of the Presidency of Madras. Reprint, Botanical Survey of India, Calcutta, pp 1916–1935

KFRI (Kerala Forest Research Institute) (2004). Resource Enhancement and Processing Of Cane & Bamboo Species Suitable For Handicrafts, Research Report No 256, ISSN No. 0970-8103

Nadaf AB, Zanan RL, Wakte KV (2011) A new endemic species of Pandanaceae from India: *Pandanus palakkadensis*. Kew Bull 66:183–186

Orwa C, Mutua A, Kindt R, Jamnadass R, Anthony S (2009). Agroforestree database: a tree reference and selection guide version 4.0. World Agroforestry Centre, Kenya, pp 1–5

Prabakaran AJ, Sujatha M (1999) *Jatropha tanjorensis* Ellis & Saroja, a natural interspecific hybrid occurring in Tamil Nadu, India. Genet Resour Crop Evol 46:213–218

Sarma SK (ed) National Bamboo Mission (2008) Hand Book on Bamboo. Cane and Bamboo Technology Center Guwahati, Assam, India, under the Ministry of Agriculture, Govt India, pp 44

WCMC (World Conservation Monitoring Centre) (1998) *Bombax insigne* var. *polystemon*. The IUCN red list of threatened species 1998: e.T33489A9782099. http://dx.doi.org/10.2305/IUCN.UK.1998.RLTS.T33489A9782099.en

Zanan RL, Nadaf AB (2012a) *Pandanus mangalorensis*: a new species of Pandanaceae from Southern India. Kew Bull 67(3):555–559

Zanan RL, Nadaf AB (2012b) *Pandanus martinianus* (Pandanaceae), a new endemic species from northeastern India. Phytotaxa 73:1–7

Zanan RL, Nadaf AB (2013) Conservation status of Indian Pandanaceae. Am J Plant Sci 4:51–56

Part III

Conservation of Wild Relatives

17 Classification of Wild Species to Facilitate Conservation and Gene Transfer

17.1 Introduction

Correct identity of the wild relatives of cultivated species with distinctive features and verified taxonomic entity is key to facilitate their successful utilization in breeding programs of cultivated species by transfer of desirable gene. Unfortunately, the issue of synonymy, i.e., multiple names for a single species entity, has been critical in the assessment of biological diversity in general and wild species diversity. A systematic study is essential to estimate wild species diversity, with an objective of reducing or solving the misunderstanding about such lacunae/gaps. The biological species concept plays a key role in this regard, though it is underappreciated in all biodiversity matters (Mallet 1995), including genetic resources, especially with those concerned with species diversity and synonymy, affecting their use. Widespread acceptance of the biological species concept (Mayr 1942) led many taxonomists in the mid-twentieth century to treat earlier names as synonyms of large polytypic species. Modern workers, perhaps increasingly motivated by the phylogenetic interests, lay greater emphasis on diagnosis of taxa and less on their reproductive limits (Cracraft 1989). Thus, a trend for reutilization of names that were earlier treated synonyms has been seen. Nearly 15% of the names of all valid species are today considered synonyms.

Measuring biological diversity with any purpose often concerns with enumerating numbers of species. An understanding of the extent of species diversity is crucial to the range of studies including bioprospecting, economic exploitation, threat, ecosystem function, and conservation and use. A core aspect of this extrapolation is the number of species already described for a given group (crop). Large discrepancies remain between estimates of described species of any group of considerable size of seed plants. Recent estimates of the number of described seed plants differ by 62% (Scotland and Wortley 2003). These widely different estimates introduce a level of uncertainty in calculation of genetic diversity. A key factor affecting estimates of described species is that of multiple names for a single biological entity, i.e., synonymy. For any group, because of synonymy, the number of published

A.K. Singh, *Wild Relatives of Cultivated Plants in India*,
DOI 10.1007/978-981-10-5116-6_17

names exceeds the actual number of described species. For seed plants, the number of published name is 1,015,000 in *Index Kewensis* but not the extent of synonymy, because many groups have not been subjected to complete (experimental) taxonomy. Govaerts (2001) estimates, therefore, imply that 58.4% of 1,015,000 binomials in *Index Kewensis* are synonyms (Scotland and Wortley 2003), and therefore, the estimates of the species are often not based on rigorous understanding of species boundaries and pattern of synonymy. The range of synonymy rates for some sample taxa lies between 67 and 88%. The mean percentage of the names that are synonyms, counting binomials only and excluding intraspecific names, is 78%. The sample taxa cover full range of issues relating to synonymy. For example, although *Strabilanthes* has many synonyms due to overzealous publication of new genera by Bremekamp (1944), the closely related *Hemigraphis* does not but still these two have a similar level of synonymy. Further, the rates of synonymy across a wide range of taxa and geography are strikingly similar (Scotland and Wortley 2003). For large species, rich groups of plants, synonymy and number of undescribed species remain the two largest constraints in understanding the species diversity and use.

In the present effort, there has been an attempt to avoid duplicates and unaccepted or unresolved species names and include only distinct genotypes with information on phylogeny and reproductive isolation. However, it can be observed that there are many species with synonyms and of doubtful identity and taxonomic entity, needing further investigation to resolve the issues of doubt and poor understanding, to facilitate conservation and use without redundancy. Synonyms are not purely errors in applying systematic classification but are also by-products of the systematic concept used. Future investigations using experimental taxonomy approach and biological concept of species will assuredly reveal many current names to be synonyms of others, now in circulation. Changing systematic concepts and improved technologies for detecting and defining species will continue to breathe new life into yesterday's entombed synonyms and taxonomy. This will resolve phylogenetic confusion and will also promote predictive and effective use of wild relatives' genetic resources in crop improvement and facilitate conservation of distinct and unique genotypes avoiding duplicates.

17.2 Experimental Taxonomy

Studies of plants, both in the field and in the laboratory, have yielded information which appears to various investigators to bear directly on the nature and delimitation of taxonomic units. The classification of organisms based on experimental facts has been termed "experimental taxonomy." Orthodox taxonomy is concerned with the convenient tabulation and grouping based on morphological similarities and dissimilarities. Experimental taxonomy fully appreciates the value of morphological differences; in fact the cytologist has disclosed a fresh field for such investigation—but it also seeks to show the causes which are responsible for these differences and ascertain their physiological, ecological, or genetical nature.

Most plants are hermaphrodites and are regularly capable of inbreeding. Some species in fact regularly inbreed to reproduce with preservations of their unique features, although many species have evolved mechanisms by which the frequency of inbreeding can be reduced and promote introgression of gene(s) from outside populations. Some species are capable of vegetative propagation, unaided by humans, and produce plantlets that can be dispersed much like seed. Still other species produce seeds asexually. Because of these observations, it became obvious that the reproductive biology of plants is integral to an understanding of species limits and of the process of speciation. This led to the development of concept of biological species, as per this, *species* are groups of interbreeding (or potentially) natural populations, which are reproductively isolated from other such groups. Establishment of this concept necessitates application of conventional (both orthodox and experimental), as well as molecular biosystematics.

The term "biosystematy" was defined by Camp and Gilly (1943) as an attempt to (1) delimit the natural biotic units and (2) to apply these units to a system of nomenclature, adequate to the task of conveying precise information regarding their essential features, defined limits, relationships, variability, and dynamic structure. Biosystematics came to be understood as including the collection of observational and experimental data on the breeding system as part of the basis for making taxonomic decisions. It would involve using following methods:

1. Field study
    (i) Sampling of population systems showing variation
    (ii) Ecological observations on population systems showing variation
    (iii) Collection of specimens for herbarium preservation and of diaspores (seeds, rhizomes, tubers, etc.) for culture in experimental garden
2. Herbarium study
    (i) Recording of range of variation within species
    (ii) Recording of geographical distribution and detection of discontinuity, if any (disjunction and variance) from specimens
3. Experimental work
    (a) *Field*:
        (i) Observation on variations under controlled environment.
        (ii) Sympatric trial through hybridization between population systems.
        (iii) Observation on vitality and vigor of hybrids.
    (b) *Laboratory*:
        (i) Cytological observation on chromosome number, karyomorphology, and homology of chromosomes in hybrids at meiosis
        (ii) Observation on miscellaneous aspects, viz., anatomy, embryology, palynology, etc., for relationship evaluation in case of non-hybridizing populations
        (iii) Perform molecular biosystematic experiments
4. Information documentation and analysis

Above findings are compiled with baseline data for geographical distribution, comparative morphology, cytology, cross-compatibility, hybrid morphology and fertility, etc., to describe the essential features of the species that shall help in the identity of the species and for further analysis to trace the phylogenetic relationship with related taxa.

17.2.1 Biochemical and Molecular Biosystematics

Biochemical and molecular techniques are available approaches for evaluating genetic diversity in plants. Seed protein has been successfully used in defining species relationships in various groups of plants, while isozymes have been widely used as molecular markers for the identification of the genetic relationships among genera, species, and varieties. In many genera, isozyme electrophoresis has been used to study the phylogenetic relationships. In the last two decades, an enormous development has taken place in biochemical and molecular systematics. The approaches in this field have been diverse, both in theory and in the operational aspects of gathering and handling data. A broad distinction should be made between the systematic studies of (a) the different classes of micro-molecules or chemical phenotypes and (b) macromolecules which are close to the core of hereditary information, i.e., the (genotypic) DNA sequence versus the (phenotypic) RNA and protein sequences.

Molecular biosystematics, although field based, uses both micro- and macromolecules with virtues of taxonomic markers, being polymorphic and conservative to resolve distinctions as well as microevolution concerning species. A precise idea of such molecules/compounds can be gathered by following further biochemical investigations using various phytochemical analysis techniques for assessing the variation, both for primary and secondary metabolites. However, modern biosystematics should not be confused with molecular taxonomy. Although related, biosystematics always resolves variation pattern in population systems first in the field, which is eventually evaluated either by in-field genetic experiments or by in-lab molecular studies.

A. *Micro-molecules (Biochemical Analysis: Chemotaxonomy)*
   1. Primary metabolites: ethanol, citric acid, lactic acid, aconitic acid, and certain amino acids
   2. Secondary metabolites:
      (a) Phenolic compounds: flavonoids, phenolic acids, coumarins, xanthones, quinones, etc.
      (b) Terpenes: monoterpenes (geraniol, menthol, pinene, camphor, carvone,. etc.), diterpenes (phytol), sesquiterpenes (farnesol), triterpene (squalene), tetraterpene (carotenoids), polyterpene (natural rubber)
      (c) Tannins: hydrolysable (ellagitannins), condensed (proanthocyanin, leucocin)
      (d) Alkaloids
      (e) Glucosinolates, etc.

B. *Macromolecules (Molecular Analysis)*

It is being performed for better understanding of variation at molecular level, which may be responsible for phenotypic variation.

1. Large molecules of starch, cellulose, and lipids, containing no hereditary information.
2. Molecules contain hereditary information, which are primary (DNA: nuclear, chloroplast, mitochondrial), secondary (RNAs (m-RNA, r-RNA, t-RNA, Sn-RNA, RNA I, Cp-RNA & Mt.-RNA), and tertiary (proteins) in nature.

Aspects Studied and Methods Used in Molecular Biosystematics

1. Protein material: ribulose-1, 5-bisphosphate carboxylase/oxygenase (RUBISCO), seed albumin, plastocyanins, cytochrome C, plastoquinone, etc.
   Methods: Immunoelectrophoresis, comparative amino acid sequencing, isoelectric focusing, electrophoresis
2. Nucleotide: material, DNA (nuclear, mitochondrial, and chloroplastidial), RNA (ribosomal)
   Methods: DNA–DNA hybridization, DNA–RNA hybridization, nucleotide sequencing

Nucleotides Useful in Biosystematics Study

1. Nuclear: rDNA -18S (encoding the smaller subunit of rRNA); ITS I and II + 5.8S to a lesser extent 26S gene (along with 5.8S genes encoding the larger subunit of rRNA), gap C gene encoding cytosolic glyceraldehyde-3-phosphate dehydrogenase (GAPDH)
2. Chloroplast: (rbcL gene of cpDNA encoding RUBISCO, matK gene within trnk introns, ndhF etc.)
3. Mitochondrial: (mt DNA-nad1b/c exon sequence, mat R sequence)
4. Ribosomal (rRNA): 5S, 16S, and 23S in prokaryotes and 5S, 5.8S, and 26S in eukaryotes

Biosystematics, being the field-based experimental taxonomy, has the potential to help in species identification and thereby biodiversity assessment. Since it has three major components, viz., morphology, genetics, and ecology, which are the best tools to work out in the respective manner the three components of biodiversity, viz., species, genetics, and ecological diversities, it is nearly perfect. Perfection in the assessment of biodiversity is certain to pave the pathway not only to recognition of species and infraspecific entities but also to trigger the optimum sustainable utilization of these bio-resources. Undoubtedly biosystematics has the potential to establish rapprochement between traditional taxonomy and molecular taxonomy.

Consequently, even at intraspecific level, it helps in further classification of variation with internationally accepted taxonomic hierarchy. The three intraspecific categories below the rank of the species are:

Subspecies: A population of some biotype providing regional appearance/composition of a species in relation to physical, chemical, genetic, and biological aspects
Variety: A population of some biotype forming local appearance/composition of a species in relation to a specific physical, chemical, genetic, and biological aspects
Forma: A category with sporadic variation in one or two characters

Biosystematics studies, in addition to resolving the taxonomic issues and species identity and entity, also lead to the following advantages:

- Delimit every natural biotic entity composing biodiversity
- Provide a more meaningful system of nomenclature
- Resolve microevolution leading to speciation
- Ascertain genetic relationships among organisms
- Overall improvement of traditional taxonomy and optimization of services to other disciplines of science
- Provide a more scientific basis for recognition of the intraspecific categories

17.3 Phylogenetic Classification

After validation of specific status avoiding duplication of taxa with alternative and synonymous names, it is desired that the species be arranged in a hierarchical series following a phylogenetic classification, which along with essential constant features also consider the evolutionary sequence, including common ancestry, evolutionary divergence, and inequality of evolutionary rates for studying relationships and evolutionary affinities. The difficulties in assessing phylogenies usually result because of parallel, convergent, or reticulate evolution. Some of the phylogenetic classifications developed acounting evolutionary divergence and affinity are the Engler and Prantl, Hutchinson, and Whittaker classification.

Phylogenetic classifications may be monophyletic, polyphyletic, or paraphyletic. Monophyletic refers to one ancestry and exists when all the included organisms in a group share a common ancestor and none of the descendants of that ancestor are excluded. When the included organisms are derived from two or more distinct ancestors, the classification is polyphyletic. When a classification does not include all the descendants of the most recent common ancestor, it is termed paraphyletic. Phylogenetic classifications should be based on monophyletic groups, whenever possible.

Cladistics is a method commonly used to determine phylogenies or ancestor–descendant relationships in monophyletic groupings of taxa using outgroup comparisons. The term is derived from clade, which refers to a monophyletic group. A branching pattern (cladogram), which depicts the divergence of the taxa based on the distribution of shared derived character states, may be constructed that may depict evolutionary relationships and genetic distances among groups/taxa.

In statistical analyses and modern computer, science programs are available designed to facilitate analysis of complex data sets. A branching network, or unrooted tree, is first constructed based on the data from the in-group and outgroup simultaneously. Then the tree is "rooted" along the branch connecting the outgroup(s) to the rest of the tree. Because our knowledge of characters is rarely, if ever, complete, we need to remember that all classifications remain hypotheses of relationships.

17.4 Classification of Species (Genetic Resources) for Better Management and Use

The main objective of the present effort is to promote conservation and use of alternative genetic resources, i.e., the wild relatives of cultivated species in the breeding programs. Efficient management is the key for successful use of plant genetic resources (PGR). There has been attempts to address this issue from time to time. Systems have been proposed for their grouping with stronger genetic base to enable effective maintenance, conservation, and sustainable use in crop improvement. This has led to the development or proposition of some guidelines for classification or grouping of the accessible genetic diversity of a target cultivated/crop species that can improve the management and simplify use. The two popular concepts that were put forward to achieve these objectives were *Gene Pool* and *Core Collection.*

To conserve total spectrum of variability avoiding duplication and with information on relationships between the components of available accessible genetic diversity, it is required to develop a framework in which both informal and formal systems can operate with minimum confusion and redundancy. To achieve this, first came into existence the concept of gene pool, proposed by Harlen and de Wet (1971) dealing with total accessible genetic diversity of a taxa, including the diversity within a cultivated species and in its possible genetically affiliated wild species. Because it was more to promote the use of accessible plant genetic resources in crop improvement through conventional breeding methods or with the aid of developing biotechnological techniques, it was primarily based on cross-compatibility relationships to bring the accessible PGR within utilization limits following an appropriate breeding approach. Later came the concept of core collection (Brown 1989) to deal with "problem of plenty" of cultivated/crop species collections, which is beyond the scope of the present effort/book.

17.4.1 Gene Pool Concept

The methods of formal orthodox taxonomy have not been very satisfactory for the classification of variability existing within cultivated plant species and among its wild relatives and from conservation and use point of view. It has been creating a lot of confusion, particularly, because of the differences of opinion among researchers, who developed these classifications in their informal ways. For example, Snowden (1935) classified 31 species of cultivated *Sorghum* alone, in addition to the wild and weedy ones that are fully genetically compatible with the domesticated sorts. Jakushevsky (1969) reduced these to nine and de Wet and Huckabay (1967) to one, thereby adding confusion from conservation and use point of view. There are an innumerable number of such examples of confusion in other taxa, including vegetables species such as *Cucumis*, regarding the specific status of species like *C. callosus*, *C. trigonous*, and *Trichosanthes* with varying numbers of species in different publications. In addition, wild taxa are different from their cultivated counterparts. They differ not only in their morphological features but also in breeding behavior that had led to their genetic and reproductive isolation. Therefore, they require special taxonomic and descriptive treatment. Unfortunately for long, there were no guidelines for consistent grouping of related taxa.

Harlen and de Wet (1971) developed the concept of "gene pool" by developing a structure for representing the total available genetic diversity in a group and characterizing it by assigning the constituent taxa to primary, secondary, and tertiary gene pools. At the intraspecific level, cultivars are grouped into races and subraces in an informal way without rigid rules.

The differences between wild and domesticated taxa or plants are well known. However, both wild and cultivated plants have undergone various types of changes and selection pressures due many factors. The result of these is an enormous amount of conspicuous variation among very closely related forms. Harlen and de Wet (1971), to provide a genetic perspective and focus on cultivated plants in relation to other components of genetic diversity, proposed the informal categories of (1) primary gene pool (GP-1), (2) secondary gene pool (GP-2), and tertiary gene pool (GP-3).

17.4.2 Primary Gene Pool (GP-1)

This corresponds with the traditional concept of the biological species. Among the components of this gene pool, crossing between plants of distinct populations is easy; hybrids are generally fertile with good chromosome pairing; gene segregation is approximately normal and gene transfer is generally easy. The biological species almost always includes spontaneous races (wild and/or weedy) as well as cultivated races. To be explicit, the species can be divided into two subspecies:

- Subspecies A to include the cultivated races
- Subspecies B to include the spontaneous races

In taxonomic terms, primary gene pool approximates to taxonomic species identical to application of biological concept of species.

17.4.3 Secondary Gene Pool (GP-2)

This includes all biological species that will cross with the cultivated/crop species and would approximately be an experimentally defined cenospecies. Gene transfer is possible, but one may have to struggle with barriers, which may be poorly or not at all developed. They may result in some hybrids being weak, affecting gametic cycle, recombination pattern, difficult to bring to maturity with recovery of desired types in advanced generations, etc. This gene pool is available for use. However, the plant breeder or geneticist should put extra effort to overcome the cross-incompatibility and recombination-barriers with application of various possible cytogenetic manipulations to establish a fertile hybrid and effective gene transfer.

In taxonomic terms, secondary gene pool in majority of cases corresponds to the next division of genus, such as sections. This depends on the evolutionary state of isolating mechanism vis-à-vis morphological divergence, which may not go hand in hand.

17.4.4 Tertiary Gene Pool (GP-3)

At this level, crosses can be made between the wild species and the crop species, but the hybrids tend to be lethal or completely sterile. Gene transfer is either not possible with known techniques or radical measures are required, such as embryo culture or grafting to obtain hybrids, doubling chromosome number or using bridging species to obtain some fertility or other biotechnological approaches. The GP-3 defines the outer limits of potential genetic reach. Since very few people have worked at GP-3 level, it is rather ill defined. But if a cross can be made at all, there is always a chance that some technique will be discovered that will make its use possible. Perhaps the most powerful tool now known for introducing genes from GP-3 into crops is using complex hybrids that may work as bridge or through application of some of the biotechnological techniques, including recombinant DNA technologies.

Taxonomically, tertiary gene pool may correspond to its extent of genus, which is taxonomically well isolated from its closest allies. It may also extend beyond the generic boundaries.

The secondary gene pool might outline groups that would be acceptable to some taxonomists as generic limits, but the tertiary gene pool may extend too far. At any rate, the GP-3 describes the extreme outer limit of the potential gene pool of a crop/cultivated species. It is not a taxonomic unit. They are simply guides for placing existing classifications into genetic perspective from use perspective.

The differentiation of polyploidy must of course be accounted. There may be no solution that would apply to all crops, but as a general guideline, it was proposed that separate gene pools be recognized for different ploidy levels. The genetic barriers due to polyploidy are not always strong and gene transfer across ploidy levels may be rather easy in some cases. The barriers are there; however, it is generally useful to indicate their presence by providing separate epithets.

The formal taxonomy has failed most conspicuously at the infraspecific level (Jeffrey 1968). The tendency to overclassify is conspicuous, and formal categories are provided to groups of cultivars that have little or no genetic integrity. The extra intraspecific categories of Jirasek (1966), for example, would provide a minimum of over 2000 classes leading to more confusion than clarification. The term "varieties" should be avoided altogether as a botanical term, because it is too easily confused

Table 17.1 A further extension of Harlan and de Wet's gene pool system (Smartt 1990)

Gene pool	Constituents	Harlan and de Wet category	Experimental taxonomic category
First order	(a) Cultigen	GP-1	Same ecospecies (biological species)
Second order	(b) Weedy form Wild counterpart		
Third order	Cross-compatible species producing ± fertile hybrids	GP-2	Same coenospecies (Syngameon)
Fourth order	Cross-compatible species producing viable but sterile hybrids	GP-3_3	Same comparium
Fifth order	Cross-compatible species producing inviable hybrids		
Sixth order	Incompatible related species		

with agronomic or horticultural varieties (cultivars) and is not especially appropriate for cultivated plants in any case. The intraspecific categories suggested were:

Species
Subspecies A: The cultivated races
1. Race
2. Subrace
3. Cultivar
4. Line, clone, genotype

Subspecies B: The spontaneous races
1. Race
2. Subrace

Race: "A group of related individuals with enough characteristics in common to permit their recognition as a group." There are ill-defined races, hybrid races, in the process of formation and complex races made up of derivatives of two to several races. For example, sorghum has so far indicated that the basic races are bicolor, guinea, caudatum, kafir, and durra. There are, however, clearly identifiable hybrid races involving most combinations of these basic races.

Subrace: The subrace is simply a convenient division of a race. It may or may not be appropriate to divide a race into subraces.

Smartt (1990) further developed the concept of gene pool with minor differentiation, applying equal value to considerations of genetic resources for purposes of classification, evaluation, documentation, and conservation (Table 17.1). He suggested that from the standpoint of germplasm resources, some amplification of this basic scheme was desirable. Harlan and de Wet (1971) have clearly recognized that the wild and domesticated components of GP-1 merit distinction; it may be quite difficult to maintain the desired degree of separation between GP-1 A (domesticated) and GP-1 B (wild) segments of this gene pool.

Additional distinction can also be drawn between domesticated and wild components of the primary gene pool. They constitute, from the point of view of the plant breeder, gene pools of first and second choice within the total gene pool of the biological species, the GP-1. The original definition of the secondary gene pool (GP-2) is perfectly satisfactory as it stands. Tertiary gene pool (GP-3) is big and could be useful. The GP-3 might well be considered to include all those taxa, which may or may not cross-fertilize with cultigen. Therefore, there are arguably grounds for subdividing GP-3 into A and B components to include viable but infertile hybrids in A and those hybrids that are not viable in B. The B division would include all species between which cross-fertilization could be demonstrated with appropriate cultigen. This would approximately equate the order of the range of gene pools with the degree of genetic isolation visualized by the breeder or which is determined by naturally evolved isolating mechanisms.

17.4.5 Factors Contributing to Use Gene Pools

The most important gene pool from breeder's point of view is GP-1 A. This gene pool presents fewer problems in utilization than any other. The factors contributing to the development of this gene pool are of importance. Wide geographic dispersion promotes the development and maintenance of genetic diversity by a combination of founder effects, hybridization, genetic drift, selection, and mutation, and together they can often produce and maintain highly distinctive genotypes in specific areas. Different mode of utilization and varying incidence of pest and pathogens are two component factors in maintaining diversity. Ecological diversity, intensity of cultivation, and diversity of usage together with moderately high population densities in production region also promote and maintain high level of genetic diversity. These factors together result in the accumulation of extensive genetic resources in cultigens to produce rich primary and secondary centers of diversity. The extent of variability in GP-1 B depends on the present range and distribution of the wild conspecific counterpart of the cultigen. This may be very restricted or very extensive or not known at all.

It is perhaps in development of the GP-2 that the greatest variation is shown between different crops. Richness of species diversity and biosystematics relationships between species of a genus influences it. A crop belonging to a genus or infrageneric taxon, which shows considerable and comparatively recent speciation, is likely to have a more extensive GP-2 than the one which developed in a small genus and/or one in which recent evolutionary diversification had not apparently occurred.

In most crops, there is some difficulty in defining the GP-3, when its extent is determined by production of hybrids, which may be "sterile, not viable or anomalous." Although it is easy to compile data on reported viable but sterile hybrids, published reports may not be sufficiently informative regarding hybrid viability. Early aborted hybrids, which are not viable, may not be reported and/or even the occurrence of fertilization may be referred when it has not actually taken place. The suggestion has been advanced that in these circumstances, the GP-3 be subdivided into A and B components to include viable but sterile hybrids in A and hybrids that are not viable in B. The B division would include all species between which cross-fertilization could be demonstrated with the appropriate cultigen. Also, on relative cross-compatibility relationship, GP-3 can be further extended within and beyond generic boundaries, as suggested by Smartt (1990) and is done by Singh and Simpson (1994) in case of *Arachis* to include cross-incompatible species in quaternary gene pool.

17.5 Perspective

It is desirable that most of the wild relatives of crop species, which because of their lesser economic importance have been neglected and mostly described using the principles of orthodox taxonomy, are further studied, because this might have created many unacceptable or unresolved names, synonyms, and confusion of identity or distinctiveness. They need to be subjected to biosystematics studies. Biosystematics have a potential to succeed in developing keys for species identification with possible morphological markers to help in the collection and assessment of total spectrum of genetic diversity, i.e., biodiversity. Since the concept of biosystematics envisages to account for variation contributed by the three major components, viz., morphology, genetics, and ecology to plant populations, it offers the best tools to work out in the respective manner for the three components of biodiversity, viz., species, genetic, and ecological diversities, and should be pursued in most cases. Perfection in assessment of species diversity is certainly not only to ensure recognition of species and infraspecific entities but also to facilitate gene pool classification and visualization of most appropriate breeding strategies based on cross-compatibility and hybrid fertility. This helps trigger optimum sustainable utilization of different types of genetic resources using appropriate traditional and modern biotechnological approaches. Undoubtedly biosystematics has the potential to establish the rapprochement between traditional taxonomy and molecular taxonomy. In view of this, greater application of biosystematics in plant genetic resources, especially while dealing with the wild relatives of cultivated/crop species, is an absolute necessity to ensure correct identification of the species to facilitate correct collection, conservation, and use. Biosystematics should come up with all its tools to use data ranging from morphology to informational homology of chromosomes and molecular information packages, for augmenting success and prosperity in the use of diverse wild relatives of crops in genetic improvement of economically important cultivated/crop species.

References

Bremekamp CEB (1944) Materials for a monograph of the Strobilanthinae. Verhandelingen der Nederlandsche Akademie van Wettenschappen Afdeeling Natuurkunde, Tweede Sectie 41(1): 1–305.

Brown AHD (1989) Core collections: a practical approach to genetic resources management. Genome 31:818–824

Camp WH, Gilly CL (1943) The structure and origin of species. Britonia 4:323–385

Cracraft J (1989) Speciation and its ontology: the empirical consequences of alternative species concepts for understanding patterns and processes of differentiation. In: Otte D, Endler JA (eds) Speciation and its consequences. Sinauer, Sunderland, pp 28–59

De Wet JMJ, Huckabay JP (1967) The origin of *Sorghum bicolor.* II. Distribution and domestication. *Evolution* 21:787–802

Govaerts R (2001) How many species of seed plants are there? Taxon 50:1085–1090

Harlan JR, De Wet JMJ (1971) Towards a rational classification of cultivated plants. Taxon 20:509–517

Jakushevsky ES (1969) Varietal composition of sorghum and its use for breeding. Bull Appl Bot Genet Plant Breed 41:148–178

Jeffrey C (1968) Systematic categories for cultivated plants. Taxon 17:109–240

Jirasek V (1966) The systematics of cultivated plants and their taxonomic categories. Preslia 38:267–284

Mallet J (1995) A species definition for the modern synthesis. Trends Ecol Evol 10:294–299

Mayr E (1942) Systematics and the origin of species. Columbia University Press, New York

Scotland RW, Wortley AH (2003) How many species of seed plants are there? Taxon 52(1):101–104

Singh AK, Simpson CE (1994) Biosystematic and genetic resources. In: Smartt J (ed) The groundnut crop, a scientific basis for improvement. Chapman and Hall, London, pp 96–138

Smartt J (1990) Grain legumes – evolution and genetic resources. Cambridge University Press, Cambridge, p 379

Snowden JD (1935) A classification of the cultivated sorghums. Royal Bot Gard, Kew No 5:221–225

18 Collection Strategies

18.1 Introduction

Despite the fact that wild relatives have been well collected in major global crops such as wheat, chickpea, tomato, soybean, and rice and have proved valuable in plant breeding efforts, their representation in total available genetic resources is very poor. For example, in rice, *Oryza nivara* was instrumental in helping to develop rice varieties resistant to grassy stunt virus, a disease that caused hundreds of millions of dollars of damage to Asian farmers in the 1970s (Khush et al. 1997), while in tomato a single wild species provided genes boosting solid content by 2.4%, estimated to be worth about US$ 250 million a year to the global tomato industry (Esquinas-Elcazar 1981), and in groundnut transfer of foliar disease resistance from *Arachis* species resulted in increasing yield worth US$ 500 million (Sasson 1996). A recent survey conducted by the International Center for Tropical Agriculture (CIAT) in coordination with the Global Crop Diversity Trust (Crop Trust) and the Royal Botanic Gardens (RBG), Kew, under the project "Adapting Agriculture to Climate Change: Collecting, Protecting and Preparing Crop Wild Relatives," revealed:

- Twenty-nine percent of the total, or 313 wild relative plant species analyzed, are completely missing from the world's gene banks.
- A further 257 (23.9%) of species are represented by fewer than ten samples having been collected for each, leaving out a substantial amount of potentially important plant diversity.
- Over 70% of the total crop wild relative species are in urgent need of collection and conservation to improve their representation in gene banks.
- Over 95% are insufficiently represented regarding the full range of geographic and ecological variation in their native distributions.
- The most critical collecting gaps occur in the Mediterranean and Near East; Western and Southern Europe; Southeast and East Asia; and South America (https://www.sciencedaily.com/releases/**2016**/03/160321123704.htm).

A.K. Singh, *Wild Relatives of Cultivated Plants in India*,
DOI 10.1007/978-981-10-5116-6_18

Castañeda-Álvarez et al. (2016) found that wild relatives of important food security crops like banana, plantain, cassava, sorghum, and sweet potato need urgent effort for collection and conservation, along with those of pineapple, carrot, spinach, and many other fruits and vegetables. Even the wild relatives of vital staples like rice, wheat, potato, and maize, which tend to be better represented in gene banks, still have significant gaps in their collections. In addition, recognizing the value of wild relatives, which carry novel gene pool with greater potential, further collections from wider range of distributional habitats, their bioprospecting, and conservation offer greater opportunities for genetic enhancement. This would increase their potential value further (Ladizinsky 1989). Moreover, the present collections of wild species maintained in gene banks worldwide lack representation from the main habitats or the distributional range and the adequate information on their potential value.

18.2 Collection

Wild relatives of cultivated species are generally restricted to a narrow specific ecogeographical region and are difficult to collect. Many of the wild relatives are inconspicuous in nature, grow in smaller populations, and often found mixed with other wild species. This makes an inexperienced explorer confused while executing collection missions (Ladizinsky 1989). Moreover, collection of plant genetic resources (PGR) means collection of the total spectrum of genetic diversity from a region needing a rigorous sampling procedure. For collection of PGR, particularly of wild relatives, a lot of spadework and background information is required, prior to proceeding for collection. Distribution maps, cross-reference with maps of earlier collections related to targeted populations, information on time of harvesting/maturity, season of collection (for vegetative propagules, tubers, bud woods, etc.), and the areas to be surveyed, and route maps are required. Boarding facilities, mode of transport, equipment, and other accessories are required and need to be worked out well in advance. Sometimes modifications are required in exploration mission as per the need. For example, rescue collections of wild relatives are generally recommended in the natural habitats, which are threatened due to their conversion into agricultural lands, urbanization, high pressure of grazing, commercial exploitation of the species, or proneness to natural calamities such as floods, drought, forest fires, etc. Some wild relatives also deserve collection on priority basis due to their confinement to specific locations. Variation needs to be collected from the areas having a large number of wild relatives rather than collecting the gene pool in total. The representative sample should at least meet the breeders' and users' requirements. Collections are most effective when there is a specific requirement for research or for crop improvement program. For collection of material widely distributed, areas having broader range of genetic diversity in wild relatives with the desired trait(s) or gene(s) should be explored. During collection trips, species occurring scanty or found depleted are given top priority but with due care, limiting genetic erosion.

The wild-related species are, by and large, distributed in specific habitats. Weedy species have wider distribution, but endemic species have narrow or restricted distribution. These can be depicted on species distribution-maps. For this, the database, regional floristic records, monographs, herbarium material, plant collection reporters, or recent publications can be consulted, and mapping is done based on associated geographical/ecological zones in terms of geographical coordinates including broader perspective areas.

To ensure the identity of the target species, the key features that distinguish the target species from its close relatives in that region is required. Additionally, information on distribution range, habitat, and effect of nonbiological factors on growth and development, with suggested seed collecting techniques, ensures high-quality collections. Herbaria can assist explorers in having an idea about the type of species diversity of the region. Herbarium and field notes may also provide useful data to help in planning of exploration, since the herbarium specimens are generally collected during flowering, with information that can be optimal for seed collection. Collectors often prefer nowadays images of live plants. However, images must be authoritatively authenticated for a species, to be of any use as field guide. Hence, images from herbarium specimens identified by an expert would be more useful. Known companion species and the preferred habitat may provide further help in the successful identification and collection of wild relatives in the field.

Collection and conservation of wild relatives in India have been discussed (Anjula et al. 2005). Several organizations, such as the European Native Seed Conservation Network (ENSCONET) has published seed collecting manual for wild species (ENSCONET 2009). Common collecting activities from preparation to final deposition into gene bank are organized in following steps:

18.2.1 Planning

This may include:

- *Authorization*: All collections must be made legally. As collection of wild species would require access to forest areas, it may require permission from the appropriate forest authorities. As per the Biological Diversity Act (2002), access to genetic resources for non-Indians in India is controlled by the National Biodiversity Authority and its constituent organizations, the State Biodiversity Board, and the local Biodiversity Management Committees, responsible for the documentation and management of native seed collecting and advice before collecting anything. Collection should be preferably participatory involving local stakeholders with their prior informed consent. It should be based on mutually agreed terms (MAT) for the sharing of genetic material, associated information, and benefit sharing on commercialization. Permission should cover collection of seeds, herbarium specimens, etc., for a wide range of species.
- *Gathering information on the species to be collected:* To obtain detailed description and key for differentiation and identification of targeted species, a literature search should be made consulting the local and regional floras, listings, databases, and monographs. It may also include visiting local herbaria and genetic

resource centers, which are useful sources of information to ascertain the precise distribution and the periodic variations in growth and development that can be correlated with climatic conditions. The collector may like to cross-check the information with other sources for possible deviation or authentication of information.

- Collectors also need to collate information from the above sources on other biological features of the targeted species such as breeding system, including apomixes, ploidy, etc. The information on the reaction of the targeted species against the common diseases or pests of cultivated species is also useful, whereas information on seed storage behavior would help in the post collection care and handling of collected genetic material.
- There can be great variation among targeted species about information regarding their geographical distribution and about their populations known with specific features. The detailed knowledge may be available with local botanists and plant ecologists. However, in the absence of such help, the literature must be searched, particularly for maps displaying the precise distribution of species. In certain cases, it may be available in electronic format for integration with geographical information systems (GIS).
- *Coordination prior to collecting:* Confirm range of distribution of the targeted species, identity of flowering, and likely appropriate date for physiologically mature seed for collecting, through literature or contact with local botanists. Such an interaction on the distribution-map of the targeted species can help in developing the most appropriate time schedule for a successful collection expedition. The planning process can be extended with the integration of GIS for plotting the species distributions, earmarking the populations, and combining with most productive route. In case of isolated regions, means of access to location, accommodation, service stations, etc., may be other factors worth considering. The meteorological data, particularly in relationship of the prevailing temperature and relative humidity conditions during the collection, can be obtained from recent weather reports. This would facilitate selecting a time, when equilibrium for seed moisture contents with outside environment can be achieved easily (Table 18.1). A contingency plan must be drawn to face the expected emergency situations. In this regard the local guides can be of great help and if possible must be procured to avoid most problems.

18.2.2 Selection of Populations/Sites

Several inherent factors, such as breeding behavior and size of the population, influenced by the biotic and abiotic stress factors, induce genetic diversity among the populations. The differences of environmental conditions prevailing in different geographic areas impose different selection pressures on the target taxon's populations and create further genetic and phenotypic differentiation. Therefore, the area to be explored can be divided into sectors based on ecogeographic information. The more distant populations growing in environmentally diverse conditions are

Table 18.1 Expected approximate moisture content in seeds of different crops at equilibrium with air having different relative humidities and 77 °F (25 °C)

	Relative humidity					
	15%	30%	45%	60%	75%	90%
Crop group/crop	Moisture content (%)					
Grains						
Paddy	5.5	8.0	10.0	12.0	14.0	17.5
Wheat	6.5	8.5	10.0	12.0	15.0	19.5
Maize	6.5	8.5	10.5	12.5	15.0	19.0
Sorghum	6.5	8.5	10.5	12.0	15.0	19.0
Oilseed						
Soybean	–	6.5	7.5	9.5	13.0	19.0
Groundnut	2.5	4.0	5.5	7.0	10.0	13.0
Cotton	–	6.0	7.5	9.0	11.5	–
Flax	4.5	5.5	6.5	8.0	10.0	15.0
Mustard	4.0	5.0	6.0	7.0	9.0	–
Vegetables						
French bean	5.0	6.5	8.5	11.0	14.0	–
Beet root	5.0	6.0	7.5	9.5	12.5	–
Bhindi (okra)	7.5	8.0	9.5	11.0	13.0	–
Brinjal (eggplant)	4.7	7.0	–	–	–	–
Cabbage and cauliflower	3.5	4.5	6.0	7.0	9.0	–
Capsicum and chilli	6.0	7.0	8.0	9.0	11.0	–
Carrot	5.0	6.0	7.0	9.0	11.5	–
Cucumber	6.0	7.0	7.5	8.0	9.5	–
Lettuce	4.0	5.0	6.0	7.0	9.0	–
Onion	6.0	7.0	8.5	10.0	12.0	–
Pea	5.0	7.0	8.5	11.0	14.0	–
Spinach	7.0	8.0	9.5	11.0	13.0	–
Tomato	6.0	7.0	8.0	9.0	11.0	–
Water melon	6.0	7.0	8.0	9.0	10.5	–

Based on Harrington and Douglas (1970)

expected to be more genetically diverse, whereas the populations of the same sector are expected to have similar morphological and adaptive traits. Other specific factors, such as contributing to genetic erosion or threat to existence, can be added to this information for greater emphasis. Maxted et al. (1995) described how the eco-geographical surveys can be used in the selection of collecting sites, whereas Dulloo et al. (2008) outlined the use of genecological zonation for selection of specific site. For greater information, see Bacchetta et al. (2008).

Using geographic information systems (GIS), detailed geographical information can be collected regarding sites, and the possible effect of environmental factors on plant populations can be estimated and accounted (Draper et al. 2003, 2004). It helps in the physical characterization of environmental features associated with the origin and existence of the samples, avoiding redundancy, increasing efficiency, and reducing cost of collection with increased sampling of the genetic diversity, facilitating rational use.

18.2.3 Number of Samples to Be Collected

Ideally, if possible made collection from every population found in the distribution range of the targeted species, which shall sample the total spectrum of genetic variability. However, rationally, ecological specialization in range of distribution and breeding system affecting gene distribution decides the number of populations that need to be sampled. For example, in outcrossing, wind-pollinated woody perennials, because of higher dispersal of pollen, the greater proportion of their gene diversity is captured within populations. Therefore, sampling of fewer populations in larger quantity may be sufficient to sample most diversity than in the case of self-pollinated annuals, where a high proportion of the gene diversity is fixed and found between populations (Hamrick et al. 1991), requiring a collection of more number of populations in smaller quantity. Fragmented distribution of the species leading to geographical and genetic isolation of populations indicates that there is likely to be genetic differentiation between the populations, because of genotype x environment interaction. But it depends on the available information, and as collection is not a regular effort, it has been advised to start with sample of minimum five populations (see Falk and Holsinger 1991) from a geographical range. But Neel and Cummings (2003) have shown that five populations would cover only 67–83% of all alleles, on an average. Therefore, except in the case of species with very few populations, five populations will not provide full coverage. Guerrant et al. (2004) recommended collection of as many as 50 populations to provide fuller coverage of variability spread out over a long period. In some cases, ecological diversity can be taken into consideration for genetic differentiation. Therefore, the greater the number of populations collected, the greater is the probability of capturing and sampling of genetic diversity.

Collection effort should cover the full extent of distribution of the species at the site. As it is difficult to differentiate between where a population ends and another one starts, it is advisable to keep the population samples separately collected based on above differentiation criteria. Further decision may be taken on the basis underlying principles of population genetics.

Collection within a population can be made until an apparent topographical or other barrier to genetic exchange is encountered. Pollen and seed dispersal method, which may vary to different extent, may impact it. Therefore, in a situation of insufficient information, the practical approach would be to establish an arbitrary of distance of, say, 10–25 km between collection sites. Such an approach has the merit, because the differences in geographical location of definite distance will be significantly meaningful to future users of the information.

Within populations, there may be genetic differences caused by ecological variation within the habitat of a species. Keeping this in mind, separation of ecotypic variation may be important. In case of the absence of information about the extent of ecotype variation, it is advisable to collect more rather than fewer samples to circumvent the situation. They can be later bulked, as the collector gains knowledge about the phenology of species, because conservation and curation of a large number of small samples creates logistic/management problems. However, it might be advisable to keep obvious distinct morphological or ploidy types separate, particularly in the case of facultative or obligatory vegetatively propagated species.

Finally, it is also important to make sure that the population from which collection made is wild and not planted or product of human intervention, such as hybridization. Besides the passport data, collection of additional information on distinctive features of the population will be useful in future conservation assessments and use.

18.2.4 Minimum Number of Plants to Be Sampled

The collectors should strive to collect genetic variability from as many individual plants as possible. Collection can be made by random selection of plants over as wide an area as possible. In case of significant variation in the habitat, a stratified sampling method can be adopted in which the population is divided into subpopulations (strata) based on physical or biological factors, and random samples are taken of each stratum, keeping seeds of each ecological type separate.

The literature on the collection of plant genetic resources provides significant guidance for collection from different geographical regions in diverse crops, including wild species (ENSCONET 2009). A significant amount of it can be captured from works on crop species (Marshall and Brown 1975). They recommended capturing of at least one copy of 95% of the alleles that occur in the target population at frequencies greater than 5%. To achieve this target, they estimated the collection/sampling of a minimum number of randomly chosen 30 plants in the case of cross-pollinators or 59 in the case of self-pollinators species. As the breeding system may not be known in case of many wild relatives, they recommended sampling of 50 plants in a population as a benchmark.

The Center for Plant Conservation in the United States recommends the sampling of between 10 and 50 plants per population (Falk and Holsinger 1991). However, guidance based on capturing single copies of alleles seems more relevant when collecting material for crop breeding than it does in the context of nature conservation. When the material is required for reintroduction, adaptation is likely to be of greatest importance, and the frequencies of alleles in the sample should closely match to those populations previously present at the site. Broadhurst et al. (2008) have highlighted the considerations this fact to be kept in mind, while collections are made for large-scale restoration. When sampling is to capture the frequencies of alleles in a big population of cross-pollinated species, the sampling of 200 individuals and five seeds from everyone was suggested by Marshall and Brown (1983).

Finally, keep in mind the genetic implications of sample size vis-a-vis breeding system. In cross-pollinated species, a few samples including many individuals and seeds per individual versus many samples including fewer individuals and seeds per individual in self-pollinated species should be applied carefully. Further, the number of seeds that should be collected and the suggested limit of the proportion of the population (20% in endangered species, to 5000) also need attention. It must be ensured that populations are distinct entities and not extension through spread of vegetative propagules.

In summary, try to collect seed from at least 50 and preferably 200 plants. This can be modified based on local/prevailing situations, which may include very small populations, annual or long-lived perennial, accessibility, time of collection, and eventual use. Therefore, seed collecting is the "art of the possible." Often, it is impractical to obtain a big sample from the wild and keeping individual plant harvests separate. However, effort should be to collect as big a sample as possible, keeping in mind the limitations of regeneration, conservation, and use.

18.2.5 Sample Size and Seeds per Plant

While sampling, the amount of seeds collected from individual plant has genetic implications, and under some circumstances (see 18.2.4), it also affects the sample size. Where natural habitats are source of genetic material for collection, conservation, and use, to ensure the survival of populations over years, which is crucial in case of annuals than perennial species, Way (2003) advised not to collect more than 20% of the total volume of mature seeds available on a collection day. Also, unless the seed quantity has been reduced to a level needing regeneration to facilitate continued conservation and use, avoid repeat collections of a species from the same site for consecutive years. To obtain the optimum sample size, Guerrant et al. (2004) advised collecting low levels of samples spread over several seasons/years. These rules may not apply to species adapted to a narrow/specialized ecological niche and threatened populations, to ensure survival/protection.

The multiplication of wild species under controlled condition may be difficult. For this reason, an attempt should be made to collect sufficient quantity of seeds to maintain a collection over longer duration avoiding regeneration/multiplication. This shall also avoid the chances of genetic erosion during multiplication, maintaining true-to-type. As a general guideline, it is advised to collect around 5000 seeds per accession. However, in the case of small populations, there can be debate even about the desirability of collection. Calculate the number of seeds collected based on 1000 seed weight. In case of endangered species, where small quantities are conserved, minimize the usage of seed in routine viability testing, or extend the interval of monitoring.

18.2.6 Sampling Approach

Sampling should be done as randomly as possible to avoid bias and capture as much diversity as possible. However, it is not easy. In case of large populations in a uniform landscape, good sampling can be achieved by adopting a systematic approach with sampling at regular intervals along a narrow section traversing through natural features of the earth's surface or in the grid of the landscape. Walking, say, three or four paces in each direction, along sections/lines and sampling, can achieve this, with repeatition of exercise covering the whole area/landscape. In case of a collection team, each member can walk on a different transect or a different bar of the grid and collect. Biased sampling should be avoided by selecting individuals based on physical appearance. As much as possible, even-number of seeds are collected from maternal genotypes. To restrict bias in the genetic sampling, the temptation to concentrate on collection of seeds from individuals with higher fecundity (or any other feature) must be avoided. In case of sampling of less than 20 plants, it is advisable to keep seeds from different plants separately. This will maximize the contribution of the maternal genotypes at regeneration/multiplication. However, having many subsamples of a collection increases the curatorial workload. In sampling, greater emphasis is to be given to the physical distance between plants.

Consideration should also be given to effects that may be created in sampling by the spatial distribution of population and within plants. There may be significant differences in the genetic and physiological maturity of seeds from different regions of the inflorescence. For instance, in Gramineae, the maturity of seeds progresses from the base of the inflorescence with older seeds (earlier flowering and pollination) to young ones at the top, whereas in Apiaceae, the outer fruits are older (have set earlier) than the inner ones. Therefore, sampling from across the seed head will be beneficial capturing physiologically mature seed, avoiding immature and old seeds.

Time of collection may have a very significant effect, which could influence the adaptation of the sample in case of reintroduction. Collecting/sampling early or late in the season will bias the genotypes conserved, both with respect to the diversity of alleles and the allelic frequencies. The variation between samples collected in different seasons may also be important from variability point of view, especially in the case of annuals or short-lived species with very dynamic populations, responding to prevailing climatic conditions. In such cases the repeated samples are the best, but they need to be kept separate to avoid mixing (fresh seed being with aged seed or influenced by climatic variation) and can be brought together for use or on regeneration. However, collectors may not have the opportunity of repeated visits, and such bias are accepted part of the collection exercise.

18.2.7 Seed Collecting

General recommendation:

- Do not collect immature and empty seeds. Pull fruit, crush or cut open to look at small number of seed.
- Avoid mixing of collections.

- Seeds should be collected in well-labelled cloth or non-glossy paper bags. Selection of bags should correspond to the nature of seed. For example, it may be time consuming to extract grass seed from cloth bags.
- Tie the cloth bag at the neck, while staple the paper bag after folding.
- Subject to the availability of equipment, calibrate the moisture contents of the seed samples at site itself. If seeds are too wet, dry them using desiccant or by loosely hanging bags containing seeds in the air (Table 18.1).
- While leaving a site after collection, make sure nothing is attached to your shoes or clothing that can result in contamination.

Checking seed maturity:

- To ensure the collection of physiologically mature seed, either use guide or morphological marker traits, such as fruit color that signals maturity.
- Avoid collecting immature fruits or seeds. But collecting slightly immature fruits and leaving them to mature in laboratory in some cases is possible and may be beneficial, for example, in case of species with explosive fruit or capsule.

Harvesting: Select appropriate techniques as per the nature of species.

- The dehiscent fruits (such as siliquas, leguminous pods, or capsules) can either be directly collected into cloth/paper bag or seeds can be extracted washed in a bowl or bucket and poured into bag after removing debris.
- Cut the branch with seed heads or inflorescence (panicles), and place them into bag, inserting head first. Seed extraction can wait up to the end of collection trip for laboratory extraction. This approach suits to several families, including Asteraceae.
- Larger fruit may be collected one after the other. The fruits with flesh may be put into punched plastic bags, ensuring aeration. Seed may be extracted either during the trip or at the end, depending on the period of the trip.
- For collection of fruit from tall trees, collect fruit either by shaking of the tree or climbing. The seeds can be shaked out onto the paper and put into bags. Do not collect fruits or seeds that have dropped on the ground.
- In case of orchids, as the seeds are too small, the fruits should not be touched, and a razor or scissor may be used to cut the pedicel, dropping the fruit into a bag. They may require further care in handling and extraction of seeds.
- In some cases, with seed head of explosive fruit, to secure good amount of seed they can be bagged and collected later. There can be some other innovative traps to catch the seeds.
- In rare cases where plant is not at seed-producing stage, the whole plant or cutting can be collected with the permission of owner and later be propagated under controlled condition for harvest of seed.

18.2.8 Information Documentation

(a) *Passport data*: Information documentation is a very important component of germplasm collection, as seed/genetic material without passport information is nearly of no use. Along with provenance data, the record of information on a number of plants and threat to the target species are important. The data should be recorded in an objective manner. The technology for data recording has evolved from collecting information on paper to handheld computers in a desired structure.

Record location by geographical coordinates on a map or through geographical positioning system (GPS) along with altitude and other administrative data. GPS receivers sometime do not work because of physical interference from trees or nonavailability of satellite signals at the bottom of deep valleys. In such situation, Google Earth or Google Maps can be used to record geographical coordinates of the site.

(b) *Identification/verification*: Identification of species needs application of instinct, knowledge, and skill. Identification is thus the art and science of recognizing a living entity and distinguishing it from others, which is gained by experience. For the identification of target species, use national, regional, or local monographs. Collection of herbarium specimens during seed collection allows the verification of collector's identifications (which help subsequent identifiers) and facilitates establishing a reference samples that can help in future unambiguous verification. Ideally, high-quality images of the plant specimens should be carried in the field to help identification.

Herbarium specimen made should represent most the identifying characters of the plant. If possible, it is advised to collect more than one specimen for duplication in other herbaria. It should be from separate plant to account for population variation, if any. While collecting herbarium, seed, etc., do not cause avoidable damage to populations, especially in case of threatened species. If it is difficult to obtain full specimen, the situation can be salvaged with collection of plant parts or with seed extraction specimen remains. As a last resort, record reference number of previous specimen. An expert must authenticate herbarium specimens, and his address and affiliation be recorded. In case of naturalized or reintroduced species, proper record should be made.

(c) *Soil samples*: Collection of soil samples is important in case of species that have symbiotic relationship with soil microorganisms, particularly in case of orchids, legumes, and trees that are efficient recycler of available nutrient. This helps in reestablishment of species. Label the soil samples with associated seed accession identity number.

18.2.9 Post Collection Care

Care of collected seed material is essential before they are processed for long-term conservation with storage in seed gene bank. If they are not handled properly against

Table 18.2 Summary of methods for handling of fruits after collection

Fruit type	Seed type	Viability maintenance	Bulk reduction
Dry dehiscent	Orthodox	Sun drying	Drying, shaking, sifting, or manual removal of fruit parts
Dry indehiscent		Sun drying	Flailing, pounding, or threshing sifting or manual removal of fruit parts
Serratinous (cones, capsules, dry compound fruits)		Sun drying	Barbecuing, shaking, sifting
Fleshy		Thin pulped (sun drying) others: de-pulping by pounding or washing	De-pulping
Immature (any type)		Moist storage	Not applicable
Dry fruit	Recalcitrant	Moist storage	Drying until dehiscence, sifting or manual removal of fruit parts
Fleshy fruits		De-pulping, moist storage	De-pulping

Based on Schmidt and Thomsen (2003)

encountered adverse conditions, it may result in loss in seed viability to a significant extent, defeating the objective of collection of genetic variability for conservation and use. For detailed information on care of fruit, seed, and vegetative propagules during and post collection and handling of germplasm, see Singh et al. (2007).

Time permitting, clean the seed samples in the field itself during collection. The easiest method for cleaning the seeds is by using a light plastic washing up bowls and drainers, where debris can be removed by sieving or by gentle blowing. The semidried seeds are collected into a paper bag or envelope. However, it is best to leave the cleaning until the seeds reached seed bank, which are equipped with specialized equipment.

Water contents, being the basis of movements of cations and anions, thereby initiate any metabolic activity at a specific temperature in any living organ or tissue. Therefore, drying of the seed or the genetic material to safe limits and processing and storage under low relative humidity (32%) and temperature (25 °C) conditions help maintain the nearly equilibrated moisture contents, contributing significantly to improved upkeep of the material during collecting and transportation.

Freshly collected fruits and seeds are susceptible to damage, primarily because they often have relatively high moisture content conducive to fungal and bacterial infection that accelerate the process of seed aging. Also, moist seeds and fruits respire, producing heat and consuming oxygen. If the oxygen is depleted because of inadequate aeration, fermentation replaces respiration. Both respiration and fermentation produce heat. High temperature in combination with high moisture content tends to accentuate respiration or fermentation leading to faster seed deterioration.

How fast is the seed deterioration depends on species, physiological quality of the seed at collecting, contamination of the seed lot with dirt, rotten fruits, twigs, etc., and the surrounding environment temperature and relative humidity. Therefore, fruits and seeds need proper post collection care for reduction or management of moisture contents through proper drying or otherwise (Table 18.2). The desired aeration of the collecting bags or transporting containers enable control of these factors so that deterioration can be restricted/limited, maintaining the high quality of the seed/genetic material collected for conservation in the gene banks.

It is advisable not to transport the whole fruit, except those cases where the separation is a cumbersome and long-drawn process. In the case of fleshy fruits, it is possible to gently macerate the fruits against a hard surface or plastic sieves in running water to remove the flesh. Thereafter, they can be either be air-dried or on silica gel. If the transportation of fleshly fruits is essential for some reasons, then place them in open or in perforated plastic bags that are either loosely spread or closed with plenty of fresh air. For transportation of seeds, always use permeable containers that allow them to air-dry. Many permeable containers are available, such as paper, plastic or cloth bags, cardboard boxes, or wicker weave baskets, bamboo storage baskets, etc.

If the transportation back to gene bank is going to take a longer time, the seeds can be dried over silica gel, dried rice, or charcoal during collecting. This becomes essential if the average outside relative humidity is greater than 50% (hygrometer can be used to assess the seed moisture and calculate equilibrium relative humidity). Silica gel is used in about 1:1 (silica gel to seed material). Falling of direct sunlight over the seed containers should be avoided as it may cause rise in temperature. Also, at night, if the temperature drops significantly, the humidity might rise within the vehicle. If possible, avoid this by putting the dry seed in available cool and dry location conditions. Similar approach can be adopted in long collection trips, with instead carrying seed material around in the vehicle, leaving the dried seed at the base camp in an air-conditioned hotel room.

18.3 Perspective

Recognizing the importance of wild relatives in general and in genetic improvement of crops, a massive national hunt for collection of wild relatives of crop species was made in India, between 1999 and 2004, under the National Agricultural Technology Project (NATP) funded by the World Bank. A total of 3030 accessions of wild relatives of crop plants were collected. This resulted in increase in germplasm collections of wild relatives of cultivated/crop species from 6406 accessions (125 species) during 1976–1999 to 16,712 (373 species) during 1999–2004 (Anjula et al. 2005). Basically, this assemblage consisted of major crop groups recorded from the subcontinent (Singh et al. 2013) based on gaps in wild species collection. However, there still appears a gap regarding the completeness of collections. Question, whether the collections have achieved the target of collecting species from most parts of the country representing all geographical and ecological areas and total spectrum of species and genetic diversity, still remains unanswered. Therefore, there is a need for compilation of all passport information available on the past collections from all sources and generated during these explorations and produce a more comprehensive updated national database on available plant species and genetic diversity in India. A thorough review of the database on available collections and made in the last decade under NATP shall help identifying geographical, ecological, and genetic diversity gaps.

There are wide gaps in information regarding wild species distribution, diversity, traits, phenology, reproductive biology, etc. The present documentation of available information in this regard is the beginning and much needs to be done to make the present effort a success with revision of status in properly structured tabular form, to facilitate greater exploration and pointed collection of priority species from priority areas, particularly in case of the species under threat and prospective candidates of desired gene(s) from appropriate locations to facilitate conservation, before they are lost forever.

For the protection of rich diversity in wild relatives, there should be dissemination of information on appropriate collection strategies. This would help use of safe collection methods, ensure non-destruction of populations, avoid negative impacts on habitats and species diversity, and seek public support to maximize economic benefits and opportunities offered, and minimize the losses. The collection opportunity can also be used to educate people regarding over-exploitation of commercially important wild species from forest/reserve areas, adoption of controlled grazing, and harvesting seed on maturation, using minimum number of plant/parts for herbarium, regeneration/perpetuation, etc., and organizing man and biosphere, and environmental education programs.

To avoid habitat loss and loss of populations through repeated collections or harvest from wild, studies should be initiated on agronomy of the selected economically important wild species for their ex situ multiplication and seed increase both under in situ and field conditions. These efforts can provide new opportunities for livelihood support to local communities.

Development of awareness regarding the importance and the threat to wild relatives at various levels, for the promotion of community involvement in management and in intensification of in situ/on-farm conservation of economically important wild species at national level, is another area needing attention. This can be achieved by providing greater community control, maybe through Biodiversity Management Committees under BDA over common property resources of the locality and special grants or awards for communities or individuals who maintain high levels of biological diversity on private land. The provisions provided under National Gene Fund envisaged in Protection of Plant Varieties and Farmers' Rights Act (PPV & FR) 2001 can be used to support such programs. These provisions can be further strengthened at the village level through assistance, integration, and corporation of Panchayat Raj System.

References

Anjula P, Bhandari DC, Bhatt KC, Pareek SK, Tomer AK, Dhillon BS (2005) Wild relatives of crop plants in India: collection and conservation. National Bureau of Plant Genetic Resources (NBPGR), New Delhi, India, p 73

Bacchetta G, Bueno Sánchez Á, Fenu G, Jiménez-Alfaro B, Mattana E, Piotto B, Virevaire M (eds) (2008) Conservación ex situ de plantas silvestres. Principado de Asturias/La Caixa, Oviedo, p 378

Biological Diversity Act 2002 and Biological Diversity Rules 2004. National Biodiversity Authority, Government of India, Chennai, India, p 57

Broadhurst LM, Lowe A, Coates DJ, Cunningham SA, McDonald M, Vesk PA, Yates C (2008) Seed supply for broadscale restoration: maximizing evolutionary potential. Evol Appl 1(4):587–597

Castañeda-Álvarez NP, Khoury CK, Achicanoy H, Bernau V, Dempewolf H, Eastwood R, Guarino L, Harker R, Jarvis A, Maxted N, VMüller J, Ramirez-Villegas J, CSosa C, Struik PC, Vincent H, Toll J (2016) Global conservation priorities for crop wild relatives. Nature Plants 2(4):16022

Draper D, Rosselló-Graell A, Garcia C, Gomes C, Sergio C (2003) Application of GIS in plant conservation programmes in Portugal. Biol Conserv 113:337–349

Draper D, Marques I, Roselló-Grael A (2004) Criaçao de um Banco de Sementes representativo da flora afectada pela Construçao de Barragem do Alqueva (II fase), Relatório Final. Jardim Botánico – Museu Nacional de História Natural. Universidade de Lisboa, Lisboa, p 147. http://www.edia.pt

Dulloo ME, Labokas J, Iriondo JM, Maxted N, Lane A, Laguna E, Jarvis A, Kell SP (2008) Genetic reserve location and design. In: Iriondo JM, Maxted N, Dulloo ME (eds) Conserving plant genetic diversity in protected areas. CABI Publishing, Wallingford

Esquinas-Elcazar JT (1981) Genetic resources of tomatoes and wild relatives. International Board for Plant Genetic Resources (IBPGR), Rome

European Native Seed Conservation network (ENSCONET) (2009) ENSCONET Seed Collecting Manual for Wild Species. Editors: Royal Botanic Gardens, Kew (UK) & Universidad Politécnica de Madrid (Spain), pp 32

Falk DA, Holsinger KE (eds) (1991) Genetics and conservation of rare plants. Oxford University Press, New York, pp 225–237

Guerrant EO Jr, Fiedler PL, Havens K, Maunder M (2004) Appendix 1. Revised genetic sampling guidelines for conservation collections of rare and endangered plants. In: Guerrant EO Jr, Havens K, Maunder M (eds) Ex situ plant conservation: supporting species survival in the wild. Island Press, Washington, DC

Hamrick JL, Godt MJW, Murawski DA, Loveless MD (1991) Correlations between species traits and allozyme diversity: implications for conservation biology. In: Falk DA, Holsinger KE (eds) Genetics and conservation of rare plants. Oxford University Press, New York

Harrington JF, Douglas JE (eds) (1970) Seed storage and packaging: application for India. National Seed Corporation and US AID, New Delhi, pp 3–22

Khush GS, Ling KC, Aquino RC, Aquiero VM (1997) Breeding for resistance to grassy stunt in rice. In: Proceeding of the 3rd International congress SABRAO. Plant Breeding Papers, Canberra, pp 3–9

Ladizinsky G (1989) Ecological and genetic considerations in collecting and using wild relatives. In: Brown AH, Frankel OB, Marshall DR (eds) The use of plant genetic resources. Cambridge University Press, Cambridge, pp 297–305

Marshall DR, Brown AHD (1975) Optimum sampling strategies in genetic conservation. In: Frankel OH, Hawkes JG (eds) Crop genetic resources for today and tomorrow. Cambridge University Press, Cambridge

Marshall DR, Brown AHD (1983) Theory of forage plant collection. In: McIvor JG, Bray RA (eds) Genetic resources of forage plants. CSIRO, Melbourne

Maxted N, Van Slageren MW, Rihan J (1995) Ecogeographic surveys. In: Guarino L, Ramanatha Rao V, Reid R (eds) Collecting plant genetic diversity: technical guidelines. CABI Publishing, Wallingford, pp 255–286

Neel MC, Cummings MP (2003) Genetic consequences of ecological reserve design guidelines: an empirical investigation. Conserv Genet 17(4):427–439

Protection of Plant Varieties and Farmers' Right Act (2001) Ministry of Agriculture. Government of India. Universal Law Publishing Co. Pvt. Ltd., New Delhi, p 127

Sasson A (1996) Biotechnologies and the use of plant genetic resources for industrial purpose: benefits and constraints for developing countries. In: Castri F di, Younes J (eds) Biodiversity, Science and Development: Towards a new Partnership. CAB International, pp 469–487

Schmidt LH, Thomsen KA (2003) The seeds processing In: Smith RD, Dickie JB, Linington SH, Pritchard HW, Probart RJ (eds) Seed conservation: turning science into practice, Kew International Workshop, 26-31 July 2001, Wakehurst Place, Royal Botanic Garden, Kew, UK, pp 281–306

Singh AK, Radhamani J, Srinivasan K (2007) Post collection care and handling of the germplasm. Proc Indian Natn Sci Acad 73(2):55–70

Singh AK, Rana RS, Mal B, Singh B, Agrawal RC (2013b) Cultivated plants and their wild relatives in India– an inventory. Protection of Plant Varieties and Farmers' Rights Authority, New Delhi, India, p 215

Way MJ (2003) Collecting seed from non-domesticated plants for long-term conservation. In: Smith RD, Dickie JB, Linington SH, Pritchard HW, Probert RJ (eds) Seed conservation: turning science into practice. Royal Botanic Gardens, Kew

19 Conservation Strategies

19.1 Introduction

Wild relatives of crops/cultivated species are essential components of natural and agricultural ecosystems/agrobiodiversity and hence are indispensable for maintaining ecosystem health and supporting genetic improvement of crop by providing alternative genetic resource. Their protective management is essential for maintaining a healthy environment and conservation to promote sustainable use in crop improvement for increasing agricultural production, ensuring food and nutritional security. Unfortunately, occurring as they do on untended lands, crop wild relatives (CWR) are more vulnerable due to changes in land use patterns, with intensification of agriculture and urbanization causing climate change. Many are at the risk of extinction. Further, CWR fall between juxtaposed agendas of agricultural and conservation. Agriculture looks at tended lands, and conservation does not focus on wild agricultural genetic resources, further compounding their vulnerable position.

The natural populations of many CWRs are increasingly at risk. Destruction and degradation of natural environment or their conversion to other land uses are threatening the habitat loss of wild species related to crops. Deforestation is the leading cause for the loss of many populations of important wild relatives of fruits, nuts, and many industrial crops. Populations of wild relatives of cereal crops that occur in arid or semi-arid lands are being severely reduced by overgrazing, thereby resulting in desertification. The growing industrialization of agriculture is drastically reducing the occurrence of CWRs within the traditional agroecosystems. Therefore, scientific and rational conservation and use of CWRs are essential elements for maintaining the environmental balance and crop improvement, to improve food and nutritional security with elimination of poverty.

A recent review of global conservation efforts of CWR by Castañeda-Álvarez et al. (2016) showed their poor representation in gene banks. Around 29% of wild relatives associated with 63 crops were completely missing from the gene banks; 24% are represented by fewer than ten samples. Over 70% of taxa require high priority further collecting to improve their representation in gene banks, and over 95%

A.K. Singh, *Wild Relatives of Cultivated Plants in India*,
DOI 10.1007/978-981-10-5116-6_19

are insufficiently represented about their full range of geographic and ecological variation in their native distributions. Most critical gaps for further collecting were found in the Mediterranean, Near East, Western and Southern Europe, Southeast and East Asia, and South America. However, improvement in conservation ensuring availability of CWR for crop improvement is required worldwide in almost all countries. Therefore, there is a need for:

- Determining the conservation status of CWR in gene banks and threats to their existence
- Management or protection approaches to conserve CWR in priority sites in cost-effective manner
- Developing long-term indicators and risk threshold levels, which the local people can use to monitor CWR for their own needs and which policymakers can use to define interventions required for their conservation and sustainable use

19.2 Probable Conservation Approaches

19.2.1 By Regulatory Mechanisms

Recognizing the importance of wild relatives of cultivated/crop species and their predominant occurrence in wild/natural habitat, conservation can be facilitated/strengthened with enforcement inhibitory approaches on collection of economically important wild plant species, which would restrict overexploitation by regular harvest from the nature out of human greed. It may require enactment of certain legislations from policymakers. Certain provisions are already in force; the Chemical Export–Import Council of Mumbai, India, is looking after the export–import of various herbal products, whereas the Government of India is implementing the Indian Forest Act (1927). Under minor forest produce category, 29 plant species have been banned from the trade outside India. In 2000, this list was further expanded with the addition of 114 medicinal plants regulating their harvest from nature, requiring legal procurement of a certificate from the concerned forest officer to accompany the export consignment for cultivated material. There are more than 30 Indian national Acts facilitating conservation of biodiversity. Some important ones are The Indian Forest Act 1927, Import and Export Act (Control) 1947, The Wildlife (Protection) Act 1972, Forest (Conservation) Act 1980, The Environment (Protection) Act 1986, Spices Board Act 1986, etc., that are helping conservation of wild relatives of domesticated plant species, particularly medicinal and aromatic plants, spices and condiments, industrial crops, etc., indirectly. The recently promulgated National Biodiversity Act 2002 enforcing control on access to genetic diversity is another deterrent providing soft protection to national biodiversity.

19.2.2 By Promoting Cultivation Practices

The number of plant species used in the traditional medicinal system goes beyond thousand (Singh et al. 2016), whereas the National Medicinal Plants Board (NMPB) has worked out a list of 960 medicinal plants that are under trade and thereby of commercial significance. However, only around 36 species are cultivated on a regular or marginal basis, meaning that most of them are harvested from wild. This includes the sister species of some important medicinal aromatic plant. Similarly, many alternative vegetables used by tribal communities, forage plants in rural areas, and ornamental (orchids) and industrial plants like bamboo and canes are harvested from nature, threatening the very existence of the populations of these species and thereby the species/genetic diversity. Therefore, one of the approaches can develop cultivation practices after bioprospecting their potential value from scientific and economic points of view. This would help provide new opportunities for livelihood support to local communities, increase their production to meet the market demand, and limit their harvest from the nature. The Government of India has established NMPB under the Ministry of Health and Family Welfare with such objectives. The NMPB has established state medicinal plant boards for different states and has identified 28 medicinal species of high demand in market/ industry for development of cultivation practices and good quality planting material for distribution to keen farmers. Therefore, efforts are being promoted for cultivation of medicinal plants so that industrial sector is served with true potent quality raw material for manufacturing of quality herbal product. This approach can be extended to other crop groups such as floriculture, agro-industrial plants, etc. However, this approach must be complemented with other approaches that would either help in conservation of selected plant material or provide protection to populations growing in diverse environments responding to changing environment and generating/evolving diversity that may help meet future requirements.

19.2.3 By Conventional Conservation Strategies

Basically, there are two conservation approaches, one dealing with strategies providing protection to self-perpetuating populations of the plant species in their natural habitats and the other by bringing them or their genetic material into protection, using diverse techniques. They have broadly classified into two:

1. *In situ conservation*: Means conservation of species/genetic diversity within their natural ecosystem and habitat
2. *Ex situ conservation*: Means conservation of components of genetic material of biological diversity outside their natural habitat

Based on habit, habitat, breeding behavior, threat perception, size of sample, and the objectives of conservation, appropriate strategy or a combination can be selected.

19.2.3.1 In Situ Conservation

In situ conservation methods are those which seek to maintain self-perpetuating populations in natural ecosystems. Therefore, in situ methods providing protection to species populations will be more suited for conservation of wild relatives of crop species germplasm. In situ conservation through genetic reserves can produce a range of social benefits, some of which are relevant to the use of germplasm in breeding programs. Maintenance of some populations of wild species within their natural ecosystems is also desirable for purposes of research and education and more importantly for preservation of biological diversity. Its importance to germplasm conservation strategies will vary greatly between gene pools. However, most protected areas do not include management genetic resources as one of their management objectives. The management needs of the populations of the CWR target species are quite specific and separate from the management of the protected area itself. For this reason, the concept of genetic reserves was introduced taking care of total spectrum of genetic variability. Maintaining stable populations of these species may entail landscape manipulations at a lower level reducing the value of the reserve for perpetuating landscape processes and structures but promote preservation of genetic structure of the target species.

In situ conservation serves the purpose of long-term survival and evolution of species in their natural habitat in association of other life forms. In situ conservation of crop wild relatives in some cases may be the most appropriate and may be the only alternative because of several reasons:

1. The member of being complex natural ecosystem, e.g., moist tropical forest tree species, which are interdependent for survival.
2. Many have high dormancy that cannot be broken by available methods.
3. Others have highly specialized breeding system (dependent on specific insect) facilitating their natural perpetuation.
4. Many more are adopted to a very narrow ecological niche and getting them out could be suicidal.

Protection of areas that are rich in such plant diversity can be one approach with establishment of biosphere reserves, national parks, wild life sanctuaries, and/or specialized species sanctuaries. But many times, they cannot meet the requirements of individual species because disturbance of habitat causes degradation of vegetation, which may disturb the stability of plant components. Thus, ecological imbalance greatly influences the survival and conservation of plant biodiversity and may result in change of composition and loss of certain species. Conservation methods involve more than simply establishing protected natural areas, which contain random or fragmentary populations of wild relatives of crops species. This would need initiation of programs for habitat conservation of the targeted species/populations and would involve a number of component activities, which are essential for efficient efforts and to secure a sustainable as well as useful alternative genetic resource for crop improvement. These activities would include:

1. Initial surveying (distribution, genotypic variability, correlation of ecological, and geographical factors: within and between)
2. Formulation of conservation objectives
3. Selection of species/populations
4. Selection of sites and size of unit in terms of minimum requirements
5. Reserve design, management (participation of local people/institutions reducing human pressure impact), and monitoring
6. Access to germplasm (regulation, documentation)
7. Linkage with ex situ conservation and use

The above activities necessitate the development of extensive new research programs dealing with sample chosen for populations of a wild species within networks of reserves, reproductive biology, geographic distribution, phenotypic variability, ecological age, physical parameters within the environment, minimum population sizes (10,000) for saving low frequency of alleles, and natural subdivisions of populations expected to be reflected through alleles in linkage groups (the total number of individuals should exceed 100 mature individuals per subpopulation). To ensure the existence of appropriately distributed conserved populations, they should be cross-checked with existing reserves and conservation areas. Dispersed and isolated areas of species habitat might be needed in addition to large reserve areas.

It is always advisable to follow complementary conservation approaches to have a balance of advantages and disadvantages to harness effective benefits. Such approach may also provide additional benefit, for example, in situ conservation may help in protection of habitat, thereby the ecosystem and the environment, in addition, to continued evolution for generation of new variability for present and future use. Therefore, recognizing the importance and risk faced by the wild relatives of cultivated/crop species, a comprehensive approach is the need of the hour.

India has traditional refuge of natural resources called sacred grooves—a traditional method of protecting biodiversity-rich areas. It would be quite effective for species, which are on the verge of extinction. Sacred grooves are seen throughout the country and vary in size from few plants to several acres and dealing mainly with plants either of medicinal or religious significance, however, may include certain ornamentals and plants of cottage industrial use. There are four regions important for grooves, Khasi and Jaintia Hills, Western Ghats, Aravalli Hills and Sarguja, and Chand and Bastar area in Central India. Maximum numbers of grooves are in Kerala, Maharashtra, Madhya Pradesh, etc. Additionally, in situ conservation of medicinal plants has been taken up by some nongovernmental organizations like Foundation for Revitalization of Local Health Traditions (FRLHT) in southern states and Tropical Botanical Garden and Research Institute (TBGRI) in Kerala.

Approximately 4.2% of the total geographical area of India has been earmarked for extensive in situ conservation of habitats and ecosystems. The Indian Council of Forestry Research and Education (ICFRE) has identified 309 forest preservation plots of representative forest types for conservation of viable representative areas of biodiversity. One hundred eighty-seven of these plots are natural forests and 112 plantations, covering a total area of 8500 ha. A program called "Eco-development" for in situ conservation of biological diversity involving local communities has been initiated. This integrates the ecological and economic features for sustained conservation of ecosystems by involving local communities. The economic needs of local communities are taken care under this program through provision of alternative sources of income and steady availability of forest and related products, including medicinal plants, alternative source of food, ornamentals, plant-based raw material resource for cottage industry, etc. To conserve the representative ecosystem, a Biosphere Reserve Program has been initiated. Eighteen biodiversity-rich areas of the country have been designated as biosphere reserves. Programs have also been launched for scientific management and wise use of fragile ecosystems. This has led to establishment of 448 wildlife sanctuaries, 85 national parks, 19 wetlands, 20 lakes, etc., in different biogeographical provinces under various conservation programs helping conservation of many naturally occurring wild species belonging to various crop groups. Gene sanctuaries have been established for conservation of wild relatives of cultivated/crop plants. For example, Nokrek National Park or Nokrek Biosphere Reserve is National *Citrus* Gene Sanctuary set up in buffer zone of 47 square kilometers to conserve *Citrus indica*. It is known for wild varieties of *Citrus* fruits that provide a gene pool for commercial *Citrus*. Barsey *Rhododendron* Sanctuary is established in Western Sikkim in Singalila Range. Deorali Orchid Sanctuary in Sikkim is renowned for its rare and extensive collection of orchids. Further strengthening of these efforts is needed with:

- Expansion of protected area network and strengthen their activities
- Expedite formulation of management plans of protected areas based on natural resources and socioeconomic profiles
- Development and implementation of socio-eco-development projects adjacent to protected areas to meet the demands, increase the income of local people, and involve them in conservation efforts
- Minimization and elimination of activities leading to loss of biodiversity, including habitat destruction, overexploitation of plants from nature, pollution, and introduction of exotics
- Reintroduction and establishment of viable populations of threatened species
- Intersectional consultation and cooperation to restrict habitat destruction and overexploitation of biological resources

19.2.3.2 Ex Situ Conservation

This refers to conservation of plant species' genetic material under protected conditions or in man-made gene bank using various techniques, as per the habitat and breeding behavior. It would include:

A. *Plant conservation in field gene bank*: It is suited for perennial wild species, including shrubs and trees. It involves maintenance/conservation of whole plant(s) in a protected area/facility. It is partially evolutionary in nature.
   1. Some examples are botanical gardens, arboreta, herbal garden, clonal repositories, etc.
   2. Selective conservation (conservation under selective pressures)

B. *Seed conservation in seed gene bank*: It is the most common economical method of germplasm conservation, as more than 1750 seed banks are established across the world for ex situ conservation of plant diversity, including closest wild relatives of crop species. It involves storage of seeds, the physical carrier of genetic information from one generation to the other at low temperatures. Predominantly orthodox seeds, which are desiccation tolerant, can be dried to low moisture level (93.9% plants, Royal Botanic Gardens, Kew 2008), and are not freeze sensitive, are stored at low temperatures. They can survive for years, under low temperature conditions (up to −18 °C), if dried to low moisture contents (5–10%). This conservation is static in nature (for a recent status, see Hay and Robert 2013).
C. In vitro *conservation of explants or organs*: It involves culture and maintenance of cells, tissues, and organs under aseptic conditions, following the principle of totipotency at hypothermic temperatures. The cultures are subjected to slow growth in glass or plastic containers to increase the subculturing intervals. It is most suited for medium-term conservation of vegetatively propagated plant species, ensuring direct availability of genetic resources in crop improvement. It is static in nature.
D. *Cryopreservation of genetic material*: Cryopreservation plays an important role in cell and tissue banking. It involves preservation of seeds, organs, sperms (pollen), or cultures by cooling to temperatures below the freezing point of water in liquid nitrogen between −150 and −196 °C following the principle of cryogenics. It involves several steps. Initially, it was evolved to facilitate conservation of recalcitrant or intermediate type of seeds that are sensitive either to desiccation or freezing or both. It is now extended for long-term conservation of all types of genetic material, including seeds, and in vitro culture for cost-effective conservation.
E. *DNA banking*: It involves storage of total genomic DNA or its components, such as DNA libraries, DNA clones, etc., under freezing temperatures.

Both in situ and ex situ conservation approaches have their advantages and disadvantages. Ex situ methods are well suited to capture and store all alleles and genotypes safely on long-term basis, while in situ conservation permits evolution to continue, allowing populations to adopt to changing environment. Ex situ conservation can also support in situ conservation by providing material for reintroduction of species and restoration of habitat. Similarly, ex situ conservation can be regularly enriched and sustained with introduction of fresh and more evolved genetic diversity from in situ conservation. Though in situ and ex situ activities are carried out independently, they can be linked and integrated for maximum benefit for both conservation and use of plant genetic resources.

Figure 19.1 illustrates the possible approaches that can be adopted for conservation of wild relatives of cultivated/crop species, either independently or in combination to complement the efforts. Wild species are very diverse group, including species with different life span, habit, habitat, and diverse reproductive biology. For conservation of such diverse plant biodiversity, a complementary comprehensive

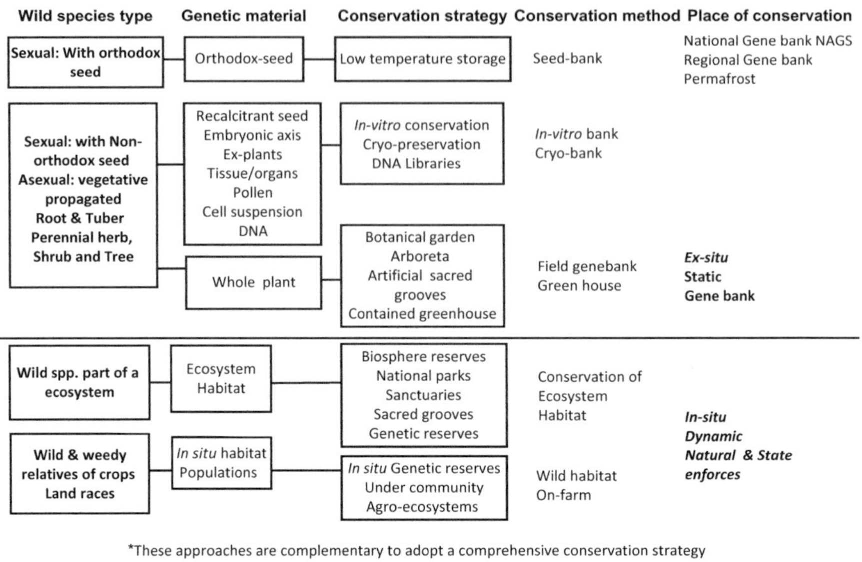

Fig. 19.1 Possible approaches for conservation of wild relatives of crop species

approach must be followed that would help maximize the benefits of different approaches to ensure optimum conservation of a species or its gene pool. Following this approach, it is very likely that a range of ex situ conservation methods would be applicable to satisfy the needs of diverse type of crop wild relatives that are habitat to a range of climatic conditions with breeding systems ranging from obligate vegetative propagation to facultative and obligate self-pollinating or cross-pollinating systems. It is quite logical to have an approach with appropriate and balanced applications of both in situ and ex situ conservation strategies. For many populations, because of their adaptation to a very narrow ecological niche, in situ conservation may be the only answer. Similarly, for conservation of specific types associated with a geographical region, in situ or on-farm conservation may be one option in addition to their ex situ conservation in seed gene bank.

Altered landscape/evolutionary gardens are an example of ex situ complementarity, wherein in situ conservation might be quite relevant to wild species conservation in natural ecosystems. There is a range of types in altered landscapes that can maintain a significant number of genotypes and provide convenient sources of germplasm with introduced scientifically determined planting of cultivars that can enrich the natural gene pool and permit additional future recombinations. Such special evolutionary gardens (with condensed genetic diversity within natural distributions of respective species) are deemed useful for more assured and/or increased source of germplasm. For many crops, some ex situ stands (field gene banks) are already maintained as "evolutionary gene banks" such as for rubber and oil palm. It is necessary to devise better strategies so that such ex situ field gene banks become more common and effective.

Within ex situ also a balance must be struck as per the need. For example, it must be assessed whether they would be best conserved in a field gene bank or under in vitro as cell, tissue, organ, pollen, and perhaps as DNA libraries or in combination thereof. Therefore, a network of complementary and comprehensive strategy is the need of the hour to ensure effective conservation and sustainable use of wild relatives of cultivated/crop species for present and future generations.

Each ex situ technology must be selected based on its merits in utility, security, complementarities, and the advantages over other available techniques. Inappropriate application of any technology will result in a failure and loss of germplasm resources. For example, application of in vitro technology in conservation without rigorous research and experimental foundation on "regeneration" of plants from morphogenic cultures, and their conservation maintaining the "genetic purity" of accession, would be futile and waste of energy and resources. This may result in endangering the germplasm because of its conservation in suboptimal conditions and with a narrow genetic base, defeating the basic principles of germplasm conservation.

For difficult plant systems, a significant part of the gene pool can initially be conserved in field gene banks, followed by research efforts for seed conservation using either conventional methods or biotechnological methods, such as in vitro conservation. These approaches are complementary to each other having balance of merits and demerits. In vitro conservation may be economical and provide improved security to germplasm and independence from climatic variation, whereas the field gene banks allow evaluation of germplasm, assisting users and continued evolution of genetic diversity of species responding to the climatic variation.

The basic objective in conservation of plant genetic resources is preservation of genetic diversity existing in the form of a functional unit called "gene(s)." In clonal species, all clonal gene combinations may not be equally important, but overtime they may become important. They are best conserved as seed (also pollen); therefore, whenever the seeds are produced in as many as root and tuber crops, they should be incorporated in conservation strategies.

Devising conservation strategies should be for efficiency. It can help in identification of research needs. For example, if an ideal strategy for a species is in vitro conservation and slow-growth method, but if it is not yet available, then this means it requires research and development. In this way, true needs of conservation with focus on efficiency, cost, and utility can be developed in the future. Therefore, because of the complementarities of the approaches, a continued dialogue between environmentalists, conservationists, breeders, and biotechnologists is required to ensure necessary stimulus for research and to identify the best approach, its adaptation to pressing problems, and realization of new biotechnological solutions in agriculture.

India maintains many botanical gardens, herbal gardens, and/or crop-specific field gene banks maintaining many wild relatives, either independently, under universities, or under some nongovernmental organizations. Central government and state governments together run and manage 33 botanical gardens that maintain the diversity of species/populations including crop wild relatives. In addition, Botanical

Survey of India and several other organizations such as the National Botanical Research Institute, Lucknow, the Central Institute for Medicinal and Aromatic Plants, Lucknow, Tropical Botanical Gardens and Research Institute, Thiruvananthapuram, and the Regional Plant Resources Center, Bhubaneswar, were initiated by the state government in 1985, and many other state departments maintain botanical gardens to help in ex situ conservation of rare, endangered, threatened species and wild relatives of crop plants. However, there is a strong need for consolidation and strengthening of the network of ex situ conservation centers, such as botanical gardens, herbal gardens, herbaria, etc.

Seed banking has been the main component of ex situ conservation [Global Strategy for Plant Conservation (GSPC) 2002 http://www.plants2020.net/]. In India, ex situ conservation of PGR of domesticated and economically important agricultural plant species including their wild relatives is the sole responsibility of National Bureau of Plant Genetic Resources (NBPGR) that operates under Indian Council Agricultural Research (ICAR) to support various national and international crop improvement programs. The National Gene Bank (NGB) at NBPGR is the largest ex situ seed repository in India and the third largest in the world (after the United States and China). It conserves collection of landraces, traditional and rare cultivars, released varieties, breeding lines, genetic stocks, and wild and weedy relatives of crop plants under long-term stores maintained at -18^0 C. A total of 186 species (7381 accessions) of wild relatives including 63 species (2364 accessions) from indigenous source have been conserved. Additionally, germplasm accessions of wild relatives have also been conserved in cryo-gene bank (59 species; 231 accessions) and in vitro repository (15 species; 46 accessions) as complementary collections.

19.3 Perspective

Decline in the population of wild relatives of cultivated/crop species is likely to continue because of several factors, such as indirect and direct effects of changing land use pattern, increased demands for food, energy production, housing, water, and development, and poorly understood potential impacts of plants on climate change and impact of nonnative species, pest, and disease. Therefore, conservation efforts must be strengthened, particularly of potentially valuable crop genetic resources with modification in the traditional approaches to plant conservation, which have mainly emphasized upon the conservation of individuals, populations, and species. Conservation of CWR includes conservation of genetic diversity of genes and their alleles, which are core.

Looking at the vastness of genetic diversity in wild species, in case of seed-bearing species, information about seed storage behavior is essential to facilitate conservation using an appropriate strategy. Nonorthodox seeds cannot be stored successfully on long-term basis using conventional seed gene bank protocols. Therefore, if the seed storage behavior of a species has not been documented, it should be determined experimentally or visualized as per the taxonomy, origin, and other seed traits.

It is expected that seed bank collections of wild species will play an increasingly important role in habitat restoration and reintroduction of species, besides their use in crop improvement. Looking at the scenario of demand of wild species for crop improvement, the International Rice Research Institute (IRRI) distributed more than 10,000 accessions of cultivated rice to users outside of IRRI in 2012, compared with less than 1000 accessions of wild rice species which are indicative of scope of improvement in facilitated access of these resources for use. The varying diversity in life span demands that they be conserved in large quantities, only in long-term storage conditions. Cryopreservation may be the only recourse to ensure the effective ex situ seed conservation of short-lived species. Sample size, monitoring, and regeneration issues in case of wild species remain researchable issues needing attention in future and can be relaxed.

Recognizing the variation in morphology of seeds of wild species, it is advisable to adopt the international standards for gene banks (FAO 2014).

One of the major impediments to the potential use of wild species germplasm, for reintroduction or habitat restoration (as well as causing difficulties for viability testing), is lack of knowledge on how to break dormancy and germinate the seed, which requires research attention. Increasing focus may also be directed to how accessions are evaluated for assessment of their genetic potential to use in crop improvement, beyond restoration and species reintroduction.

References

Biological Diversity Act (2002) and Biological Diversity Rules 2004. National Biodiversity Authority, Government of India, Chennai, India, p 57

Castañeda-Álvarez NP, Khoury Colin K, Achicanoy Harold A, Vivian B, Hannes D, Eastwood Ruth J, Luigi G, Harker Ruth H, Andy J, Nigel M, Müller Jonas V, Julian R-V, Sosa Chrystian C, Struik Paul C, Holly V, Jane T (2016) Global conservation priorities for crop wild relatives. Nature Plants 2:16022. doi:10.1038/nplants.2016.22

FAO (2014) Genebank Standards for Plant Genetic Resources for Food and Agriculture. Rev. ed. Rome (Revised edition)

Forest (Conservation) Act (1980) Amended (1988, 2004) Government of India, Ministry of Environment & Forest, New Delhi

Hay FR, Robert RJ (2013) Advances in seed conservation of wild plant species: a review of recent research. Conserv Physiol 1(1):cot030. doi:10.1093/conphys/cot030

Import and Export (Control) Act (1947) Gazette of India, 1947, Part V, p 86

Royal Botanic Gardens Kew (2008) Seed Information Database (SID). Version 7.1. http://data.kew.org/sid/

Singh J, Lavania UC, Singh S (2016) Chapter 3. Indian traditional and ethno medicines from antiquity to modern drug development. In: Singh RJ (ed) Genetics resources and chromosome engineering of medicinal plants, Volume 6 medicinal plants. CRC Press, Boca Raton, pp 53–86

The Environment (Protection) Act (1986) Enacted by Parliament of India, January, Ministry of Environment, Forest & Wildlife, Government of India, New Delhi

The Indian Forest Act (1927) Act No.16 of 1927, Enacted by Central Legislative Assembly, 21 Sept., 1927 (As modified upto the 5th June, 1951)

The Spices Board Act (1986) No.10 of 1986, Government of India, Ministry of Commerce, Notification No. 122 (E) dated 26th February 1987

The Wildlife Protection Act (1972) Act No. 53 of 1972, Enacted by Parliament of India, Sept., 1972 Amended several times (1982-2013)

Breeding Strategies for Use of Wild Relatives

20

20.1 Introduction

As the list of desirable characteristics exhausts in the known natural variability of the primary gene pool of crop species, increasing attention has been paid to introgression of characteristics from related species in the genera and beyond, into the existing genotypes of cultivated species. The term "wide hybridization" is generally used to designate hybridization between widely diverse plants to generate new genetic combinations from where desired recombinants can be selected. However, the establishment of hybrid with appropriate fertility to carry on with desirable recombinants into subsequent generations is not always easy. It requires information on genomic relationships between the wild species to be used and the cultivated species for manipulation of successful hybridization, hybrid fertility, and genomic recombination. It may have problems of genetic incompatibility, genomic incompatibility, ploidy barriers, crossability barriers, hybrid fertility, etc., which may interfere either with the establishment of hybrids or with the transfer of useful alien genes from wild to cultivated species through normal chromosome pairing, genetic recombination, and segregation cycle. Some of the genetic, genomic, and ploidy barriers in cross-compatible species can be overcome through cytogenetic manipulations, while crossability barrier encountered in hybridization due to sexual incompatibility or hybrid breakdown due to pre- and/or post-zygotic events can be overcome through manipulation of pre-fertilization and post-fertilization events with certain hormonal treatments and embryo rescue from unsuccessful crosses.

For successful alien gene transfer, identification of a potentially suitable species and breeding strategy, based on information on genomic constitution of wild species and its relationship with the cultivated species, are prerequisites for establishment of fertile hybrids and subsequent successful interspecific gene transfer. It would involve several steps discussed below.

A.K. Singh, *Wild Relatives of Cultivated Plants in India*,
DOI 10.1007/978-981-10-5116-6_20

20.2 Characterization and Evaluation of Wild Species

As transfer of genes from wild species would not be an easy process and may require a long-drawn research project with several steps to achieve the final goal, the first step would be to evaluate the wild species of the target crops against main yield-reducing factors. It may be for the characteristics of potential value in general and for the target traits for which the variability is lagging in cultivated species. Therefore, a characterization and evaluation program of wild species should be initiated in target crops to develop information regarding the essential features (taxonomy) and regarding the potential value of different wild relatives for lacking variability, by screening them on priority against such features. This would need listing of characteristics required, development of effective screening methods for identification of species/accessions with desirable variability, and information documentation on desirable traits in a database.

The other component of characterization of wild species is required at the genomic level, using both traditional cytogenetic techniques of genome analysis, involving karyomorphological studies designating different genomes, cross-compatibility relationships, and analysis of chromosome pairing in meiotic cycle of resultant hybrid between the wild and cultivated species, tracing genomic relationships. This should be extended to molecular techniques of genome characterization with greater resolution power, such as genome sequencing (DNA), genome mapping, in situ hybridization, etc., to trace phylogenetic relationship between wild and cultivated species to facilitate visualization of most appropriate breeding method for gene transfer/introgression.

20.3 Identification of Potentially Valuable Species

The reports of useful genetic variability in wild species have generated interest in collection, characterization, and evaluation of wild species. Internationally accepted screening techniques have been developed in most crop species for evaluation of accessions against major biotic and abiotic stresses to facilitate exchange of information and the desired genetic material. This wide interest in use of wild species has resulted in generation of lots of information on the potential value of the various wild species in different crops available in literature. Therefore, a thorough literature search and screening of the available accessions of wild species for various desirable characteristics and development or strengthening of a database listing wild species with various desirable traits is the need of time, to facilitate their conservation and utilization in the breeding program as per the need.

20.4 Prioritization of Species

As discussed above, the wild species that are to be used in the breeding program may have different levels of genetic distinctness from the cultivated species, affecting their cross-compatibility with cultivated species and probable success in genetic introgression. Therefore, based on cross-compatibility relationships, different approaches may be required to be resorted for utilization of wild species. In case of species that are component of primary gene pool as a member of same biological species, there should not be much problem in transfer of genes, and conventional breeding methods should be sufficient to effect gene transfer. The biological species would include wild and weedy landraces of cultivated species that might have been classified as separate subspecies or botanical varieties, falling in group B of primary gene pool.

In case of the species that are part of the secondary gene pool and defined as cenospecies, to overcome sterility barriers, unique and imaginative methods should be employed for manipulation of genome and chromosome pairing for effecting recombination and thereby the transfer of genes.

In case of species belonging to tertiary gene pool, which represents the outer boundary of the gene pool, gene transfer is not expected to be possible by the known techniques and radical methods, such as embryo culture, doubling of chromosome, and the use of bridge species, and biotechnological technique such as genetic transformation may be required.

The convenience with which the gene transfer can be affected may be one of the criterias to decide the priority species to be used in (pre-) breeding program. The other criterias as discussed above would be lack of availability of specific variability in cultivated species for a specific or ensuing constraints and wild species being the only source.

20.5 Breeding Options for Interspecific Gene Transfer

To overcome hybridization and sterility barriers and to effect desired recombination/transfer of gene through sexual or asexual means, following breeding options may be considered for exploitation of wild species.

20.5.1 Conventional Cytogenetic Manipulations

20.5.1.1 Direct Hybridization

1. *Same chromosome number:* In case of same chromosome numbers, direct hybridization is an easiest route for incorporation of wild species germplasm.
2. *Different chromosome number:* In case of different chromosome numbers, difficulties are encountered because of difference in ploidy levels between wild and cultivated species, and the following four basic methods are employed.

(a) Direct hybridization at different ploidy levels producing sterile hybrids
(b) Raising the ploidy level of wild species followed by hybridization at the same ploidy level as that of the cultivated species
(c) Doubling chromosome number of the wild species or of their hybrids before hybridization and resynthesizing chromosome number equal to the cultivated species
(d) Reducing the chromosome number of cultivated species to the level of wild species before hybridization and resynthesizing chromosome number equal to cultivated species, i.e., using the approach of doubled haploids to create inbred lines

Transfer of desirable features from wild species has been and can be attempted both in diploid and polyploid crop species following the above conventional breeding options involving ploidy manipulations (Singh et al. 1990). In diploids, such as sugar beet, direct interploidy hybridization and hybridization with induced autotetraploids have resulted in transfer of fragment of chromosomes or genes for characters like cyst nematode (Jung et al. 1986). In *Ribes* ssp. (currants and gooseberries), genetic introgression was achieved by using autotetraploid and amphidiploid of wild species. In polyploid crops, direct hybridization and hybridization involving autopolyploids, amphidiploids, and haploids as per the ploidy level of crop and/or of the wild species or the bridge species and their hybrids, followed by backcrossing, have been used most frequently in several crops, such as clovers, *Trifolium* spp. (Parrot and Smith 1986), alfalfa, *Medicago sativa*, etc. (Barnes et al. 1977). Similarly, in polybasic allopolyploids like *Brassica* and segmental allopolyploids like groundnut, direct hybridization and hybridization with autopolyploid and amphidiploid have been successfully used for regaining hybrid fertility and introgression of clubroot resistance in case of *Brassica* (Johnston 1974) and foliar disease resistance in case of groundnut (Singh et al. 1991). Haploid breeding has been used for gene introgression in potatoes. In potatoes dihaploids are produced from tetraploid *S. tuberosum* and hybridized with diploid wild species with selection of desired recombinants at the diploid level, followed by resynthesis of cultivated polypoid.

In all above methods, hybrid production followed by backcrossing with recurrent cultivated species parent cultigen to regain the agronomic features through recombination breeding is the course, if the objective is production of alien gene introgression lines. The cycle of backcrossing may result in production of predominant cultivated species like interspecific derivatives/populations on which selection is applied for desired recombinants. In this regard, the strategy for application of selection in early generation or advance generation would depend on genetics of traits. The selected desired recombinant would need to be advanced to uniformity and stability. Finally, they are to be evaluated in adaptation trials for assessment of agronomic potential.

Hence, plant breeding is dependent on meiotic recombination for the generation of new genotypes that can be selected for improved phenotypes. Low frequency and often uneven distribution of meiotic crossing over has been a major constraint to efficient and timely crop improvement. Using wild species for the transfer of advantageous gene(s) in crop improvement programs inevitably requires many years of backcrossing to break up linkage drag or knowledge/methods for manipulating the recombination frequency and distribution in these hybrids.

These approaches can result in synthetic amphidiploid lines and alloplasmic lines (developed by exchanging the cytoplasm through interspecific and intergeneric hybridization followed by successive backcrossing to the nuclear genome's donor species) with advantages of CMS and useful agronomic traits, such as herbicide resistance, modification of flavor and nutrient content, and morphogenetic potential (Namai 1987), disomic or monosomic alien chromosome addition lines, and disomic and monosomic alien chromosome substitution lines. Therefore, these approaches can provide materials that are useful both for developing practical breeding strategies/material and for studying the genetic effects of individual chromosomes on plant traits (Kaneko and Bang 2014).

Many disomic alien chromosome substitution lines (DASLs) have been bred in wheat cultivars. In these lines, one member of a chromosome pair is replaced by a homoeologous chromosome from wild-related species, e.g., *Secale*, *Hordeum*, and *Aegilops* (Khush 1973). Such DASLs can be used to study the genetic effects of individual chromosomes and to estimate the number of genes that control a given trait and their linkage relationships. These lines can be generally developed in the progeny after self-fertilization of a monosomic alien chromosome substitution line (MASL). In this breeding system, the MASLs can be generated only when the two corresponding chromosomes are homoeologous and complement each other, especially within diploid species. DASLs are known to occur spontaneously during successive backcrossing after interspecific hybridization. Several addition lines have been produced in wheat with improved quality of disease resistance, earliness, protein content, etc., but none of these have been accepted as commercial variety. Substitution lines can be produced only in species where extensive cytogenetic and genetic knowledge has accumulated.

Utilization of chromosomal translocation has also been used for gene transfer in case of wheat, tobaccos, etc., through various manipulations like irradiation where the desired gene is situated on nonhomoeologus chromosome.

20.5.1.2 Bridge Crosses

This method has been used through crosses with a species common in cross-compatibility relationship between wild donor and recipient cultivated species. Bridge crosses have been used in wheat, *Cucurbita*, and *Solanum* with traditional steps to effect desired recombination.

20.5.2 Biotechnological Approaches Overcoming Hybridization Barriers

The wild germplasm belonging to secondary and tertiary gene pools have several kinds of hybridization barriers. They either operate at the time of fertilization or after fertilization, depending upon the degree of reproductive isolation.

Pre-fertilization barriers include failure of pollen germination, pollen tube entry into the stigma, and growth of the pollen tube carrying the male gametes through the style. Many crosses may show unilateral incompatibility, that is, pollination is effective only in one direction, while the reciprocal shows strong pre-fertilization barriers, though success generally is greater when the wild species is used as female parent (Shivanna 1996). There are several techniques to overcome the pre-fertilization barriers, particularly the application of a variety of growth regulators at the respective stages of fertilization to achieve normal growth and fertilization and to initiate seed production (Sastri 1984). Abortion of embryo causes post-fertilization barriers leading to formation of shriveled seeds or rudimentary seeds with embryo, because the endosperm degenerates after fertilization restricting nutritive supply to embryo. Embryo culture has been advantageously used for establishment of hybrids overcoming the post-fertilization barriers. This technique has been widely used for obtaining interspecific or intergeneric hybrids promoting the growth of embryo that includes ovary culture, ovule culture, etc. Successful application of this technique has resulted in production of hybrids in case of *Phaseolus*, *Trifolium*, *Gossypium*, *Cucurbita*, *Lycopersicon*, *Hordium*, *Triticum,* etc. In vitro fertilization has been effectively used in raising several intergeneric hybrids (Zenkteller 1990). On establishment of hybrids, they may also involve various other cytogenetic manipulations to improve chromosome pairing and thereby genetic recombination and hybrid fertility.

20.5.3 Asexual Biotechnological Approaches

Biotechnological approaches (Tester and Langridge 2010) can allow for the transfer of genes beyond taxonomic boundaries, from far more distantly related wild relatives of crop species into breeding programs.

20.5.3.1 Somatic Hybridization

To overcome the fertilization barriers, though technique like embryo rescue is the choice because of simplicity and operational ease, biotechnological approaches such as fusion of protoplast of two distant species and its regeneration into hybrid plant have been one option successfully employed in *Brassicas* (Prakash et al. 2009).

20.5.3.2 Genetic Modification/Transformation

The recent developments in biotechnology have made it possible for transfer of genes from tertiary gene pool and beyond the taxonomic boundaries through the asexual method of genetic modification/transformation. The genetic modification (GM) involves incorporation of the genes through *Agrobacterium*-based genetic transformation system or particle bombardment, to transfer isolated DNA, etc., between organisms and produce a modified organism containing recombinant DNA (Stewart 2004; Brown and Caligari 2008). This technique is also being used to identify and transfer single or small groups of genes of economic value into commercial varieties from diverse sources. However, there has been considerable controversy and concern about genetic escape of such genes and potential ecological disruption associated with genetically modified organism and the GM technology. This has been due to the use of transgenesis, which involves exchanges of genes which do/could not occur in nature between a bacterium or an eukaryote and a plant.

However, the basic tenets of these techniques suit to exploitation of wild relative of crop/cultivated species in genetic improvement of cultivated species with transfer of specific genes, overcoming the problem of the genetic drag of undesirable traits along with desirable traits in interspecific breeding. In conventional breeding, getting rid of undesirable linked traits requires a multiple backcrossing cycle to break the linkage. This technique of GM has been referred as cis-genesis. Therefore, cis-genesis is the use of GM technology to transfer genes, which occur naturally and could also be transferred to the crop using slower conventional breeding methods. The use of cis-genesis would accelerate the process of transfer of genes from wild relatives of crops/cultivated species into crops and is likely to have far lower risks or no risk, as the transferred genes already occur in the wild, and chances of ecological disruption appear much less likely. For many crops, particularly vegetatively propagated ones, cis-genesis can be used directly for improvement of existing varieties, which have been shown to be safe for use in the market.

Improved techniques for genome analysis (e.g., genome/gene sequencing, marker-assisted selection, next-generation sequencing, and transcriptomics) have further facilitated application of this shortcut process. This approach involves use of genomic-based resources, map-based cloning, and the analysis of quantitative trait loci, gene isolation, and genetic modification.

In these biotechnological approaches, much of the success of genetic engineering depends on locating useful genes to clone and then transferring those genes to genetic lines in a breeding program. Nearly the same way, as conventional plant breeding depends on genes found in germplasm, it depends on genes/alleles found in wild species. It will involve complete genome sequence of cultivated species, followed by high-throughput resequencing of related genomes, including wild-related species genome. Greater allelic diversity has been found among the wild species than among the cultivated species. Such approach will allow screening of crop wild relatives for important genetic diversity much more efficiently and in greater volume than before. Next-generation sequencing, by way of this high-throughput targeted resequencing, helps screening thousands of samples of germplasm for interesting gene variants and making them available for use in conventional breeding. For example, there are now many candidate genes shown to be involved in some way with drought tolerance, and large-scale resequencing will allow us to identify these genes and all the variants that can be found in the wild species related to many crops. Researchers in the Generation Challenge Program (www.generationcp.org) have identified allelic diversity in candidate genes for drought tolerance in cereal crops and a group of hormone-related genes for stress and ripening. Similar approaches can be extended for targeting other stress genes in wild relatives of crop species to meet the challenges of climate change and engineering new cultivars. These developments would speed up the identification of useful gene variants for production of new varieties, though help from conventional breeding will still be needed. Genome sequencing has now become affordable and is available in many crops, such as rice, maize, potato, tomato, etc., and efficient gene isolation methods such as map-based cloning and allele mining are available for using cloned indigenous genes.

It has been possible to construct autonomous minichromosomes in yeast, which behave like normal chromosomes by way of segregation at meiosis, delivering multiple genes to daughter cells. This has been possible in maize too (Carlson et al. 2007; Yu et al. 2007; Sabelli et al. 2009), presenting the possibility of engineering plant processes and the ability to select segments of the genome from a wild relative known to influence disease resistance or drought tolerance and to transfer them directly into breeding material. In future, biotechnological approaches will assist transfer of sets of genes or even complexes of genes conferring tolerance to drought, salt, and temperature from wild relatives.

20.5.4 Other Methods

In addition to the above methods, a variety of other methods have been used either to facilitate production of hybrids or to facilitate the chromosome clearing and genetic recombination. Similarly, additional methods have been applied to utilize wild species germplasm through manipulation of sporophytic or gametophytic incompatibilities through the use of mentor pollen (mixture of treated compatible pollen loosing fertilizing capacity, but retaining the capacity to stimulate incompatible pollen for fertilization). However, these methods have not been very successful and therefore rarely used in the crop improvement program.

20.6 Merit and Demerits of Various Breeding Approaches

20.6.1 Conventional Breeding

Merits: Suited to the species/genetic resources falling within the limits of biological species, including spontaneous wild or weedy races of cultivated species and the subspecies and botanical varieties. Hybridization will be easy; hybrids are generally fertile with good chromosome pairing, and the gene recombination and segregation will be approximately normal; therefore, gene transfer will be easy.

Demerits: There would be a dilution effect on gene transfer, and hence obtaining stable and widely adopted genotypes is difficult. These approaches do not provide access to distant genetic resources.

20.6.2 Sexual Biotechnological Breeding

Merits: Suited to cenospecies that will cross with the crop/cultivated species and shall help overcoming problems of hybrid fertility, chromosome pairing, and genetic recombination. Provide access to cross-compatible species of secondary gene pool, consisting of maximum usable genetic variability, and thereby to most genes of the gene pool.

Demerits: Needs expertise to overcome the barriers to hybridization using advanced techniques, such as tissue culture for establishment of hybrids. Hybrids may be weak and difficult to bring to maturity, or hybrids may be sterile restricting progress to subsequent generations, and percentage of recovery of desired recombinants in advance generation may be very low.

20.6.3 Asexual Biotechnological Breeding/Genetic Modification/Transformation

Merits: Provide access to species that are nearly or cross-incompatible, representing the outermost boundary of the gene pool and beyond, but representing higher level of desirable features, for example, nonhost reaction for a biotic stress. It facilitates specific gene(s) transfer, overcoming the problem of the genetic drag of undesirable traits along with desirable traits of interspecific breeding. As it would surpass the backcrossing cycle, and accelerate the slow process of gene transfer through conventional breeding. Avoid incorporation of toxic components from wild species. Additionally, the possibility of marker-free plant transformation without bacterial antibiotic gene as selection markers enables creation of clean cisgenic plants without transgenes.

Demerits: As it surpasses the conventional interspecific hybridization, it requires expertise or collaborative support for use of advanced techniques and technologies for production of hybrids, such as somatic hybridization, genome analysis using molecular genome sequencing, genome mapping, gene isolation, and genetic transformation.

20.7 General Steps of Interspecific Gene Transfer

Irrespective of the options to be followed to affect the gene transfer from alien germplasm, the following common steps would be required, integrating the conventional breeding:

- *Production of interspecific hybrids*: It would involve establishment of interspecific or intergeneric hybrids using appropriate techniques/manipulation as desired.
- *Backcrossing to produce stable interspecific derivatives*: This would involve backcrossing of the hybrids with recurrent cultivated species parent to regain all the agronomic/economic traits retaining the incorporated desirable gene from the wild species.

- *Raising of hybrid populations*: Interspecific hybrid derivatives incorporating desirable features may need to be advanced with sufficient population to apply selection pressure under both natural and artificial conditions.
- *Screening of interspecific derivatives for desirable traits*: The stable cultivated species like interspecific hybrid populations would need to be screened for desirable traits under field conditions to select stable genotypes.
- *Advancement of selected interspecific to uniform derivatives*: The selected interspecific genotypes would need to be advanced until they become stable and uniform without any further segregation for agronomic traits and the incorporated desirable features.
- *Testing of interspecific derivatives in diverse agroclimatic conditions*: It has been observed that wide hybridization often results in development of genotypes with wider adaptability. Also, under conventional breeding approach, not being a directed process controlling genetic recombination, it may result in recombinants of several other nontargeted desirable features. Therefore, testing of interspecific derivatives under diverse agroclimatic conditions may result in identification of genotypes with desirable recombinants, other than the target traits with wider adaptability.
- *Use of interspecific derivatives in conventional breeding as genetic resource*: In terms of normal conventional breeding, the stable interspecific derivatives produced after incorporating desirable features from the wild species are result of pre-breeding efforts, and these derivatives may function as genetic resource in the conventional breeding program, particularly resistance breeding program.

20.8 Perspective

Despite increasing concern about food security and the wider recognition of the potential of genes found in wild sources to improve crop performance in many aspects, a relatively small amount of efforts has focused on generating knowledge about these resources both from basic and applied points of view. This reflects a lack of effort in refining our understanding regarding the genetic information about this valuable component of genetic resources to guide planning on their conservation and use in the breeding program. It is desirable to encourage research in this direction, followed by information dissemination.

Baring few major crop species, genomic information is very poor or altogether lacking in majority of the cultivated species. Therefore, in majority of cultivated species, there is a need for genome analysis using conventional biosystematic approach, leading to genome identification, characterization, ascertaining taxonomic entity, and developing information on phylogenetic relationship between the wild species and particularly with cultivated species, to facilitate identification of appropriate breeding strategies to affect desired genetic introgression.

The efforts on evaluation of wild species against priority traits should be a continuing process as per the demand for assessment of genetic potential of wild species and identification of priority species based on economic importance of desirable traits associated and feasibility study on ease of transfer of traits.

As the biotechnological approaches, such as cis-genesis, are providing new avenues in plant breeding using cloned indigenous genes, there is a need to generate full genome sequences in major cultivated/crop species and prospective priority wild relatives with traits of immediate significance. This will enable efficient gene isolation, using methods such as map-based cloning and allele mining. It will enable precision breeding incorporating desired traits directly, more quickly and efficiently, in an effective way, engineering desired cultivars, or improving the existing promising cultivars.

References

Barnes DK, Bingham ET, Murphy RP, Hunt OJ, Beard DF, Skrdla WH, Teuber LR (1977) Alfalfa germplasm in the United States: genetic vulnerability, use, and maintenance. USDA-ARS Tech. Bull. 1571. USDA-ARS, Hyattsville, MD

Brown J, Caligari P (2008) An introduction to plant breeding. Blackwell Scientific Publications, Oxford

Carlson SR, Rudgers GW, Zieler H, Mach JM, Luo S, Grunden E, Krol C, Copenhaver GP, Preuss D (2007) Meiotic transmission of an in vitro–assembled autonomous maize minichromosome. PLoS Genet 3(10):1965–1974. doi:10.1371/journal.pgen.003017

Johnston TD (1974) Transfer of disease resistance from *Brassica campestris* L. to rape (*B. napus* L.) Euphytica 23:681–683

Jung C, Wehling P, Lbptien H (1986) Electrophoretic investigations on nematode-resistant sugar beets. Pl Breed 97:39–45

Kaneko Y, Bang SW (2014) Interspecific and intergeneric hybridization and chromosome engineering of Brassicaceae crops. Breed Sci 64:14–22

Khush GS (1973) Cytogenetics of alien addition and substitution. In: Cytogenetics of Aneuploids. Academic, New York, pp 238–248

Namai H (1987) Inducing cytogenetical alterations by means of interspecific and intergeneric hybridization in brassica crops. Gamma Field Symp 26:41–89

Parrot WA, Smith RR (1986) Recurrent selection for 2n pollen formation in red clover. Crop Sci 26(6):1132–1135. ISSN 0011-183X

Prakash S, Bhat SR, Quiros CF, Kirti PB, Chopra VL (2009) *Brassica* and its close allies: cytogenetic and evolution. Plant Breed Rev 31:21–187

Sabelli PA, Hoerster G, Lizarraga LE, Brown SW, Gordon-Kamm WJ, Larkins BA (2009) Positive regulation of minichromosome maintenance gene expression, DNA replication, and cell transformation by a plant retinoblastoma gene. Proc Natl Acad Sci 106:4042–4047

Sastri DC (1984) Incompatibility in angiosperms, significance in crop improvement. Adv Appl Biol 10:71–111

Shivanna KR (1996) Incompatibility and wide hybridization. In: Chopra VL, Praksh S (eds) Oilseeds and vegetable brassicas: Indian perspective. Oxford and IBH Publishing Co, New Delhi, pp 77–102

Singh AK, Moss JP, Smartt J (1990) Ploidy manipulations in interspecific gene transfer. Adv Agron 43:199–240

Singh AK, Stalker HT, Moss JP (1991) Cytogenetics and use of alien genetic variation in groundnut (*arachis*) improvement. In: Tsuchiya T, Gupta PK (eds) Chromosome engineering in plants: genetics, breeding, evolution Part B. Elsevier Science Publishers, Amsterdam, pp 65–77

Stewart CN (2004) Genetically modified plant: environmental impacts of genetically modified plants. Oxford University Press, Oxford

Tester M, Langridge P (2010) Breeding technologies to increase crop production in a changing world. Science 327:818–822

Yu W, Han F, Birchler JA (2007) Engineered minichromosomes in plants. Curr Opin Biotechnol 18:425–431

Zenkteller M (1990) *In vitro* fertilization of ovules of some species of Brassicaeae. Plant Breed 105:221–228

Future Perspectives

21

21.1 Introduction

To face challenges of food and nutritional security of ever-growing human population and the challenges of the present scenario of climate change and to improve over the narrow genetic base of crops that has caused around 75% crop losses in the twentieth century would require engineering of new sustainable and productive cultivars in the twenty-first century. This would need access to wider genetic diversity of genetic resources and application of advanced breeding technologies to accelerate the process of crop improvement. This should include increasing use of wild genetic diversity, particularly those of the wild relatives of crop/cultivated species in crop improvement. However, it is globally recognized that biodiversity is at risk from multiple threats, including climate change. The genetic diversity contained in plant genetic resources of a cultivated species, particularly among wild species that are related to crops, is under threat. It is threatening the opportunities and the ability to respond to the new stresses in the changing agricultural environment. Therefore, there is strong reason to understand and develop knowledge about the genetic value of the wild relatives, to embark upon their collection, conservation, characterization, and evaluation, promoting their use in breeding program. This will enable facilitated access to new and more evolved genetic variation, in addition to the one found in the plants of the crop/cultivated species, against the various biotic and abiotic stresses, for nutritional quality and other desired traits.

India is one of the mega-centers of biodiversity and contains a wide range of crop/cultivated species wild relatives spread over many plant families and in diverse geographical regions of the country (Singh et al. 2013; Singh 2017). The collection, conservation, characterization, and evaluation of the wild relatives of crop species are required to strengthen the plant breeding programs of Indian National Agricultural Research Systems (NARS) with additional genetic resources. As no country is self-sufficient in crop genetic resources, the collection and conservation of crop wild relatives in India shall also contribute to global wealth of crop genetic resources, for access as per the internationally agreed frameworks developed in

A.K. Singh, *Wild Relatives of Cultivated Plants in India*,
DOI 10.1007/978-981-10-5116-6_21

recent past, such as the International Treaty on Plant Genetic Resources for Food and Agriculture (ITPGRFA) 2001, operational from 2007, and the Nagoya Protocol, 2010, of Convention on Biological Diversity (CBD 1992), operational from 2012.

There have been several publications inventorying the wild relatives of crop species found in India, particularly the most recent "Cultivated Species and Their Wild Relatives in India- An Inventory," published under the auspices of Protection of Plant Varieties and Farmers' Rights (PPV & FR), India (Singh et al. 2013), and the present publication. Also, there are publications raising concerns about the rarity of systematic use of wild relatives in crop improvement, and most efforts are being done in an ad hoc manner, reviewing the examples of interspecific hybridization in some species (Warschefsky et al. 2014). Therefore, such efforts need to be continued at definite intervals to provide an updated comprehensive listing of wild species as per their importance and that of cultivated species, updating knowledge on their potential genetic value, threat to existence, gaps in collections, conservation, and characterization to promote systematic use of wild diversity in desired direction. Warschefsky et al. (2014) have also proposed a future framework in this direction, considering present perspective, which is further extended here.

21.2 Most Are Threatened: Action Plan for Protection/Conservation of Important Ones

Global changes have been causing irretrievable losses of wild habitat and thereby of wild biological diversity, including wild relatives of crop/cultivated species, which have been reported to be under threat universally (Ford-Lloyd et al. 2011; Maxted et al. 2012), including in different ecologies of India (Singh 2004). Many have become rare due to loss of their habitat or due to fragmentation and degradation of the wild habitats. Genetic erosion and loss of wild species caused by change in land use pattern, major hydroelectric dam, infrastructure developments, and intensification of modern agriculture with priority crops exemplify these threats. Many wild species are also under threat, because of their direct harvest from nature for associated economic potential, such as medicinal or pharmaceutical value. This scenario needs to change either with policies restricting such activities or through legislations, providing protection and/or re-habitation to wild species populations in neighboring or similar environment/habitat elsewhere. Additionally, the management plan of the protected area with CWR does not include specific prescriptions that will favor the conservation of individual target species. Hence, integration of prescriptions that promote growth and development of targeted wild species would be another need.

This would need collaboration and active participation of local government or non-government institutions, which can help build local consensus and educate the local people about the diverse opportunities and benefits that can be harnessed by protecting wild relatives of cultivated both in situ and ex situ. Collaborative efforts would be more effective and productive than going alone. Also, it may demand urgent efforts for immediate exploration and collection mission of wild relatives of crop/cultivated species to capture their important genetic diversity, before they are lost forever. This has also become more essential because wild relatives of crops did not get due attention in early germplasm collections.

21.3 Awareness Development Regarding the Importance of Wild Species

To succeed in collection and conservation of wild relatives of crop/cultivated species, there is a need of a program involving both at national and international communities for developing awareness about the importance of wild species in saving the degradation of environment and harnessing other benefits and opportunities they offer. They can help in containing the adverse effect of environmental pollution causing climate change (see Chapter 3). The wild species also offer opportunities in providing alternative source of food, forage, herbal medicine, etc., besides the leading role they can play in genetic improvement of cultivated/crop species as donor of genetic resistance to various stresses, providing increased resilience and production with lesser inputs. As said earlier, in this regard, support from locals can help build consensus, for example, if these wild species are recognized as being the major contributors to ecosystem services of significant value, such as carbon sequestration metigating global warming. This perhaps would boost the efforts substantially and would be more effective than lone efforts.

21.4 Building Comprehensive Collection of Wild Relatives of Indian Origin/Indigenous to India

Despite diverse difficulties, national efforts to collect crop wild relatives in a systematic manner would be the first step towards facilitating the availability of these species for use in the crop breeding program, to engineer climate resilient crop varieties that can meet the challenges and demands of twenty-first-century agriculture. Collections must cover as much of the geographical and ecological range of wild relatives as possible to cover the extent of adaptive variation, capturing most genetic diversity. The collections must be made along with full information regarding site of collection in relation to variation in environmental and physical features, such temperature range, precipitation, soil type, companion species, and reaction to prevailing stresses. This task helps to fill both physical and genetic gaps in future germplasm collections, if left. Also, it places the emphasis on under-sampled regions and maximizing the range of adaptive variation present in the available collections across diverse ecological axes (Dempewolf et al. 2014).

21.5 Continued Conservation

As most wild species are habitat to wild conditions, where they have acquired their unique features, it will be more appropriate to focus more on their in situ conservation, which has the advantage of being a part of a dynamic system, enabling the species continuously responding to changing environment and evolving itself accordingly. It is estimated that CBD has not been able to meet its 2010 target of biodiversity conservation. In India, conservation efforts do not specifically target wild relatives of crops. In most regions, in situ conservation of these resources is very minimum or nil, as most protected areas have been established with objective of overall conservation of biodiversity, without any emphasis on specific protection or preservation of habitat of crop genetic resources. Additionally, the wild relatives of cultivated species have also been facing genetic risks due to genetic introgression from cultivated forms while in the case of plant species with medicinal or pharmaceutical value or of forage or ornamental value, from their direct overharvesting from the nature. Few of these issues are easy to deal with. However, identification, notification, and prioritization of the threatened crop wild relatives are essential for attracting research and conservation attention.

There have been biodiversity conservation efforts in India, with establishment of biosphere reserves, national parks, or other conservation areas. However, as said above mainly with the objective of conservation of biodiversity at large with environment/climate as such, and not the preservation of crop genetic resources. There is very little effort on conservation of specific species habitat or of those group related with crops/cultivated species, to establish facilities like gene sanctuaries, genetic reserves, etc. Therefore, there is an urgent need to initiate a national effort in this direction.

As no country is self-sufficient in plant genetic resources, there is an inherent interdependence in case of most cultivated/crop species. There is a need for more intense international collaboration to face the challenges of growing global food and nutritional insecurity and of climate change. This will require greater resilience of crops against some of the ensuing stresses, such as higher temperature regimes, drought, etc. To meet the adaption requirement of each crop/cultivated species, the breeding system, and component carrier of genetic material, it is now important that national, regional, and global strategies should be developed and implemented in a complementary manner, including both in situ and ex situ conservation, to facilitate conservation of whole gene pool of a taxa, including wild relatives, ensuring their global availability for use in crop improvement. It will require identification of priority wild taxa and sites containing these taxa for establishment of genetic reserve, which will be a protected area in collaboration of local communities. It should be addressed by various national and international initiatives, including the International Union for Conservation of Nature's Crop Wild Relative Specialist Group (www.cwrsg.org), a Global Environment Framework project (in situ conservation of crop wild relatives through enhanced information management and field application; www.cropwildrelatives.org./index.phb?id=3261), the recently inaugurated CWR global portal, and the Food and Agriculture Organization of the United Nations' initiative to establish a global network for the in situ conservation of crop wild relatives species diversity (Maxted and Kell 2009). This becomes more essential as we know that 17.5% of the worldwide vascular plant genera encompass food and agricultural crops. The genetic improvement of the crops, particularly under changing climate, depends on the genetic diversity of available plant genetic resources, which are arguably inadequately conserved and poorly used, particularly in relation to crop wild relatives. Therefore, at global level also, the international community needs to reinitiate and support systematic cooperative efforts to collect crop wild relatives. Such cooperative efforts have severely gone down with the enforcement of the provisions of CBD, providing sovereign ownership to the nation on their bio-resources.

21.6 Promotion of Biotechnological Approaches in Combination of Conventional Breeding

Improving genetic knowledge, interspecific hybridization techniques, and the molecular technologies will continue to improve our capacity to incorporate the desirable traits found in CWR. If there is desire to increase the use of crop/cultivated species wild relative plant genetic resources, then there is a need to fully exploit the conventional and emerging novel approaches or combinations of technologies to increase the rate of gene discovery within naturally occurring genetic variation. This shall accelerate their introduction into new varieties of crops producing novel combinations of genes. Therefore, there is a need of maximizing the use of newly emerging biotechnological tools coupled with traditional plant breeding. Improving cooperation requires not only the exchange of technical expertise but also the willingness among scientists to cooperate. Further, it will make possible that genetic resources procured through sustained collection and conservation of wild relatives of crop/cultivated species are made available with desired basic genomic information (needs to be addressed with some urgency) along with necessary techniques required for their deployment.

21.7 Developing Basic Information on Phylogeny and Genomics

With available systematic methods, it has been possible to generate significant biosystematics information on wild relatives of major crop/cultivated species. These efforts have paid rich dividends in several crops, such as rice, potato, tomato, etc. However, the same is lacking for many other cultivated and crop species of marginal value and their wild relatives and needs to be extended to resolve taxonomic issues and ascertain identity and entity of the wild species along with their essential features and phylogenetic relationship with cultivated species. This shall facilitate identification of appropriate breeding strategies, enabling their use in crop improvement through introgression of desired traits. Also, it appears that such basic research along with information on useful traits will not lead to increased use of these species, provided other barriers, such as crossing and recombination problems, and associated agronomic problems are not addressed.

With development in molecular genetics techniques, it is now increasingly possible to generate information on full genome sequences for crop species and their wild relatives. Information on genome sequences for many of cultivated/crops species, of both major (rice, maize, soybean) and minor (chickpea, pigeonpea) crop species, and for their wild relatives have been published. These efforts need be expanded, and as introduction to this effort, hundreds of accessions from the target wild species are sequenced to construct functional subsets of crop wild relative diversity. In this regard, low-cost genotyping can facilitate generation of genetic data, in cost-effective manner. This can help identification of population genetic indices for prioritization of genotypes needing full resequencing. Combined with sampling strategies that emphasize population-level coverage and focusing on high-frequency alleles within individual sampled populations, selection of subsets of sampled accessions enriched for adaptive alleles can be achieved (Warschefsky et al. 2014).

21.8 Creation and Phenotyping of Purpose-Driven Hybrid Population Sets

Advanced backcross introgression populations (Tanksley and McCouch 1997) and nested association mapping (McMullen et al. 2009) are synergistic for trait discovery and use in conventional breeding programs. In these programs, one can overcome the barriers created by natural phenomena, growth habit, pod shattering, etc., or otherwise, thereby preventing the use of wild germplasm in breeding, and dissect the genetic basis of adaptive traits. However, these populations must be carefully phenotyped using standardized phenotyping procedures in replicated trials, for a range of high-priority traits related to crop production, such as ability to tolerate changing climatic conditions, resistance to emerging pest and disease threats, etc. In this regard, international and national partnerships can facilitate phenotyping under a wide range of conditions that would ensure both high-power of trait–genomic associations and their relevance to different kinds of crop production environments. Because of these reasons, the pre-breeding populations can facilitate direct use in conventional breeding programs and can be maintained under participatory breeding networks.

21.9 Establishing a Predictive Network of Genotype–Phenotype Associations

Identification of genes and genomic regions in wild species that can improve yield and resilience in the crop can be facilitated by a genotype–phenotype map for crops and their wild relatives. The association observed between genomic and environmental variation in natural populations, i.e., collections, combined with information on trait–genotype associations established through phenotyping of cultivated x wild interspecific derivatives or introgression lines (pre-breeding population), will help in the identification of agronomically valuable alleles with great precision. This would help initiate their deployment in crop improvement programs.

21.10 Deployment of Identified Phenotypes into Crop Breeding Pipelines

For deployment of identified desirable phenotypes in crop breeding program, one needs to select high-value genome intervals from wild species for improvement of elite crop genotypes. For this advanced backcross introgression (ABI) lines, preconstructed as a library of partially overlapping introgressed segments in elite cultivated genomes can provide a ready-to-go breeding resource once these high-value wild genome intervals are identified. Creation of such ABI germplasm resources in advance of (or in parallel to) trait discovery will speed the delivery of wild traits into elite backgrounds, whereas the immortal nature of such a resource ensures that subsequent discovery of new traits can benefit from a preexisting ABI pipeline for trait(s) delivery. Such ABI libraries should ideally be created with multiple elite genotypes that together encompass traits for the primary agro-climatic zones of the crop under consideration.

21.11 Overcoming Other Problems

In many ways, perennial crops are a challenge. Because of their perennial nature, they present unique challenges to crop improvement efforts. The extended immaturity stage of such individuals makes repeated backcrossing an extremely time-consuming effort. The predominant outcrossing nature prevents production of inbred lines. However, identifying genomic regions containing adaptive loci by examining the correlation between environmental variables and allele frequencies across a geographic range and taking advantage of their cultivation system, by vegetative/para sexual means, such as grafting, provide both breeding material and novel rootstock material in crop improvement to overcome this lacuna.

Crop wild relatives with adoption to limited ecological niche may pose a different challenge for utilization in breeding, because of their lack of adaptation to a variety of environments. In plants with more limited or narrower distributions, this pattern tends to be further weaker (Hereford 2009). However, despite this in the crops like maize and chickpea (Moeller et al. 2007; Abbo et al. 2003), wild relatives with limited ranges still demonstrated variation among populations consistent with local adaptation (Pyhajarvi et al. 2013). Even if the wild relative range is extremely limited and local adaptation is minimal, interspecific hybridization with geographically diverse or distant types would likely yield some expansion in the range of adaptation and genetic variation.

Thus, there is much to be done to facilitate collection and conservation and promote the use of wild genetic resources in crop improvement. Basic biosystematic information desired to be developed to avoid taxonomic confusion vis-à-vis cultivated species and to facilitate identification and collection of genetically distinct species with genetic variation evolved based on biological and physical factors/features. Genome analysis is to be performed using both conventional and molecular methods to trace genetic divergence of the targeted wild relatives vis-à-vis cultivated species and to locate the genetic loci of important traits. This shall help in identification of appropriate breeding strategies for transfer of genes conferring desired traits in a more predictive, effective, and directive manner. In priority species, information on genomic constitution and potential traits with pre-breeding efforts needs to be initiated for production of backcross introgression populations or readily available interspecific derivatives with desired traits, if possible with associated mapping. These areas, if attended to in future perspective on use of crop wild relatives, shall be able to create the desired impact in utilization of wild relatives as alternative genetic resources in genetic improvement of crops/cultivated species, engineering production of desired cultivars. Besides, their bioprospecting may provide new source of food and other by-products and opportunities to mitigate the adverse effect of pollution, causing climate change.

The appendixes provided at the end of the book, as supplementary tables, provide a bird's-eye view on some of the major aspects of crop/cultivated species wild relatives. Appendix 1 lists the representative wild relatives found in different biogeographic regions of India; Appendix 2 the wild relatives of cultivated species that have been recorded to be under threat, with status as per IUCN Red List Categories; and Appendix 3 with the list of representative wild relatives of crop plants with reported important desirable traits.

References

Abbo S, Berger J, Turner NC (2003) Viewpoint: evolution of cultivated chickpea: four bottlenecks limit diversity and constrain adaptation. Funct Plant Biol 30:1081–1087

Convention on Biological Diversity (1992) Secretariat of the Convention on Biological Diversity (2005) www.cbd.int/doc/handbook/cbd-hb-all-en.pdf

Dempewolf H, Eastwood RJ, Guarino L, Khoury CK, Müller JV, Toll J (2014) Adapting agriculture to climate change: a global initiative to collect, conserve, and use crop wild relatives. Agroecol Sustain Food Syst 38:369–377

Ford-Lloyd BV, Schmidt M, Armstrong SJ, Barazani O, Engels J, Hadas R, Hammer K, Shelagh PK (2011) Crop wild relatives—undervalued, underutilized and under threat? Bioscience 61:559–565

Hereford J (2009) A quantitative survey of local adaptation and fitness trade-offs. *Am Nat* 173:579–588

International Treaty on Plant Genetic Resources for Food and Agriculture (2001) Food and agriculture organization of the United Nations. Via delle Terme di Caracalla, Rome, p 25

Maxted N, Kell SP (2009) Establishment of a global network for the *In Situ* conservation of crop wild relatives: status and needs. FAO, Rome

Maxted N, Kell S, Ford-Lloyd B, Dulloo E, Toledo A (2012) Toward the systematic conservation of global crop wild relative diversity. Crop Sci 52:774–785

McMullen MD, Kresovich S, Villeda HS, Bradbury P, Li H, Sun Q, Flint-Garcia S et al (2009) Genetic properties of the maize nested association mapping population. Science 325:737–740

Moeller DA, Tenaillon MI, Tiffin P (2007) Population structure and its effects on patterns of nucleotide polymorphism in teosinte (*Zea mays* ssp. *parviglumis*). Genetics 176:1799–1809

Nagoya Protocol (2010) Nagoya Protocol on Access to Genetic Resources and the Fair and Equitable Sharing of Benefits Arising from their Utilization to the Convention on Biological Diversity. A supplementary agreement to the 1992 Convention on Biological Diversity (CBD), Nagoya, Japan

Pyhajarvi T, Hufford MB, Mezmouk S, Ross-Ibarra J (2013) Complex patterns of local adaptation in teosinte. Genome Biol Evol 5:1594–1609

Singh AK (2004) Endangered economic species of Indian Desert. Genet Resour Crop Evol 51(4):371–380

Singh AK (2017) Revisiting the status of cultivated plant species agrobiodiversity in India: an overview. Proc Indian Natn Sci Acad 83(1):151–174. doi:10.16943/ptinsa/2016/v82/48406

Singh AK, Rana RS, Mal B, Singh B, Agrawal RC (2013) Cultivated plants and their wild relatives in India– an inventory. Protection of plant varieties and. Farmers' Rights Authority, New Delhi, India

Tanksley SD, McCouch SR (1997) Seed banks and molecular maps: unlocking genetic potential from the wild. *Science* 277:1063–1066

Warschefsky E, RVarma P, DouglasR C, Jbvon WE (2014) Back to the wilds: tapping evolutionary adaptations for resilient crops through systematic hybridization with crop wild relatives. Am J Bot 101(10):1791–1800

Erratum to: Spices and Condiments

Anurudh Kumar Singh

Erratum to: Chapter 11 in: A.K. Singh, *Wild Relatives of Cultivated Plants in India*, https://doi.org/10.1007/978-981-10-5116-6_11

The original version of the book was inadvertently published with an incorrect image for Fig 11.2 (b) in page 139. This image has now been replaced with the correct image as given below.

The updated online version of this chapter can be found at
https://doi.org/10.1007/978-981-10-5116-6_11

A.K. Singh, *Wild Relatives of Cultivated Plants in India*,
DOI 10.1007/978-981-10-5116-6_22

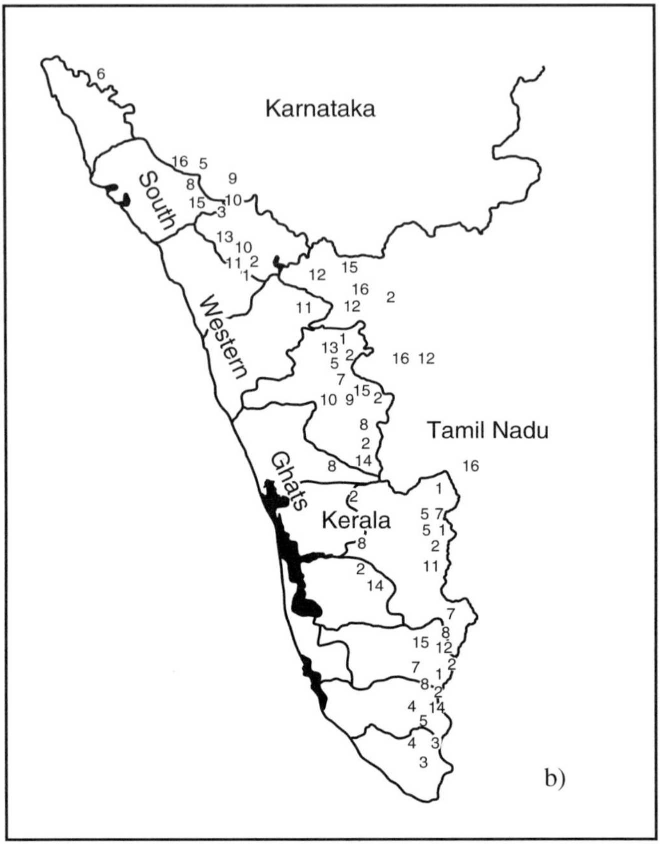

In addition, the author corrections have also been incorporated in the Appendix II.

Glossary

There exist diverse definitions and connotations associated with terminology used in plant genetic resource. This short glossary does not present a standardized formulation for the use of these terms but describes how certain key terms have been used in the present book.

1. **Agriculture**: Agriculture is the science or practice of farming, including cultivation of the soil for growing of crops and rearing of animals to produce food and for other human needs. It evolved by man interactions with bio-resources and the landscape with suitable modifications in them for economic exploitation.
2. **Agrobiodiversity**: An evolutionary divergent but highly interrelated component of biodiversity dealing with variation of plants, animals, fish, insects, microbes, avian, etc., used directly or indirectly for food and agriculture. It comprises the diversity of genetic resources (varieties, breeds) and species used for food, fodder, fiber, fuel, and pharmaceuticals. It also includes the diversity of non-harvested species that support production (soil microorganisms, predators, pollinators) and those in the wider environment support that agroecosystems (agricultural, pastoral, forest and aquatic) as well as the diversity of agroecosystems (after FAO).
3. **Alternate source of food**: Alternative sources of food like new plant species, used by tribal communities that most people do not think of as edible and economical.
4. **Backcrossing**: It is crossing of a hybrid with one of its parents (recurrent parent) or an individual genetically like it, to achieve offspring with a genetic identical or closer to that.
5. **Biogeographical Realms**: These are large spatial regions within which ecosystems share a broadly similar biological evolutionary history
6. **Biological Diversity Act**: Promulgated in 2002 to regulate, access, conservation and sustainable use of biodiversity, protection of associated community

The original version of this back matter was revised.
An erratum to this back matter can be found at https://doi.org/10.1007/978-981-10-5116-6_22

A.K. Singh, *Wild Relatives of Cultivated Plants in India*,
DOI 10.1007/978-981-10-5116-6

knowledge, secure sharing of benefit on commercial use, conservation of rich areas, and protection and rehabilitation of threatened species, involving states.

7. **Biological diversity**: It refers to variation of all living organisms, their genetic material, and the ecosystems of which they are a part. It is described at three levels: genetic, species, and ecosystem diversity.
8. **Biological species**: It is the most widely accepted species concept. It defines species in terms of interbreeding. Ernst Mayr defined a species as "the groups of interbreeding natural populations that are reproductively isolated from other such groups."
9. **Biosphere reserves**: These are areas of terrestrial and coastal ecosystems which promote the conservation of biodiversity (encompassing all species) with its sustainable use. They are internationally recognized within the framework of UNESCO's Man and Biosphere (MAB) program and nominated by national governments.
10. **Biosystematics**: The study of living organisms based on observational and experimental data on the breeding system for classification of biological units into taxa, making taxonomic decisions, based on relationships, variability, and dynamic of interrelationships.
11. **Biosystematy**: It deals with attempt to (1) delimit the natural biotic units and (2) to apply these units to a system of nomenclature.
12. **Biotype**: A population of individuals with identical genetic constitution. It may be homozygous or heterozygous.
13. **Carbon sequestration**: A natural or artificial process by which carbon dioxide is removed from the atmosphere and held in solid or liquid form. Here, it refers to capturing of CO_2 from atmosphere through biological process of plants, the photosynthesis, and store on long-term basis.
14. **Cenospecies**: It refers to the closely related independent species, capable of interbreeding and thereby gene exchange.
15. **Center of diversity**: A geographical area where a plant species first developed its distinctive properties (in farmers' fields or in the wild). A primary center of diversity is the region of true origin (often referred to as the center of origin), and secondary centers of diversity are regions of subsequent spread of a crop.
16. **Center of domestication**: Region where a plant species was first brought into protective use followed by centralized propagation from its local wild progenitors or where agricultural use of a species first originated.
17. **Center of origin**: The center of origin is a geographical area where a plant species is either domesticated or is considered to have first appeared in cultivation from their wild progenitors developing distinct features. The concept of center of origin was developed by NI Vavilov and has been subsequently modified.
18. **Cisgenes**: It refers to those natural indigenous genes, isolated from the crop/cultivated plant species or from cross-compatible or otherwise species that are part of gene pool.
19. **Cis-genesis**: The use of isolated cisgenes coding for desirable traits and incorporating them into crop/cultivated using recombinant DNA/GM technology.

20. **Congeneric species**: It refers to those organism/plant species that belong to the same genus, but form two or more different species.
21. **Conservation of biological resources**: It is concerned with phenomena that affect the maintenance, loss, and restoration of biodiversity and the science of sustaining evolutionary processes that engender genetic, population, species, and ecosystem diversity.
22. **Conservation of plant genetic resources**: Refers to the collection, maintenance, and preservation of intra- and inter-genetic variation of a species (the representative sample of the genetic variation) used in food and agriculture.
23. **Conspecific species**: It refers to those distinct sympatric organism/plant species that are distributed/inhabit the same geographic region.
24. **Convention on Biological Diversity**: It is an international treaty adopted at the Earth Summit in Rio de Janeiro in 1992 and enforced in UN on 29 December 1993, with three main goals, (i) conservation of biological diversity, (ii) sustainable use of its components, and (iii) fair and equitable sharing of benefits arising from use of genetic resources.
25. **Cytoplasmic male sterility**: The complete or partial failure of an individual to produce functional male gametes (pollen). Extrachromosomal hereditary components determine it.
26. **Domestication**: Taming of population of organisms to harness desirable/economic traits.
27. **Domiculture**: Concentrated propagation of plant biodiversity of economic value in a plot by the human community/society as a part of management of overall landscape and economic exploitation of targeted species.
28. **Ecosystem**: The totality of environment comprising living things together with their nonliving habitat.
29. **Ecotype**: A type evolved in response to a habitat, or compatible with a habitat, or a group of biotypes especially adapted to a specific environmental niche.
30. **Evaluation**: Recording of performance of a collections/accessions for desirable traits that are important for genetic enhancement and addition of value about specific feature.
31. **Ex situ conservation**: Means conservation of components of genetic material of biological diversity outside their natural habitat.
32. **Experimental taxonomy**: The classification of organisms based on experimental facts has been termed "experimental taxonomy."
33. **Forma**: Lowest category of species with sporadic variation in one or two characters.
34. **Gene center**: Generally, the geographical region of a species or genus, often associated with its origin and maximum variability (genetic variation).
35. **Gene flow**: It is the movement of genes from one population to another population, more specifically, the movement of different alleles between the populations of the same species. It creates diversity within a gene pool of a species.
36. **Gene Pool**: It represents the total accessible genetic diversity of a taxa, including the diversity within a cultivated species and in its possible genetically affil-

iated wild species (within genus and beyond), for introgression of genes at a given time. It has been classified into:

(a) *Primary gene pool*: Consisting of wild and weedy species/races of cultivated species that are freely cross-compatible producing nearly fertile hybrids
(b) *Secondary gene pool*: Consisting of wild relatives, which are cross-compatible (despite ploidy/genomic differences), but produce hybrids with reduced fertility
(c) *Tertiary gene pool*: Species that are weakly cross-compatible and conventionally do not produce hybrids with cultivated species
(d) *Quaternary gene pool*: The species that are cross-incompatible with cultivated species and form the peripheral limits of a genera

37. **Gene sanctuaries**: Conservation of germplasm under natural conditions, referring to an area where plants of specific species are conserved by protecting the area from—human interference.
38. **Gene sequencing**: It may refer to DNA sequencing or a comprehensive variant of it. Whole genome sequencing is a laboratory process that determines the complete DNA sequence of an organism's genome at a single time, i.e., the precise order of nucleotides within a DNA molecule, and involves any technology that is used to determine the order of the four bases—adenine, guanine, cytosine, and thymine in a strand of DNA.
39. **Genetic diversity**: The genetic variation within a population and among the populations of a species is generally referred to as the genetic diversity.
40. **Genetic drag**: It refers to transfer/incorporation of undesirable traits along with desirable traits in interspecific breeding, because of their tight genetic linkage. Conventionally, repeated backcrossing to recurrent parent is used to break such linkages, but now recombinant cis-genesis DNA approach can overcome this by direct transfer of desired genes surpassing backcrossing.
41. **Genetic engineering**: Genetic manipulation, by which an individual, having a new combination of inherited properties, is established.
42. **Genetic erosion**: It is loss of existing genetic diversity available in form of species, varieties, strains, etc.
43. **Genetic markers**: There are three types of genetic markers, (a) morphological markers based on phenotypic traits variation; (b) biochemical markers, which are called isozymes, including allelic variants of enzymes; and (c) DNA markers (or molecular markers), which reveal sites of variation in DNA.
44. **Genetic resources**: The genetic variability available in gene pool of a species useful for enhancing/improving genetic potential of a cultivated species about agronomic features, resilience against stresses, nutritional traits, etc. over present levels.
45. **Genetic variation**: Variation in the genetic constitution of individuals/species due to the contribution of segregating genes and gene interactions.
46. **Genome**: The complete set of DNAs, including all its genes or genetic material present in a cell or organism. It contains all the information involved in building and maintenance of an organism

47. **Genomics**: The study of genes and their function. It is a branch of molecular genetics concerned with the structure, function, evolution, and mapping of genomes. It deals with study of all the nucleotide sequences, including structural genes, regulatory sequences, and noncoding DNA segments in the chromosomes of an organism.
48. **Germplasm collections**: Genetically distinct sample of crop species collected and maintained at genetic resources center for conservation and use.
49. **Germplasm**: The living genetic resources, in case of plants, mostly seeds, or tissue that is maintained for plant breeding, preservation, and other research uses. It represents the sum of the genetic variable material available for a species.
50. **Habitat**: The sum of environmental condition in a specific place that is occupied by a plant or plant community wherein exchanges occur between the plants and the resources they utilize.
51. **In situ conservation**: Conservation of biodiversity within their natural ecosystems and/or habitats of a species where they have originated and evolved naturally.
52. **In situ/on-farm conservation**: Conservation of crop/cultivated species genetic diversity in the form of varieties, on farm or in the agroecology where they have developed their distinctive features.
53. **Introgression**: The incorporation of genes of one species (donor) into the gene pool of another species (recipient) by hybridization and repeated backcrossing with recipient species. By this process, the recipient species become more variable, displaying certain characters of donor species.
54. **Landraces**: Cultigens that are highly heterogeneous, but with enough characteristics in common to permit their recognition as group.
55. **Molecular biosystematics**: Study using simple molecular biological approaches to sample and analyze variation at biochemical and molecular DNAs level.
56. **National park**: A large protected area used for biodiversity conservation purposes. Often it is a reserve of natural, seminatural, or developed land.
57. **Native species**: A species, which is a part of the original flora of the area.
58. **Natural reserves**: A large protected area of importance for wildlife, flora, fauna, or geological interest reserved and managed for conservation and research.
59. **Naturalized wild species**: Refers to exotic species, introduced centuries/millenniums back and got acclimatize to the extent to appear indigenous and have even evolved new species, subspecies, and botanical varieties.
60. **Orthodox taxonomy**: It deals with classification and naming of organisms and the convenient tabulation and grouping based on morphological similarities and dissimilarities, to indicate natural relationships.
61. **Passport data**: It refers to the information documentation on a germplasm collection, such as location, physical/climatic conditions provenance data dealing with genetic background/pedigree information, reaction to prevailing stresses and threat, etc.

62. **Phylogeny**: The developmental and evolutionary history of group of organisms or species or genus.
63. **Phytoremediation**: It refers to the use of living green plants for in situ removal, degradation, or containment of contaminants in soils, sludges, sediments, surface water, and groundwater.
64. **Plant genetic resources**: Genetic material of plants, including modern cultivars, landraces, and wild relatives of crop plants, of value as a genetic resource for present and future generations of breeders to facilitate genetic improvement.
65. **Population**: A group of individuals belonging to different biotypes. The genotypically heterogenous population represents the basic evolutionary unit (gene pool) from which new types may arise through mutation, genetic recombination under the influence of natural pressure resulting selection/differentiation.
66. **PPV & FR Act**: The Protection of Plant Varieties and Farmers' Rights Act passed in 2001 to establish an effective system for protection of plant varieties and the rights of the farmers and plant breeders and to encourage development of new varieties of plants.
67. **Race**: An intraspecific category, primarily a population or aggregate of populations with recognizable characteristics.
68. **Recombinant DNA technology**: Refers to the technology that uses enzymes to cut and paste DNA sequences of interest together from two different species. The recombined DNA sequence is inserted into a host organism to produce genetically modified (GM) organism with new genetic combinations that are of value to science, medicine, agriculture, and industry.
69. **Red Data Book**: The book contains list/inventory of species whose continued existence is threatened.
70. **Renewable bioenergy**: Plant biomass, which has stored solar energy in the form of chemical energy/organic material and can produce energy in the form of biofuel.
71. **Sampling strategies**: Sampling methods that are followed during germplasm collection with primary emphasis to capture or sample the available genetic diversity.
72. **Selection**: The choice of certain individuals based on distinctiveness for the propagation/conservation and use from a mixed population where individuals vary in their characters. The variation is produced from nonrandom differential reproduction, which leads to individuals of different genotypes being represented unequally by their progeny in latter generations of a population of self-propagating units. Such individuals may not survive natural selection.
73. **Semi-Arid region**: A region with by highly variable and unpredictable rainfall, which is below potential evapotranspiration. Climatic conditions characterized it to be a region with intermediate climates between desert and humid regions. Dominated by short or scrubby vegetation of either grasses or shrubs.
74. **Species diversity**: Refers to the number of species represented in each community and to the evenness of species' abundances.

75. **Species endemism**: Plant species associated with a region. By extension, this term is used to refer to species which are found only in that region.
76. **Species**: A group of potentially interbreeding natural populations which normally are reproductively isolated from other such groups and/or show common characteristics.
77. **Subspecies**: A population of some biotype providing regional appearance/composition/differentiation of a species in relation to physical, chemical, genetic, and biological aspects.
78. **Taxon**: Taxonomic unit of any rank.
79. **Taxonomic characters**: Any observable feature of the plant that can be used for comparison and grouping/classification.
80. **Taxonomy**: A branch of biology engaged in the classification of organisms, especially according to their natural relationships. It covers the laws of and principles of such classification.
81. **The International Treaty on Plant Genetic Resources for Food and Agriculture (ITPGRFA)**: The International Treaty on Plant Genetic Resources for Food and Agriculture, popularly known as the International Seed Treaty, is a comprehensive international agreement in harmony with Convention on Biological Diversity, which aims at guaranteeing food security through the conservation, exchange and sustainable use of the world's plant genetic resources for food and agriculture, as well as the fair and equitable benefit sharing arising from its use. It implements a *Multilateral System* of access and benefit sharing, among those countries that ratify the treaty, for a list of 64 foods and forage crops (the genera and species are listed in Annex 1 to the Treaty).
82. **Threatened species**: Species categorized into various levels of threats as per IUCN categories are listed below:
    (a) *Extinct (EX)*: A taxon is extinct when there is no reasonable doubt that the last individual has died. As indicated by exhaustive surveys.
    (b) *Extinct in the Wild (EW)*: A taxon is extinct in the wild when it is known only to survive in cultivation, in captivity, or as a naturalized population (or populations) well outside the past range. As indicated by exhaustive surveys.
    (c) *Critically Endangered (CR)*: A taxon is critically endangered when the best available evidence by quantitative analysis indicates a reduction in the number of mature individual's due to continued decline, fluctuation or fragmentation of populations, and extent of occurrence. Thereby facing an extremely high risk of extinction in the wild.
    (d) *Endangered (EN)*: A taxon is endangered when the best available evidence indicates a species facing a very high risk of extinction in the wild because of either/or due to continued decline, fluctuation or fragmentation of populations, and extent of occurrence.
    (e) *Vulnerable (VU)*: A taxon is vulnerable when the best available evidence indicates that it is vulnerable, because of either continued decline, fluctuation, or fragmentation of populations and extent of occurrence.

(f) *Near Threatened (NT)*: A taxon is near threatened when it does not qualify for critically endangered, endangered, or vulnerable now but is close to qualifying for or is likely to qualify for a threatened category in the near future.

(g) *Least Concern (LC)*: A taxon is least concern when on evaluation, it does not qualify for critically endangered, endangered, vulnerable, or near threatened and abundantly available.

(h) *Data Deficient (DD)*: A taxon is data deficient when available information is inadequate to make a direct, or indirect, assessment on risk to its distribution and/or population status.

83. **Trans-domestication**: The hypothesis, as per which a wild plant species brought from an exotic destination, is domesticated elsewhere in a foreign land/country.
84. **Variation**: Divergence in the characteristics of an organism caused by the environment or by differences in its genetic constitution. The occurrence of phenotypic differences between individuals due to heritable differences (traceable genotypic differences) or due to differences in external conditions (phenotypic and non-heritable).
85. **Variety**: A population of some biotype forming local appearance/composition of a species in relation to a specific physical, chemical, genetic, and biological aspects
86. **Weed**: A volunteer plant which is adapted to disturbed or open habitats, it may be an ability to take advantage of human disturbances.
87. **Wide hybridization**: A term generally used to designate hybridization between widely diverse organism/plants (distinct species from same genera or different) to generate new genetic combinations from where desired recombinants can be selected.
88. **Wild and weedy relative of crop/cultivated plants**: The uncultivated species that are genetically related to a crop species, including the progenitors from same genera, as well as cross-compatible wild species from the same or other closely related genera.
89. **Wildlife sanctuaries**: A wildlife refuge, which is a naturally occurring sanctuary providing protection for a species from competition and other threats.

Appendices

Appendix I

The wild relatives of crop/cultivated plant species are distributed as per their origin, evolution, and adoption to natural environment suited to their growth and development found in different biogeographical regions. The most commonly accepted 10 + 1 biogeographic regions of Indian Subcontinent listed below, inhabit respective representative wild relatives of cultivated plants species.

Representative wild relatives of cultivated species found in different biogeographic zones of India

Biogeographical zones	Wild relatives of cultivated species recorded
Trans-Himalayan zone (Ladakh and Adjacent Areas)	*Allium carolinianum*, *A. chitralicum*, *A. gilgitcum*, *A. rubellum*, *Amaranthus spinosus*, *Cicer macracanthum*, *C. microphyllum*, *Fagopyrum cymosum*, *Hippophae salicifolia*, *H. tibetana*, *Hordeum brevisubulatum*; syn. *H. turkestanicum*, *H. spontaneum*, *Lactuca dolichophylla*, *Lepidium latifolium*, *Populus ciliata*, *P. gamblei*, *P. euphratica*, *P. jacquemontiana* var. *glauca*, *P. laurifolia*, *Salix acmophylla*, *S. denticulata*, *S. elegans*, *S. fragilis*, *S. sclerophylla*, *S. wallichiana*, *Trigonella emodi*, and *T. podperae* (27)

(continued)

A.K. Singh, *Wild Relatives of Cultivated Plants in India*,
DOI 10.1007/978-981-10-5116-6

Biogeographical zones	Wild relatives of cultivated species recorded
Himalayan zone (Northwestern and Eastern Himalayas)	**Western and Northwestern Himalaya**: *Abelmoschus manihot*, *A. moschatus*, *A. tuberculatus*, *Aegilops tauschii*, *Allium altaicum*, *A. ampeloprasum*; syn. *A. porrum*, *A. cernuum*, *A. chinense*, *A. fistulosum*, *A. schoenoprasum*, *A. senescens*, *A. stracheyi*, *A. tuberosum*, *Avena barbata*, *A. fatua* ssp. *fatua*, *A. sterilis* ssp. *ludoviciana*, *Cajanus mollis*, *C. scarabaeoides*, *Chenopodium album*, *C. ambrosioides*, *C. botrys*, *C.foliosum*, *C. giganteum*, *C. glaucum*, *C. hybridum*, *C. murale*, *Cicer microphyllum*, *Cucumis hardwickii*, *Dioscorea hispida*, *Elymus himalayanus*, *E. dahuricus*, *E. dentatus*, *Fagopyrum acutatum*, *F. cymosum*, *F. tartaricum* (also cultivated), *Hordeum aegiceras*, *H. brevisubulatum*, *H. murinum* ssp. *glaucum*, *H. spontaneum*, *Indigofera heterantha*, *Lactuca serriola*, *Linum perenne*, *L. strictum*, *Luffa graveolens*, *Malus baccata*, *M. pumila*, *Mentha* x *piperita*, *M. arvensis*, *M. spicata*, *Oryza rufipogon*, *Prunus jacquemontii*, *P. jenkinsii*, *P. tomentosa*, *Ribes glaciale*, *Rosa sericea*, *R. webbiana*, *Rubus fruticosus*, *R. hypargyrus*, *R. lanatus*, *R. molucannus*, *R. niveus*, *R. nutantiflorus*, *Rumex acetosella*, *R. patientia*, *R. vesicarius*, *Saccharum filifolium*, *S. narenga*, *Salix tetrasperma*, *Setaria viridis*, *Solanum incanum*, *S. xanthocarpum*, *Taxus wallichiana*, *Sorbus aucuparia*, *S. lanata*, *Trichosanthes himalensis*, *T. multiloba*, *T. tricuspidata*, *Trigonella cachemeriana*, *T. emodi*, *T. fimbriata*, *T. gracilis*, *T. podperae*, *Triticum sphaerococcum*, *Vigna trilobata*, *V. umbellata*, *V. vexillata* var. *vexillata*, and *Ziziphus oxyphylla* (87)
	Eastern Himalaya: *Actinidia strigosa*, *Albizia kalkora*, *Allium angulosum*, *A. wallichii*, *Amomum aromaticum*, *A. dealbatum*, *A. subulatum*, *Amorphophallus bulbifer*, *Areca triandra*, *Artocarpus chaplasha*, *Avena fatua*, *Brassica trilocularis*, *Cajanus elongatus*, *C. grandiflorus*, *C. mollis*, *C. scarabaeoides*, *C. villosus*, *Camellia caudata*, *C. drupifera*, *C. irrawadiensis*, *C. kissi*, *C. sinensis*, *C. taliensis*, *Cinnamomum bejolghota*; syn. *C. obtusifolium*, *C. caudatum*, *C. glanduliferum*, *C. impressinervium*, *Citrus aurantium*, *Coffea benghalensis*, *Cucumis hystrix*, *Curcuma amada*, *C. aromatica*, *C. caesia*, *C. montana*, *Digitaria cruciata*, *Dioscorea deltoidea*, *D. hamiltonii*, *D. hispida*, *D. kamoonensis*, *D. lepcharum*, *D. pentaphylla*, *D. prazeri*, *D. wallichii*, *D. wattii*, *Duchesnea indica*, *Eleusine indica*, *Elymus tangutorum*, *E. thoroldianus*, *Eriobotrya dubia*, *E. hookeriana*, *E. petiolata*, *Fragaria vesca*, *Garcinia hombroniana*, *G. xanthochymus*, *Glycine soja*, *Hordeum agriocrithon*, *H. spontaneum*, *Kaempferia sikkimensis*, *Lactuca cooperi*, *Lepidium capitatum*, *Luffa graveolens*, *Malus baccata*, *M. sikkimensis*, *Mangifera khasiana*, *M. sylvatica*, *Momordica macrophylla*, *Musa acuminata*, *M. balbisiana*, *M. cheesmani*, *M. mannii*, *M. nagensium*, *M. sikkimensis*, *M. velutina*, *Myrica esculenta*, *Neoluffa sikkimensis*, *Oryza meyeriana*, *O. minuta*, *Phoenix acaulis*, *P. rupicola*, *Pimpinella urceolata*, *Piper betleoides*, *P. hamiltonii*, *P. laxivenum*, *P. mungpooanum*, *P. ovatistigmum*, *P. peepuloides*, *P. sikkimense*, *P. sonadense*, *P. tenuirameum*, *P. wallichii*, *Prunus arborea*, *P. bracteopadus*, *P. jenkinsii*, *Pyrus pashia*, *P. serotina*, *Rhus griffithii*, *Ribes acuminatum*, *R. glaciale*, *Rubus ghankantus*, *R. hypargyrus*, *R. lineatus*, *R. moluccanus*, *R. reticulatus*, *R. senchalensis*, *R. sikkimensis*, *R. wardii*, *Saccharum filifolium*; syn. *Erianthus filifolius*, *S. longisetosum*, *S. ravennae*, *S. sikkimense*, *S. williamsii*, *Solanum kurzii*, *S. spirale*, *Sorbus himalaica*, *S. microphylla*, *S. vestita*, *Trichosanthes cordata*, *T. tricuspidata*, *T. wallichiana*, *Vigna clarkei*, *Vitis himalayana*, *V. lanata*, *V. parviflora*, *Zingiber capitatum*, *Z. chrysanthum*, *Z. clarkei*, and *Z. zerumbet* (129)

(continued)

Biogeographical zones	Wild relatives of cultivated species recorded
Northeast zone [Brahmaputra Valley (Assam), Meghalaya and Northeast Hills]	*Alocasia cucullata, A. macrorrhiza, Amomum aromaticum, Amorphophallus bulbifer, Brassica trilocularis*; syn. *B. rapa* var. *trilocularis, Areca triandra, Camellia caudata, C. kissi, Canavalia gladiata, C. virosa, Cinnamomum bejholgota*; syn. *C. obtusifolium, C. glanduliferum, C. paniculatum, C. pauciflorum, Citrus assamensis, C. hystrix, C. indica, C. jambhiri, C. latipes, C. medica, Coffea fragrans, C. jenkinsii*; syn. *Nostolachma jenkinsii, C. khasiana, Colocasia fallax, C. mannii, Corchorus capsularis* (wild form), *C. pseudo-olitorius, Cucumis hardwickii, C. hystrix, Curcuma aeruginosa, C. amada, C. aromatica, C. comosa, C. latifolia, C. soloensis, C. sylvatica, Digitaria cruciata, Dioscorea alata, D. decipiens, D. hamiltonii, D. hispida, D. lepchanum, D. pentaphylla, D. prazeri, D. pubera, D. trinervia, D. wattii, Diospyros lotus, Dolichos falcatus, Echinochloa crus-galli, Elaeocarpus floribundus, Eleusine indica, Erianthus ravennae, Fragaria nilgerrensis, Garcinia cowa, G. spicata, G. sopsopia, Gossypium arboreum, Hibiscus furcatus, Indigofera dosua, I. heterantha* (Himalyan indigo), *Leersia hexendra* (Manipur), *Luffa aegyptiaca, L. graveolens, Malus baccata, Mangifera khasiana, M. sylvatica, Miscanthus nepalensis, M. nudipes, M. taylorii, M. wardii, Momordica macrophylla, M. subangulata, Mucuna bracteata, Musa acuminata, M. assamica, M. balbisiana, M. cheesmani, M. flaviflora*; syn. *M. thomsoni, M. itinerans, M. mannii, M. nagensium, M. paradisiaca, M. sikkimensis, M. superba*; syn. *Ensete superba, M. velutina, Myrica esculenta, Naranga fallax*; syn. *Saccharum longifolium, Nicotiana excelsior, Oryza granulata, O. rufipogon, Phoenix acaulis, P. dactylifera, P. robusta, P. rupicola, Piper attenuatum, P. cornilimbum, P. hamiltonii, P. khasianum, P. makruense, P. meeboldii, P. oldhamii, P. peepuloides, P. sylvaticum, P. wallichii, Prunus arborea, P. cornuta, P. cerasioides, P. jenkinsii, P. napanlensis, P. undulata, P. wallichii*; *Psilanthus bengalensis*; syn. *Coffea bengalensis, Pyrus khasiana, P. pashia, P. pyrifolia, Rosa gigantea, R. longicuspis, Rubus ellipticus, R. moluccanus, R. paniculatus, R. rosifolius, Rumex dentatus, Saccharum longisetosuma, S. procerum, S. rufipilum, S. wardii, Sclerostachya fusca, Setaria pallide-fusca, Solanum khasianum, S. kurzii, S. nigrum, S. spirale, S. torvum, Sorbus aucuparia, S. vestita, Syzygium assamicum, Trichosanthes bracteata, T. cordata, T. cucumerina, T. himalensis, T. khasiana, T. tomentosa, T. wallichiana, Vigna clarkei, V. trilobata, Zingiber capitatum, Z. cassumunar, Z. intermedium, Z. rubens*, and *Z. spectabile* (150)
Desert zone (Thar and Kutch)	*Acacia nilotica* (wild form), *Aegle marmelos* (wild form), *Amaranthus spinosus, Citrullus colocynthis, Commiphora wightii* (wild form), *C. caudata, Crotalaria burhia, Cucumis prophetarum, Ipomoea cairica* var. *semineglabra, Momordica balsamina, Moringa concanensis, Psoralea corylifolia, Saccharum spontaneum, Salvadora persica, Sorghum bicolor* (weedy forms), *Sorghum halepense, Withania coagulans*, and *Ziziphus nummularia* (18)
Semi-Arid zone (Gujarat, parts of Northwestern Madhya Pradesh, Haryana, Punjab)	*Corchorus olitorius, Grewia bicolor, G. flavescens, Indigofera coerulea* var. *monosperma, I. cordifolia, Salvadora oleoides, Sesamum alatum, Sesbania concolor, Solanum nigrum, S. purpureilineatum, Sorghum halepense, Trigonella occulta, T. uncata, Vinga khandalensis, Ziziphus nummularia* (jharber), and *Z. williamsii* (16)

(continued)

Biogeographical zones	Wild relatives of cultivated species recorded
Gangetic Plains (upper, middle, and lower Gangetic Plains)	*Abelmoschus cancellatus*, *A. manihot*, *A. tuberculatus*, *Allium stracheyi*, *Amaranthus polygamus*, *A. spinosus*, *Amorphophallus bulbifer*, *Asparagus curillus*, *A. sarmentosus*, *Brassica quadrivalvis*, *Cajanus scarabaeoides*, *C. volubilis*; syn. *C. crassa*, *Carissa spinarum*, *Carthamus oxyacantha*, *Carum villosum*, *Chenopodium album*, *C. murale*, *Cichorium intybus*, *Coccinia indica*; syn. *C. grandis*, *Coix agrestis*, *C. aquatica*, *C. lacryma-jobi*, *Corchorus trilocularis*, *C. acutangulus*, *Cucumis setosus*, *Curcuma amada*, *C. ferruginea*, *C. leucorhiza*, *C. rubescens*, *Dioscorea deltoidea*, *D. hamiltonii*, *D. kalkapershadii*, *Echinochloa colonum*, *E. crus-galli*, *Ficus palmata*, *Grewia asiatica* (wild form), *G. optiva*, *Hibiscus surattensis*, *Indigofera caerulea*, *I. gangetica*, *I. thothathri*, *Ipomoea aquatica*, *Lactuca remotiflora*, *Lathyrus aphaca*, *Lepidium draba*, *Luffa echinata* var. *longistylis*, *Malva sylvestris*, *Momordica balsamina*, *M. cochinchinensis*, *M. subangulata* var. *renigera*, *Oryza rufipogon*, *O. sativa* var. *spontanea*; syn. *O. perennis*, *Panicum notatum*, *Paspalum scrobiculatum*, *Pennisetum orientale*, *Phoenix paludosa*, *P. robusta*, *P. sylvestris*, *Piper sylvaticum*, *Polyalthia suberosa*, *Prunus rufa*, *Psilanthus bengalensis*; syn. *Coffea bengalensis*, *Saccharum arundinaceum*; syn. *S. procerum*, *S. bengalense*, *S. longisetosum*, *S. narenga*, *S. ravennae*, *S. spontaneum*, *S. williamsii*, *Sclerophyllum coarctatum*; syn. *Oryza coarctata* (tetraploid), *Sclerostachya fusca*, *Setaria sphacelata*, *Solanum incanum*, *S. indicum*, *S. surattense*, *Syzygium heyneanum*, *Taxus wallichiana* var. *chinensis*, *Trichosanthes bracteata*, *T. cucumerina*, *Trigonella corniculata*, *T. obcordata*, *T. occulta*, *T. polycerata*, *Urena repanda*, *Vigna prainiana*, *V. radiata* var. *sublobata*, *Vitis latifolia*, *Ziziphus oenoplia*, and *Z. oxyphylla* (90)
Indian Peninsula (Central Highlands: Malwa Plateau, Bundelkhand, Chota Nagpur and Central and Deccan Plateau)	**Central Highland**: *Abelmoschus crinitus*; syn. *A. cancellatus*, *A. ficulneus*, *A. manihot* ssp. *tetraphyllus* var. *megaspermus*, *A. manihot* ssp. *tetraphyllus* var. *pungens*; syn. *Hibiscus pungens*, *A. tuberculatus*, *Acacia donaldii*, *Alocasia macrorrhizos*, *Alysicarpus monilifer*; syn. *Hedysarum moniliferum*, *Amaranthus spinosus*, *A. viridis*, *Cajanus cajanifolius*, *C. scarabaeoides*, *C. sericeus*, *Coccinia indica*; syn. *C. grandis*, *Coix aquatica*, *Colocasia esculenta*, *Corchorus fascicularis*, *C. tridens*, *Cucumis callosus*, *C. hardwickii*, *C. setosus*, *Cucurma longa*; syn. *C. domestica*, *C. angustifolia*, *Dioscorea bulbifera*; syn. *D. sativa*, *D. glabra*, *D. wightii*, *Diospyros chloroxylon*, *D. melanoxylon* (wild form), *D. peregrina*, *D. sylvatica*, *D. tomentosa*, *Echinochloa crus-galli*, *Eleusine indica*, *Grewia damine*, *G. tenax*, *G. villosa*, *Indigofera pulchella*, *I. tinctoria*, *Madhuca indica* (wild form), *Momordica balsamina*, *M. dioica*, *M. subangulata* var. *renigera*, *Mucuna capitata*, *Narenga porphyrocoma*, *Oryza rufipogon*, *Phoenix robusta*, *P. sylvestris*, *Rhynchosia bracteata*, *R. minima*, *R. rufescens*, *Saccharum spontaneum*, *Sesamum laciniatum*, *Solanum torvum*,*Sorghum cernuum* var. *yemense*, *S. controversum*, *S. halepense*; syn. *S. miliaceum*, *S. nitidum*, *Spondias pinnata*, *Syzygium cumini* (wild form), *Trigonella corniculata*, *T. occulta*, *Vigna aconitifolia*, *V. dalzelliana*, *V. hainiana*, *V. mungo* var. *sylvestris*, *V. radiata* var. *setulosa*; syn. *V. sublobata*, *V. trilobata*, *Zingiber capitatum*, *Z. cassumunar*; syn. *Z. purpureum*, *Z. mauritiana*, (wild form), *Z. nummularia*, and *Z. xylopyra* (72)

(continued)

Biogeographical zones	Wild relatives of cultivated species recorded
	Central and Deccan Plateau: *Abelmoschus ficulneus*, *A. manihot* ssp. *tetraphyllus* var. *megaspermum*, *Boehmeria platyphylla*, *Canavalia stocksii* (variant of *C. ensiformis*), *Capparis decidua*, *Chionachne koenigii*, *Cichorium intybus*, *Corchorus antichorus*; syn. *C. depressus*, *C. urticaefolius*, *Crotalaria* spp. (27), *Eleusine indica*, *Grewia tenax*, *G. villosa*, *Indigofera deccanensis*, *I. glandulosa* var. *sykesii*, *Linum mysorense*, *Malva rotundifolia*, *M. subangulata*, *M. sylvestris*, *Momordica cymbalaria*, *Panicum hippothrix* (grains are cooked like rice), *P. psilopodium*, *P. trypheron*, *Phoenix robusta*, *P. sylvestris*, *Sesamum laciniatum*, *Setaria glauca*, *S. pallide-fusca*, *S. pumila*, *Solanum nigrum*, *Sorghum deccanense*, *S. stapfii*, *Vigna hainiana*, *Vigna trilobata*, and *V. trilobata* var. *pusilla* (34 + 26 *Crotalaria*).
Eastern Ghats [Northeastern Ghats: Koraput, Bastar (Dandakaranya) and parts of Andhra Pradesh; South-central Eastern Ghats: Parts of Andhra Pradesh and Tamil Nadu]	*Abelmoschus angulosus*, *A. cancellatus*, *A. crinitus*, *A. ficulneus*, *A. moschatus*, *Allium porrum*, *Amaranthus dubius*, *A. spinosus*, *A. tenuifolius*, *Amorphophallus campanulatus*, *Boehmeria platyphylla*, *Cajanus cajanifolia*, *C. scarabaeoides*, *C. sericeus*, *C. volubilis*, *Canavalia stocksii*; syn *C. cathartica*, *Carissa inermis*, *Cissus vitiginea*, *Coleus forskohlii*, *Corchorus antichorus*, *Crotalaria perfoliata*, *C. shevaroyensis* *Cucumis hystrix*, *C. melo* var. *agrestis*, *C. pubescens*, *Cucurma amada*, *C. angustifolia*, *C. pseudomontana*, *C. montana*, *C. zedoaria*, *Cymbopogon flexuosus*, *Dioscorea bulbifera*, *D. hamiltonii*, *D. hispida*, *D. intermedia*, *D. oppositifolia*, *D. pentaphylla*, *D. puber*, *D. wallichii*, *D. wightii*, *Diospyros melanoxylon*, *D. peregrina*, *D. racemosa*, *D. tomentosa*, *Echinochloa crus-galli*, *Eleusine indica* (wild form), *Grewia abutilifolia*, *G. hirsuta*, *G. tiliaefolia*, *G. tenax*, *G. villosa*, *Indigofera caerulea*, *I. pulchella*, *I. trifoliata*; syn. *I. barberi*, *Jasminum angustifolium*, *J. auriculatum*, *J. scandens*, *Lablab niger* var. *lignosus*, *Luffa graveolens*, *L. umbellata* (*acutagula*), *Malva rotundifolia*, *M. sylvestris*, *Mangifera sylvatica*, *Momordica balsamina*, *M. cochinchinensis*, *M. subangulata*, *M. tuberosa*; syn. *M. cymbalaria*, *Musa ornata*, *Oryza coarctata*; syn. *Sclerophyllum coarctatum* (tetraploid), *O. granulata*, *O. jeyporensis*, *O. meyeriana*, *O. nivara* (annual); syn. *O. rufipogon*, *O. officinalis* subsp. *malampuzhaensis*, *O. sativa* var. *plena*, *Panicum hippothrix*, *P. typheron*, *Phoenix humilis* var. *pedunculata*, *P. loureiroi* var. *pedunculata*; syn. *P. robusta*, *P. pusilla*, *Phyllanthus fischeri*, *P. narayanswami*, *Sesamum alatum*, *S. laciniatum*, *S. prostratum*, *Solanum erianthum*, *S. incanum*, *S. indicum*, *S. nigrum*, *S. surattense*, *S. torvum*, *S. viarum*, *Sorghum miliaceum*; syn. *S. halepense*, *Syzygium alternifolium*, *S. zeylanicum*, *Trichosanthes bracteata*, *T. cucumerina*, *T. cordata*, *T. himalensis*, *T. multiloba*, *T. occulta*, *Vigna pilosa*, *V. umbellata*, *Vitis pallida*; syn. *Cissus adnata*, *V. pedata*; syn. *Cayratia pedata*, *V. repanda*; syn. *Cissus repanda*, *V. setosa*, *V. woodrowii*; syn. *Cissus woodrowii*, *Ziziphus oenoplia*, and *Z. rugosa* (108).

(continued)

Biogeographical zones	Wild relatives of cultivated species recorded
Western Ghats (Northwestern Ghats: Konkan and Southwestern Ghats: Malabar Coast)	**North Western Ghats**: *Abelmoschus angulosus*, *A. ficulneus*, *A. manihot*, *Cajanus lineatus*, *C. sericeus*, *Canavalia lineata*, *Cinnamomum goaense*, *Coffea wightiana*, *C. crassifolia*, *Cucumis ritchiei*, *C. setosus*, *Curcuma inodora*, *C. pseudomontana*, *C. purpurea*, *C. reclinata*, *Dolichos bracteatus*; syn. *Sphenostylis bracteata*, *Echinochloa colonum*, *Garcinia malabarica*, *G. morella*, *G. talbotii*, *Mangifera sylvatica*, *Momordica tuberosa*; syn. *M. cymbalaria*, *Oryza meyeriana*; syn. *O. indandamanica*, *Panicum hippothrix*, *Pennisetum orientale*, *Piper argyrophyllum*, *P. galeatum*, *P. hookeri*, *P. hymenophyllum*, *P. trichostachyon*, *Vigna dalzelliana*, *V. khandalensis*; syn. *V. grandis*, *Zingiber neesanum*; syn. *Z. macrostachyum*, and *Z. purpureum* (34)
	South Western Ghats: *Abelmoschus angulosus*, *A. manihot* ssp. *tetraphyllus*, *Amorphophallus bonoccordensis*, *A. commutatus*, *A. hohenackeri*, *A. mysorensis*, *A. nicolsianus*, *A. smithsonianus*, *Artocarpus gomezianus* ssp. *zeylanicus*, *A. hirsutus*, *Cajanus candollei*, *C. lineatus*, *Carissa spinarum*; syn. *C. paucinervia*, *Cinnamomum filipedicellatum*, *C. heyneanum*, *C. macrocarpum*, *C. malabatrum*, *C. riparium*, *C. travancoricum*, *C. wightii*, *Colocasia esculenta*, *Corchorus pseudo-olitorius*, *Crotalaria clarkei*, *C. digitata*, *C. grahamiana*, *Curcuma aromatica*, *C. aurantiaca*, *C. caesia*; syn. *C. malabarica*, *C. coriacea*, *C. decipiens*, *C. ecalcarata*, *C. haritha*, *C. karnatakensis*, *C. kudagensis*, *C. nilamburensis*, *C. neilgherrensis*, *C. oligantha*; syn. *C. cannanorensis*, *C. raktakanta*, *C. reclinata*, *C. thalakaveriensis*, *C. vamana*, *Dioscorea hamiltonii*, *D. hispida*; syn. *D. daemona*, *D. intermedia*, *D. oppositifolia*, *D. pentaphylla*; syn. *D. jacquemontii*, *D. spicata*, *D. tomentosa*, *D. wallichii*, *D. wightii*, *Dolichos uniflorus*, *Ensete superbum*, *Eugenia singampattiana*, *Fragaria nilgerrensis*, *Garcinia morella*, *G. travancorica*, *G. wightii*, *G. xanthochymus*, *Jasminum angustifolium*, *J. flexile*, *J. malabaricum*, *J. mesnyi*, *Linum mysorense*, *Luffa umbellata* (*acutangula*), *Momordica dioica*, *M. sahyadrica*, *Musa superba*, *Myristica dactyloides*, *M. malabarica*, *Olea glandulifera*, *Oryza meyeriana*, *O. officinalis*; syn. *O. officinalis* ssp. *malampuzhaenensis*, *Piper argyrophyllum*, *P. barberi*, *P. galeatum*, *P. hapnium*, *P. hookeri*, *P. pykarahense*, *P. schmidtii*, *P. silentvalleyensis*, *P. trichostachyon*, *Pueraria tuberosa*, *Sesamum laciniatum*, *S. malabaricum*, *S. mulayanum*, *S. radiatum*, *Solanum anguivi*; syn. *S. indicum*, *S. erianthum*, *S. incanum*, *S. nigrum*, *S.pubescens*; syn. *S. torvum*, *S. viarum*, *Syzygium arnottianum*, *S. beddomei*, *S. malabaricum*, *Trichosanthes cucumerina*, *T. nervifolia*, *T. tricuspidata*; syn. *T. bracteata*, *T. villosula*, *T. wallichiana*, *Vigna bournaea*, *V. pilosa*, *V. vexillata* var. *wightii*, *Vitis pedata*, *V. repanda*; syn. *Cissus repanda*, *Zingiber cernuum*, *Z. neesanum*; syn. *Z. macrostachyum*, *Z. purpureum*, *Z. roseum*, *Z. wightianum*, and *Z. zerumbet* (112)
Coastal zone (Sand dunes of West and East Coast)	*Caesalpinia pulcherrima*, *Canavalia cathartica*, *C. maritima*, *Cassia siamea*, *C. tora*, *Casuarina equisetifolia* (wild forms), *Cissus quadrangularis*, *Citrullus colocynthis*, *Cocos nucifera* (wild forms), *Crotalaria pallida*, *C. retusa*, *C. verrucosa*, *Ipomoea aquatica*, *I. carnea* ssp. *fistulosa*, *I. pes-caprae*, *Ixora arborea*; syn. *I. parviflora*, *Phoenix paludosa*, *Physalis minima*, *Saccharum spontaneum*, *Uniola paniculata*, and *Ziziphus williamsii* (21)

(continued)

Biogeographical zones	Wild relatives of cultivated species recorded
Islands (Andaman and Nicobar and Lakshadweep Islands)	*Amorphophallus carnosus*, *A. longistylus*, *Bauhinia nicobarica*; syn. *B. stipularis*, *Bombax insigne* var. *andamanica*, *B. insigne* var. *polystemon*, *Canavalia turgid*, *Dioscorea nummularia*, *D. rogersii*, *D. vexans*, *Dipterocarpus andamanicus*, *Ficus andamanica*, *Garcinia andamanica*, *G. cadelliana*, *G. calycina*, *G. kurzii*, *Grewia calophylla*, *Hornstedtia fenzlii*, *Ixora capitulifera*, *I. hymenophylla*, *Jasminum cordifolium*, *J. andamanicum*, *J. multiflorum*, *J. unifoliolatum*, *Mangifera andamanica*, *M. camptosperma*, *Manilkara littoralis*, *Nypa fruticans*, *Oryza indandamanica*, *Phyllanthus andamanicus*, *Piper sarmentosum*, *Polyalthia parkinsonii*, *Pterocarpus dalbergioides*, *Syzygium andamanicum*, *S. kurzii*, *S. manii*, *S. polyanthum*, *S. samarangense*, *Terminalia bialata*, *T. catappa*, *T. manii*, *T. procera*, *Vanilla andamanica*, and *Vigna marina* (44)

Source: Singh (2017)

Appendix II

Exploration, surveys, and bio-perspective assessment has led to identification around 1500 plant species to be under threat in India. This includes more than 1100 higher plants, including the wild relatives of crop/cultivated species. The rarity and the threat to these species have been caused by urbanization, developmental activities, destruction of habitats by anthropogenic factors like over grazing, mining, etc. Botanical Survey of India have published five volumes (3 published and 2 in press) of the Red Data Books of Indian Plants listing 1182 species, 623 listed published earlier. In 2001, a tentative list of species needing conservation was also drawn (Archive 2001) to facilitate both in situ and ex situ conservation. Further investigations, in many plant species has confirmed the threat perception with IUCN categories. Based on survey and the investigations in this direction, the Ministry of Environment and Forest, Government of India releases press release on the plant species under threat to draw public/scientific attention and develop awareness and support for conservation of plant species, particularly those of economic value, including wild relatives of crops/cultivated species. An attempt is being made to list the representative crop/cultivated species for a bird's-eye view.

Representative Wild Relatives of Crop/Cultivated Species Reported to Be Under Threat

Wild relatives of cultivated species	Crop group[a]	IUCN category/threat
Abutilon bidentatum var. *major*	ORN	R
Abutilon ranadei	ORN	CR
Acacia campbelii	AGFOR	VU
Acacia donaldii	AGFOR	DD
Acer caesium	ORN	VU
Acer oblongum var. *membranaceum*	ORN	CR
Acer osmastonii	ORN	EN
Aconitum balfourii	ORN/M	DD

(continued)

Wild relatives of cultivated species	Crop group[a]	IUCN category/threat
Acorus calamus	M & AP	EN
Aconitum heterophyllum	ORN	EN
Aconitum kashmiranum	ORN	EN
Aconitum violaceum	ORN	VU
Aegle marmelos var. *mahurensis*	F & NT	R
Albizia thompsonii	AGFOR	R
Albizia thompsonii var. *galbana*	AGFOR	R
Allium auriculatum	VEG	EN
Allium loratum	VEG	EN
Allium roylei	VEG	EN
Allium stracheyi	VEG	VU
Anogeissus sericea var. *nummularia*	IND CR	R
Arnebia benthamii	M & AP	CR
Asparagus jacquemonti	VEG	DD
Asparagus rottleri	VEG	DD
Barleria gibsonioides	ORN/M	R
Barleria prionitis ssp. *dicantha*	ORN/M	DD
Barleria stocksii	ORN/M	Rediscovered
Bauhinia variegata	ORN	DD
Berberis affinis	M & AP	R
Berberis apiculata	M & AP	R
Berberis kashmirian	M & AP	R
Berberis lambertii	M & AP	VU
Berberis pseudoumbellata	M & AP	DD
Berberis royleana	M & AP	DD
Bombax insigne	COTIN	CR
Boswellia ovelifoliolata	M & AP	EN
Boswellia serrata	M & AP	Eroded
Brachystelma laevigatum	ORN	VU
Brachystelma pauciflorum	ORN	EN
Butea monosperma var. lutea	M & AP	EN
Brachystelma volubile	ORN	DD
Cajanus cajanifolius	GL	DD
Cajanus lineatus	GL	DD
Calamus brandisii	ORN/IND	R
Calamus inermis	ORN	VU
Camellia siangensis	COMC	DD
Canavalia stocksii	VEG	R
Capparis cinerea	F & NT	DD
Capparis diversifolia	F & NT	VU
Capparis grandis	F & NT	DD
Capparis pachyphylla	F & NT	VU
Carum villosum	SP & CON	EN
Cenchrus rajasthanensis	FOR	DD
Ceropegia attenuata	FOR	R
Ceropegia bulbosa	FOR	VU

(continued)

Wild relatives of cultivated species	Crop group[a]	IUCN category/threat
Ceropegia fantastica	FOR	EN
Ceropegia fimbriifera	FOR	VU/R
Ceropegia intermedia	FOR	DD
Ceropegia maccannii	FOR	EN
Ceropegia odorata	FOR	EN
Ceropegia pusilla	FOR	R
Ceropegia spiralis	FOR	R
Chlorophytum borivilianum	M & AP	CR
Chlorophytum malabaricum	M & AP	R
Cinnamomum filipedicellatum	SP & CO	EN
Cinnamomum heyneanum	SP & CO	DD
Cinnamomum perrottetii	SP & CO	VU
Cinnamomum riparium	SP & CO	VU
Cinnamomum travancoricum	SP & CO	VU
Citrus indica	F & NT	EN
Citrus macroptera	F & NT	EN
Coffea crassifolia	COMC	R
Commiphora wightii	M & AP	R
Crinum brachynema	ORN	CR
Crinum eleonorae	ORN	EX
Crinum woodrowii	ORN	R/CR
Crotalaria bourneae	FIB	DD
Crotalaria clarkei	FIB	R
Crotalaria clavata	FIB	EN
Crotalaria digitata	FIB	EN/R
Crotalaria fysonii var. *glabra*	FIB	EN
Crotalaria grahamiana	FIB	R
Crotalaria longipes	FIB	EN
Crotalaria medicaginea var. *rigida*	FIB	
Crotalaria paniculata	FIB	NT/LC
Crotalaria willdenowiana galbrifoliata	FIB	R
Curcuma caesia	SP & CO	EN
Curcuma coriacea	SP & CO	EN
Curcuma decipiens	SP & CO	R
Curcuma pseudomontana	SP & CO	VU
Curculigo orchioides	M & AP	EN
Cycas beddomei	ORN	EN
Cycas circinalis	ORN	EN
Cymbidium aloifolium	ORN	NT
Cymbidium whiteae	ORN	EN
Cymbopogon flexuosus	FOR	EN
Cyperus dwarkensis	FOR	R
Cypripedium himalaicum	ORN	EN
Dactylorhiza hatagirea	ORN/M	Population decline
Dalbergia congesta	AGFOR	EN
Delphinium malabaricum var. *ghaticum*	ORN	VU

(continued)

Wild relatives of cultivated species	Crop group[a]	IUCN category/threat
Dendrobium pauciflorum	ORN	EN
Dichanthium armatum	FOR	R
Dioscorea deltoidea	VEG	EN
Dioscorea wightii	VEG	R
Diospyros barberi	F & NT	VU
Diospyros holeana	F & NT	VU
Diospyros paniculata	F & NT	VU
Dipcadi ursulae var. *longiracemosa*	ORN	
Elaeocarpus munronii (Rudraksha)	INCR	NT
Elaeocarpus recurvatus	INCR	VU
Ephedra gerardiana	M & AP	EN
Ephedra foliata	M & AP	Threatened
Embelia ribes	M & AP	CR
Eugenia argentea	ORN	EN/EX
Eugenia cotinifolia ssp. *codyensis*	ORN	EN
Eugenia discifera	ORN	EN
Eugenia indica	ORN	EN
Euphorbia katrajensis	INCR	R
Ficus andamanica	F & NT	EN
Ficus angladei	F & NT	CR
Garcinia kingii	F & NT	DD
Garcinia talbotii	F & NT	Threatened
Garcinia travancorica	F & NT	VU
Garcinia wightii	F & NT	VU
Grewia damine	F & NT	Declined
Grewia gamblei	F & NT	EN
Grewia pandaica	F & NT	CR
Hildegardia populifolia	FIB	CR/VU
Impatiens johnii	ORN	EN/EX
Impatiens macrocarpa	ORN	CR
Impatiens neo-barnesii	ORN	EN
Impatiens orchioides	ORN	VU
Impatiens tenella	ORN	EN
Impatiens nilagirica	ORN	CR
Indigofera barberi	INCR	R
Indigofera coerulea var. *monosperma*	INCR	R
Indigofera constricta	INCR	R
Indigofera thothathri	INCR	VU
Ipomoea clarkei	VEG/ORN	EN
Ixonanthes khasiana	INCR	VU
Ixora lawsonii	ORN	EN
Ixora longibracteata	ORN	DD
Jasminum adenophyllum	ORN	EN
Jasminum strictum	ORN	CR
Jasminum wightii	ORN	R
Lactuca benthamii	VEG	EN

(continued)

Wild relatives of cultivated species	Crop group[a]	IUCN category/threat
Lactuca cooperi	VEG	EN
Lactuca filicina	VEG	EN
Lactuca undulata	VEG	EN
Livistonia jenkinsiana	ORN	EN
Madhuca bourdillonii	AGFOR	EN
Madhuca diplostemon	AGFOR	EN
Magnolia gustavii	ORN	CR
Mallotus philippensis	INCR	R
Mangifera andamanica	F & NT	VU/NT
Mangifera khasiana	F & NT	DD
Michelia kisopa	ORN	DD
Michelia punduana	ORN	R
Mucuna pruriens	FOR	DD
Myristica dactyloides	SP & CO	VU
Myristica malabarica	SP & CO	VU
Nardostachys grandiflora (jatamansi)	M & AP	CR
Nepenthes khasiana	M & AP	EN
Nymphaea pygmaea	ORN	EX
Oryza nivara	C	DD
Panax pseudoginseng	M & AP	VU/CR
Paphiopedilum insigne	ORN	EN
Paphiopedilum spicerianum	ORN	EN
Paphiopedilum venustum	ORN	EN
Paphiopedilum villosum	ORN	VU
Pandanus mangalorensis	INCR	CR
Pandanus martinianus	INCR	EN
Pandanus palakkadensis	INCR	CR
Pandanus unguifer	INCR	EN
Phoenix rupicola	COMCR	NT
Phyllanthus narayanswami	M & AP	EN
Phyllanthus talbotii	M & AP	R
Picrorhiza kurroa	M & AP	EN
Pimpinella katrajensis	INCR	R
Pimpinella tirupatiensis	INCR	EN
Pinus gerardiana	INCR	R
Piper barberi	SP&CON	CR
Piper mullesua	SP&CON	VU
Piper pykarahense	SP&CON	DD
Podophyllum hexandrum	M & AP	EN
Portulaca oleracea	VEG	Declined
Prunus himalaica	F & NT	R
Pterocarpus santalinus	M & AP	EN
Pueraria tuberosa	M & AP	NT
Rauvolfia beddomei	M & AP	Threatened
Rauvolfia micrantha	M & AP	Degraded
Rauvolfia serpentina	M & AP	VU

(continued)

Wild relatives of cultivated species	Crop group[a]	IUCN category/threat
Rhododendron arboreum ssp. *arboreum*	ORN	Needs attention
Rhododendron concinnoides	ORN	VU
Rhododendron elliottii	ORN	DD
Rhododendron formosum	ORN	VU
Rhododendron johnstoneanum	ORN	DD
Rhododendron santapaui	ORN	EN
Rhododendron subansiriense	ORN	VU
Rhododendron wattii	ORN	EN
Rubus almorensis	F & NT	DD
Rubus fockei	F & NT	DD
Rhynchosia beddomei	GL	R
Rhynchosia velutina	GL	VU
Rubia himalayensis	INCR	VU
Salix obscura	AGFOR	Rediscovered
Salvadora oleoides	F/AGFOR	Threatened
Salvadora persica	F/AGFOR	Threatened
Santalum album	INCR	VU
Saussurea bracteata	M & AP	R
Saussurea clarkei	M & AP	R
Saussurea costus	M & AP	CR
Saussurea involucrata	M & AP	R
Saussurea obvallata	M & AP	VU
Shorea tumbuggaia	INCR	EN
Sterculia khasiana	M & AP	EX
Sterculia urens	M & AP	EN
Syzygium alternifolium	F & NT/M	EN
Syzygium andamanicum	F & NT	CR
Syzygium benthamiana	F & NT	DD
Syzygium beddomei	F & NT	R
Syzygium bourdillonii	F & NT	EN
Syzygium caryophyllatum	F & NT	EN
Syzygium chavaran	F & NT	EN
Syzygium gambleanum	F & NT	E
Syzygium manii	F & NT	CR
Syzygium occidentalis	F & NT	DD
Syzygium palghatense	F & NT	EX/EN
Syzygium stocksii	F & NT	DD
Syzygium travancoricum	F & NT	EN/CR
Taxus wallichiana	M &AP	EN
Tecomella undulata	INCR	Threatened
Terminalia arjuna	M &AP	DD
Terminalia pallida	M &AP	EN
Tinospora sinensis	M &AP	VU
Vanda coerulea	ORN	R/CR
Vanda wightii	ORN	EX/RD
Vanilla andamanica	SP & CO	CR

(continued)

Wild relatives of cultivated species	Crop group[a]	IUCN category/threat
Vanilla wightiana	SP & CO	R
Vigna dalzelliana	GL	Threatened
Vigna khandalensis	GL	R
Withania coagulans	M & AP	Threatened
Zingiber cernuum	SP & CO	DD
Zingiber purpureum var. *palamaunsis*	SP & CO	EN
Ziziphus truncata	F & NT	DD
Ziziphus williamsii	F & NT	EN

[a]C = cereals, GL = grain legumes, OS = oilseed, FIB = fiber, FOR = forage, VEG = vegetable, F & NT = fruit and nuts, SP & CO = spices and condiments, COMCR = commercial crops, M & AP = medicinal and aromatic plants; ORN = ornamentals, AGFOR = agroforestry, INCR = industrial crops

Source: CAMP Workshop (1998), Chadburn (2012a, b, c, d, e, f), MoEF (2010), Nayar and Sastry (1987–1990), Rawat (2008), Singh (2004, 2015), Ved et al. (2015), Walter and Gillett (1998), WCMC (1998)

References

CAMP Workshops on Medicinal Plants, India (January 1997) (1998) *Myristica malabarica*. The IUCN Red List of Threatened Species 1998: e.T31219A9615327

Chadburn H (2012a) *Vigna khandalensis*. The IUCN Red List of Threatened Species 2012: e.T19892969A20163711

Chadburn H (2012b) *Acacia donaldii*. The IUCN red list of threatened species 2012: e.T19891442A20125493

Chadburn H (2012c) *Albizia thompsonii*. The IUCN red list of threatened species 2012: e.T33644A20121746.

Chadburn H (2012d) *Crotalaria paniculata*. The IUCN Red List of Threatened Species 2012: e.T19893123A20049586.

Chadburn H (2012e) *Dalbergia congesta*. The IUCN red list of threatened species 2012: e.T19892181A20055198

Chadburn H. (2012f). *Vigna dalzelliana*. The IUCN red list of threatened species 2012: e.T19892280A20163346

MoEF, Ministry of Environment and Forests (2010) English press release. Threatened plants

Nayar MP Sastri ARK (eds) (1987, 1988, 1990) Red data book of Indian plants. vol 1, 1987, p 1371; vol 2, 1988, pp 1–273; vol 3, 1990, pp 1–280, Botanical Survey of India, Calcutta

Rawat GS (ed) (2008) Special habitats and threatened plants of India. ENVIS bulletin: wildlife and protected areas, vol 11(1). Wildlife Institute of India, Dehradun, pp 239

Singh AK (2004) Endangered economic species of Indian Desert. Genet Resour Crop Evol 51(4):371–380

Singh AK (2015) Agricultural biodiversity heritage sites and systems in India. Asian Agri-History Research Foundation, Secunderabad. pp 467

Ved et al. (2015) The IUCN Red List of Threatened Species 2015

Walter KS, Gillett HJ (eds) (1998) 1997 IUCN Red List of Threatened Plants. Compiled by the World Conservation Monitoring Centre. IUCN – The World Conservation Union, Gland, Switzerland and Cambridge, UK. lxiv + p 862

WCMC (1998) *Syzygium andamanicum*. The IUCN red list of threatened species 1998: e.T32621A9718483. http://dx.doi.org/10.2305/IUCN.UK.1998.RLTS.T32621A9718483.en

Appendix III

The germplasm collections of wild relatives in major crop species have been evaluated against main yield reducing factors, particularly the biotic and abiotic stresses and for nutritional traits in food crops and yield and quality-related traits in others. This has resulted in identification many desirable traits in these wild species. However, the possible genetic diversity that may exist within a species is yet to be studied in greater details in most of cases. This has created an expectation of the specific desirable trait(s) in all accessions of that species, which in many cases may not be true.

A great variation has been observed in identification wild relatives of crop/cultivated species under various crop groups with desirable traits. A large number desirable traits have been discovered in wild relatives of field and horticultural crops of food and forage value. However, the same is not true for other groups, such as spices and condiments, floriculture, medicinal and aromatic plants, agroforestry and cottage industry crops, and others, because of several inherent lacunas or constraints, such as:

1. Many are yet to reach crop status and mostly cultivated in gardens and homestead gardens.
2. Many of them are perennial in nature with difficulties in their evaluation for desirable features.
3. Many have complex breeding system.
4. Crop or genetic improvement activities in many are still in infancy.
5. There has been lack of efforts in developing screening methodologies for/against various traits.
6. Lack of efforts toward evaluation for desirable traits.

Recognizing these constraints, a list of representative's wild relatives with desirable features is being produced herewith in summary.

Representative wild species related to crop/cultivated species with potential useful traits

	Wild-related species	Identified desirable trait(s)
Cereals		
1.	*Aegilops tauschii*	Source of disease and pest
2.	*Avena barbata*	Source of disease resistance (virus, rust, mildew) and quality
3.	*Avena fatua*	Resistant to drought, diseases (virus, rust, mildew), and high yield
4.	*Avena sterilis*	Resistant to diseases (virus, rust, mildew), herbicides, and high grain yield
5.	*Elymus dahuricus*	Source of salt tolerance (wheat and barley)
6.	*Hordeum spontaneum*	Sorce for quality, yield, disease resistance, cold, salt, and waterlogging tolerance
7.	*Eleusine compressa*	Drought tolerance (finger millet)

(continued)

	Wild-related species	Identified desirable trait(s)
8.	*Oryza granulata*	Tolerant to drought, shade, and aerobic soil, immune to bacterial leaf blight (BLB) and resistance to brown plant hopper (BPH), yellow stem borer
9.	*Oryza minuta*	Source of resistance to BLB, green leaf hopper (GLH), whitebacked plant hopper (WBPH) and BPH
10.	*Oryza nivara*	Source of grassy stunt virus and sheath blight (SB) resistance
11.	*Oryza officinalis*	Source of BPH, GLH, and BLB resistance
12.	*Oryza rufipogon*	Source of cytoplasmic male sterility (CMS), quality, yield, metal and salt tolerance, and SB resistance
13.	*Panicum turgidum*	Drought and salt tolerant
14.	*Paspalum scrobiculatum*	Drought and salt resistant
15.	*Pennisetum orientale*	Drought resistant, prolonged green
16.	*Triticum turgidum/dicoccoides* (wild tetraploid wheat)	High protein content
Grain legumes		
1.	*Cajanus albicans*	Source of high seed protein (HSP) and sterility mosaic virus (SMV)
2.	*Cajanus cajanifolius*	Source of nuclear male sterility (NMS) and (CMS), soil salinity
3.	*Cajanus crassus*	Resistance to SMV
4.	*Cajanus lineatus*	Source of cleistogamy and CMS
5.	*Cajanus mollis*	Source of highest seed protein (HSP) content
6.	*Cajanus platycarpus*	Resistance to *Phytophthora* blight (PB), soil salinity
7.	*Cajanus scarabaeoides*	Source of CMS, HSP and dwarfism, and resistance to pod borer (PB), SMV
8.	*Cajanus sericeus*	Source of HSP, CMS, and resistance to both PB and SMV
9.	*Cicer microphyllum*	Drought resistance
10.	*Glycine soja*	Source for resistance to yellow mosaic virus (YMV), adaptability, cold tolerance, and short season
11.	*Macrotyloma sar-garhwalensis*	Source of high protein content
12.	*Vigna acontifolia*	Drought resistant
13.	*Vigna mungo* var. *silvestris*	Hard seed coat
14.	*Vigna radiata* var. *setulosa*	Source of drought and buchiids resistance
15.	*Vigna vexillata*	Resistant to cowpea weevil
Oilseeds		
1.	*Brassica tournefortii*	Source of cytoplasmic male sterility (CMS)
2.	*Carthamus oxyacantha*	Highly drought tolerant
3.	*Lepidium latifolium*	Source of cold tolerance
4.	*Linum perenne*	Cold tolerance
5.	*Linum strictum*	Source of fiber strictness
6.	*Sesamum alatum*	Resistance to phyllody disease

(continued)

	Wild-related species	Identified desirable trait(s)
7.	*Sesamum laciniatum*	Resistance to diseases and pests of sesame
8.	*Sesamum malabaricum*	Source of cytoplasmic sterility and powdery mildew in sesame
9.	*Sesamum mulayanum*	Source of resistance to phyllody, powdery mildew, and wilt
Fiber yielding plants/crops		
1.	*Boehmeria macrophylla*	Fiber quality
2.	*Boehmeria platyphylla*	Source of fiber strength
3.	*Corchorus aestuans*	Source of resistance to stem rot
4.	*Corchorus depressus*	Source of drought/heat tolerant
5.	*Corchorus fascicularis*	Source of drought/heat tolerant and adaptability
6.	*Corchorus pseudo-olitorius*	Immune to fungal diseases, stem rot, root rot, black band, soft rot, and anthracnose
7.	*Corchorus tridens*	Source high protein (vegetable/fodder)
8.	*Gossypium arboreum*	Source for resistance pest and diseases and tolerance to drought and CMS
9.	*Gossypium herbaceum*	Resistance to biotic and drought stresses, source of adaptability and yield, and quality traits and CMS
Forage crops		
1.	*Cenchrus setigerus*	Source of drought resistant
2.	*Cenchrus rajasthanensis*	Source of drought resistant
3.	*Chrysopogon aciculatus*	Tolerant to heavy grazing
4.	*Dichanthium annulatum*	Tolerant to heavy grazing and lodging
5.	*Diplachne fusca*	Source of salt tolerance
6.	*Eragrostis curvula*	Extremely drought tolerant
Vegetables		
1.	*Abelmoschus angulosus*	Source of resistance to yellow vein mosaic virus (YVMV), mites, and tolerance to low temperatures and light frost
2.	*Abelmoschus caillei*	Source of resistance to YVMV and shoot and fruit borer
3.	*Abelmoschus crinitus*	Source of resistant *Cercospora* blight
4.	*Abelmoschus manihot tetraphyllus var. pungens*	Resistant to enation leaf curl virus
5.	*Abelmoschus manihot*	Source of resistant to YVMV
6.	*Abelmoschus tuberculatus*	Tolerant to YVMV and fruit borer
7.	*Allium roylei*	Source of resistance to powdery mildew, leaf blight, etc.
8.	*Canavalia cathartica*	Source of nutritional traits
9.	*Canavalia rosea*	Highly salt tolerant
10.	*Citrullus colocynthis*	Source for drought tolerance and pest resistance
11.	*Cucumis callosus*	Resistance to fruit fly and leaf-eating caterpillars and tolerance to drought
12.	*Cucumis hardwickii*	Source of cold tolerant and resistance to powdery mildew
13.	*Lablab purpureus* var. *typicus*	Drought hardy
14.	*Momordica cochinchinensis*	Rich in nutrients

(continued)

	Wild-related species	Identified desirable trait(s)
15.	*Solanum anguivi* (*indicum*)	Source of resistance to *Fusarium* wilt
16.	*Solanum melongena* Linn var. *incanum*	Source of variation for phenolic content, *Fusarium* wilt, and frost and drought resistance
17.	*Solanum melongena* Linn var. *insanum*	Source of resistant to bacterial wilt, shoot, and fruit borer
18.	*Solanum nigrum*	Source of resistant to potato bacterial wilt
19.	*Solanum sisymbriifolium*	Source of resistance to *Verticillum* wilt resistance, *Meloidogyne* spp. and eggplant pest, such as aphid, shoot, and fruit borer
20.	*Solanum torvum*	Source of resistance to *Verticillum* wilt, bacterial wilt, pest *Meloidogyne* spp., root stock for eggplant
21.	*Solanum virginianum* (*xanthocarpum*)	Source of resistance to bacterial wilt
22.	*Trichosanthes dioica* (wild type)	Source of resilience to stresses
Fruits and nuts		
1.	*Citrus assamensis*	Source for waterlogging tolerance
2.	*Citrus cavaleriei*	Source of cold tolerance and graft stock
3.	*Citrus hystrix*	Source of *Citrus* pest resistance
4.	*Citrus indica*	Source of *Citrus* disease resistance
5.	*Citrus jambhiri*	Cold hardiness, source of drought resistance, also used as rootstock
6.	*Citrus karna*	Source of graft stock
7.	*Citrus latipes*	Source cold tolerant, potential disease resistance, and graft stock
8.	*Citrus limonia*	Most used source of rootstock and *Citrus* disease resistance
9.	*Citrus maderaspatana*	Source of graft stock
10.	*Citrus reshni*	Gene sources for salt tolerance in graft stock
11.	*Citrus reticulata*	Source of graft stock and winter hardiness
12.	*Citrus trifoliata*	Source of dwarfing, graft stock, hardiness, and disease resistance
13.	*Diospyros lotus*	Gene sources for graft stock
14.	*Ficus palmata*	Source of graft stock for fig
15.	*Fragaria daltoniana*	Frost tolerant strawberry
16.	*Fragaria nilgerrensis*	Source of quality
17.	*Garcinia hombroniana*	Source of rootstock
18.	*Garcinia sopsopia*	A suitable rootstock for mangosteen
19.	*Garcinia xanthochymus*	Source of graft stock for cultivated mangosteen
20.	*Grewia oppositifolia*	Tolerant to frost
21.	*Malus baccata*	Source of resistance to diseases, cold tolerance and graft stock
22.	*Malus sikkimensis*	Source for disease resistance and dwarfing gene
23.	*Mangifera indica*	Source of graft stock
24.	*Mangifera sylvatica*	Potential source of graft stock for mango
25.	*Manilkara zapota*	For variability in fruit size
26.	*Musa balbisiana*	Source of improved vigor and tolerance to biotic and abiotic stresses

(continued)

	Wild-related species	Identified desirable trait(s)
27.	*Prunus cerasoides*	Source of disease resistance and graft stock for sweet cherry
28.	*Prunus cornuta*	Source of disease resistance for sweet cherry and as graft stock
29.	*Prunus prostrata*	Used as graft stock and potential source of disease resistance
30.	*Pyrus pashia*	Common rootstock for Asian pear
31.	*Pyrus polycarpa*	Field resistant to powdery mildew and fire blight
32.	*Pyrus pyrifolia*	Preferred rootstock for pear
33.	*Ribes glaciale*	Immunity to rust caused by *Cronartium ribicola*
34.	*Rubus ellipticus*	Shade tolerant
35.	*Syzygium cumini*	Variability used in genetic improvement
36.	*Vitis lanata*	Source of late ripening and resistant to disease
37.	*Ziziphus nummularia*	Genetic source of rootstock
Spices and condiments		
1.	*Alpinia galangal*	Source of aromatic rootstock
2.	*Cinnamomum travancoricum*	Source of antifungal activities
3.	*Cinnamomum curvifolium*	Source of high essential oil
4.	*Curcuma ecalcarata*	Source of pinocembrin and piperitenone
5.	*Curcuma leucorhiza*	Source of edible starch
6.	*Curcuma sylvatica*	Source of highest concentration (320 mg/100 g) of biologically active peptide turmerin
7.	*Myristica beddomei*	Fruit pericarp is a rich source of nutritional compounds
8.	*Myristica malabarica*	Source of good rootstock for grafting the true nutmeg
9.	*Piper attenuatum*	Source of highest percentage of crotepoxide
10.	*Piper galeatum*	Source of bold fruits
11.	*Piper hamiltonii*	Source of resistance to betelvine blackfly
12.	*Piper nigrum* (INGR 8100)	Source of proliferating spikes
13.	*Piper thomsonii* (INGR 8009)	Source of sex change from male to bisexual plant
14.	*Zingiber cassumunar*	Source of fungitoxic action against *Rhizoctonia solani*, the damping off pathogen
15.	*Zingiber rubens*	Source of cold hardyness
16.	*Zingiber zerumbet*	Sources of ginger disease resistance
Commercial crops		
1.	*Camellia oleifera*	Source of edible seed oil
2.	*Camellia taliensis*	Gene resources for tea improvement
3.	*Miscanthus nepalensis*	Potential ornamental
4.	*Narenga porphyrocoma*	Source of temperature tolerance and resistance to red rot disease of sugarcane
5.	*Phoenix paludosa*	Source for poor soil tolerance
6.	*Saccharum arundinaceum*	Source of drought and disease resistance
7.	*Saccharum spontaneum*	Source of vigor, cold tolerance, hardiness, and disease resistance and yield
8.	*Sclerostachya fusca*	Source of waterlogging resistance, heavy tillering, and earliness in sugarcane

(continued)

	Wild-related species	Identified desirable trait(s)
Medicinal and aromatic plants		
1.	*Albizia procera*	Drought tolerant
2.	*Alpinia malaccensis*	Potential ornamental plant
3.	*Cymbopogon khasianus*	Source of essential oil and methyl eugenol
4.	*Plectranthus hadiensis* var. *tomentosus*	Source of bioactive phytochemicals, especially terpenoids
Ornamental plants for floriculture		
1.	*Barleria grandiflora*	Potential regular ornamental
2.	*Begonia griffithiana*	Potential regular ornamental
3.	*Hedychium coccineum*	Early bloomer
4.	*Hedychium marginatum*	Potential regular ornamental
5.	*Jasminum parkeri*	Potential regular ornamental for rockeries
6.	*Rhododendron arboreum*	Source of variability as parent in hybrids
7.	*Rosa macrophylla*	Source of variability
Agroforestry trees		
1.	*Acacia jacquemontii*	Source of high temperature and excessive drought tolerance
2.	*Acer caesium* ssp. *caesium*	Source of seed dormancy
3.	*Populus euphratica*	Source of drought and salinity tolerance
4.	*Salix wallichiana*	Frost resistant
Cottage industry plants and others		
1.	*Bambusa pallida*	Source of cane and quality fiber
2.	*Dendrocalamus hamiltonii*	Source of palatable shoots
3.	*Indigofera glandulosa*	Source of high protein
4.	*Millettia pinnata*	Drought resistant
5.	*Morus alba*	Drought tolerant
6.	*Morus macroura*	Source of resistant to drought, salinity, and frost

Source: Analysis of reviewed literature

MIX
Papier aus verantwortungsvollen Quellen
Paper from responsible sources
FSC® C105338

If you have any concerns about our products,
you can contact us on
ProductSafety@springernature.com

In case Publisher is established outside the EU,
the EU authorized representative is:
Springer Nature Customer Service Center GmbH
Europaplatz 3, 69115 Heidelberg, Germany

Printed by Libri Plureos GmbH
in Hamburg, Germany